Fertilization

Fertilization

Second edition

Frank J. Longo

Department of Anatomy
The University of Iowa, USA

Taylor & Francis
Taylor & Francis Group

LONDON AND NEW YORK

Published by Taylor & Francis
2 Park Square, Milton Park, Abingdon, Oxon, OX14 4RN
270 Madison Ave, New York NY 10016

Transferred to Digital Printing 2009

First edition 1987

Second edition 1997

© 1987, 1997 F.J. Longo

Typeset in 10/12 Palatino by Acorn Bookwork, Salisbury, Wilts

ISBN 0 412 56350 9

A catalogue record for this book is available from the British Library

Library of Congress Catalog Card Number: 96–86753

Contents

    events of fertilization**   **225**
       Ascaris type of fertilization   225
       Sea-urchin type of fertilization   231
       Comparison of the *Ascaris* and sea-urchin types of
       fertilization   236

15 **Manipulation and *in vitro* fertilization of human and other
    mammalian gametes**   **238**

    References   245
    Index   305

Preface to the first edition

E. B. Wilson stated in the preface to the third edition of the *The Cell in Development and Heredity*, 'Every writer must treat the subject from the standpoint given by those fields of work in which he is most at home; and at best he can only try to indicate a few of the points of contact between those fields and others.' The aim of this book is to provide an overview of structural and functional aspects of fertilization processes in a manner that would be helpful not only to specialists in the field but also to investigators in related disciplines and to advanced undergraduate and graduate students in the biological sciences. Fundamental descriptive accounts at the light and electron microscopic levels of observation have been combined with analytical studies – physiological and biochemical investigations of fertilization in invertebrates and vertebrates. A comparative approach to fertilization is presented and, although a variety of animals are referred to, additional space is purposely given to organisms that have been, and continue to be, popular research material.

The text does not pretend to be comprehensive and admittedly does not cover all aspects of the field or areas related to the general subject. Historical reviews and technical details are presented only to the extent necessary to formulate an orientation to and a perspective on individual topics. Areas such as fertilization mechanisms in plants, gamete development and the role of accessory reproductive structures on fertilization, particularly in mammals, have been included only to a limited extent in an effort to keep the book within the bounds of an overview.

It is not possible to list or refer to even a major portion of the vast literature on fertilization. In this book, for the most part, each topic is referenced with reviews and/or original research articles that pertain to several different organisms and include earlier and more contemporary investigations of the subject. It is anticipated that this approach may aid the reader in exploring the depth and nuances of a particular topic.

In any venture such as this credit is due to a number of individuals who have been generous with their time and efforts. Because my own views of fertilization processes have come in large measure from previous mentors, I acknowledge their presence in these pages, particularly Everett Anderson. Sincere thanks and grateful appreciation is expressed to Becky Hurt, Tena Perry and Vicki Fagen for typing the manuscript; to Frederic So and Shirley Luttmer for their assistance in the organization of the references; to Julie Longo, Brenda Robinson and Paul Reimann for their artistic help in the preparation of the illustrations, and to Jo Ann Barnes for her generous and conscientious editorial assistance. Kitty, Julie, Joe, Crista, Thad, Jude, Gabrielle and Gian Carlo are also owed special thanks for many special favors.

Preface

In the 10 years since the first edition of this overview of fertilization was published, the number of investigations relating to fertilization events in animals has continued to expand, making it difficult, if not virtually impossible, for a single individual to keep abreast of such a field. This tremendous productivity is not only due to a desire to know and understand our 'origins' but also reflects: (1) the inherent qualities of the gametes that make them useful paradigms for answering specific questions in cell and molecular biology; (2) the application of techniques of cellular and molecular biology in helping to solve outstanding questions of fertilization, development and cell biology; (3) the realization that an understanding of such a process as fertilization is fundamental to the improvement of existing and the elaboration of new means of conception and contraception; (4) the development of *in vitro* methods of fertilization in humans, as well as other mammalian species; and (5) the improvement of strains of plants and animals.

The major thrust of this edition is basically the same as the first, i.e. an overview of fertilization events in a wide variety of animal species that have been and continue to be important models for our understanding of mechanisms of this event in the life history of a biparental organism. Because sperm and eggs are specialized cells, having the capacity when combined to generate a new individual, the emphasis of what is presented here is, by its very nature, primarily concerned with the cell biology of fertilization – that is, an exploration of the organelles and mechanism by which the gametes interact with one another.

All chapters of the first edition have been extensively rewritten and include, whenever possible, aspects of the following organisms: echinoderms, mollusks, ascidians, fish, amphibians and mammals. A separate chapter concerning *in vitro* fertilization in humans and other mammals has been included because of the tremendous growth of this area and because of the unique features involving *in vitro* fertilization in mammals. Chapters have been expanded or completely rewritten, taking into account developments in areas such as receptor–ligand interactions, signal transduction mechanisms and changes isolated sperm nuclei undergo when cultured in egg cytosol. Moreover, aspects concerning sperm and egg development have been added or expanded in order to provide further insights into and understanding of mechanisms and concepts of fertilization.

As in the first edition, this volume considers to only a limited extent or not at all a number of important aspects relating to fertilization, such as hormonal regulation of sperm and egg development and fertilization, accessory structures of both the male and female reproductive tracts and fertilization (mating) events in a number of important animal groups including protozoa, coelenterates, platyhelmenthes, nematodes, aves and reptiles, as well as members of the plant kingdom. It is recognized that integration of such areas and organisms into this volume would be extremely worthwhile but such an undertaking was deemed beyond the scope of the volume intended.

In an endeavor such as presented here appreciation needs to be given to individuals who have helped in their own special ways. Vicki Fagen, Tena Perry, Chris Ihle and Jennifer Satterfield are acknowledged and thanked for typing the manuscript. Many of the illustrations presented in the first edition have been retained and additional ones have been incorporated as well. Thanks are expressed here and in the figure legends to individuals who generously contributed their illustrations to this volume. Original drawings were done by Julie Longo and sincere appreciation for her efforts are acknowledged here. I have had the benefit of a number of supporters in many special ways. Kitty, Julie, Joe, Crista, Jude, Thad, Gabrielle and Gian Carlo are thanked for their support and for the stability and the focus they have provided me in all my endeavors. This volume is dedicated to them and all the investigators in the fields of cell and developmental biology that I have had the pleasure of sharing research problems of mutual interest. They helped to make work play.

Frank J. Longo
Iowa City, IA

General considerations of fertilization: a definition 1

The interaction of the spermatozoon and the egg initiates a series of transformations involving the nuclear and cytoplasmic components of both gametes. These transformations constitute the process of fertilization, which commences with the interaction and subsequent fusion of the gametes and culminates in the association of the corresponding groups of chromosomes derived from two pronuclei, one of maternal and the other of paternal origin (Wilson, 1925). In almost all cases investigators have pointed out that the essential aspects of fertilization are: (1) the association of the maternal and paternal genomes – biparental heredity – and (2) the activation of both the sperm and the egg – a series of events which alters the metabolism of both gametes and leads to the cleavage and differentiation of the fertilized egg or zygote. This mode of bisexual reproduction emerged during evolution and has been maintained in most metazoans (Margulis et al., 1985). Although the reasons for this are not clearly understood, fertilization may accelerate the rate of adaptation while avoiding the accumulation of detrimental mutations in a changing environment (Michod and Levin, 1987).

Meiosis, the cell division that reduces by half the number of chromosomes, is similar to and derived from mitosis (Margulis et al., 1985). The resultant cells, sperm and eggs, with half of the chromosomes' complement, fertilize, thereby restoring the diploid number of chromosomes and initiating the process of development. According to Margulis et al. (1985) this process flourished not because of its tendency to mix genes from separate sources and to generate genetic variation, but because it became fixed in the life cycles of a rapidly evolving group – the first animals.

Although isogamy, reproduction resulting from the union of two gametes that are identical in size and structure, occurs in some groups, particularly the protozoa, heterogamy, the union of two gametes of differing size and structure, is generally the rule among most groups of animals. Bell's (1988) discussion of the evolution of germ cells provides interesting insights into the basis of gamete dimorphism. Using the Volvocales as a model system, he shows that it is better to produce small gametes, as more can be generated from a given mass of material. However, if the viability of a zygote increases with size, small gametes will continue to be an advantage because they can be produced in great numbers, whereas large gametes will also be favored because they give rise to large and highly viable zygotes. The net effect of these two factors may then create a selection for gamete dimorphism.

Activation of the egg can be initiated by processes other than fertilization. Parthenogenetic activation by chemical and/or physical stimuli may lead to complete development or to the initiation of processes that simulate fertilization, but in most cases does not give rise to viable offspring. In either case, these observations indicate that the egg is endowed with the essential machinery and information to initiate developmental processes when suitably stimulated. Furthermore, activation changes that occur at fertilization and parthenogenesis do not require immediate gene action, implying that the genetic activity required for establishing the fertilization response occurs during oocyte maturation. Metabolic changes in the egg evoked by fertilization ultimately affect new gene expression and differentiation during later stages of embryogenesis. Hence, fertilization in the scheme of a biparental organism serves as a point of transition between gamete and embryonic development.

In addition to the processes of fertilization that characterize most bisexual organisms, there are groups of unisexual animals that have evolved unique mechanisms for perpetuation of their species (Bogart *et al.*, 1989).

1. Some genera of female lizards reproduce by parthenogenesis.
2. Females of some species of fish reproduce by gynogenesis, i.e. sperm from a sympatric male activates the egg but does not contribute its genome to the embryo.
3. In hybridogenesis maturing eggs of some fish and frogs eliminate an entire genome which is retrieved in fertilization by the sperm of a sympatric male.

The loss of biparental reproduction in such cases has been the subject of controversy (Margulis *et al.*, 1985).

Investigations of fertilization date well before the turn of the century under the leadership of cytologists such as van Beneden, Flemming, Strasburger, Boveri and Wilson whose observations on germ cells were closely affiliated with theoretical writings of Nageli, Weissmann, Hertwig, Roux and de Vries (Wilson, 1925). The remarkable observations of these investigators and their formulations provided the intellectual framework for the chromosome theory of inheritance. More contemporary research on fertilization tends to serve a dual role, one being its application to fertility control and the other as a model system to study basic processes of cells in general.

Much of the research in fertilization has employed the gametes of invertebrates, particularly echinoderms, mollusks and ascidians, as well as nonmammalian vertebrates such as frogs. The reasons for the popularity of these animals include their availability and minimal requirements for maintenance. In addition, because both sexes are separate and relatively large quantities of gametes can be obtained, which can be fertilized externally and develop in synchrony, specific processes of fertilization may be analyzed using a wide variety of techniques. Consequently, important insights have been gained into the causal chain of events that occur during fertilization in these animals, which are relevant to the study of fertilization in higher organisms. In contrast, the number of eggs available from mammals is generally small and, because they normally undergo internal fertilization, working with the gametes of such organisms is much more involved. Despite numerous difficulties, there have been considerable advances in our knowledge of fertilization in mammals, and in some ways we have greater insights into processes of mammalian fertilization than in invertebrates and lower vertebrates (Yanagimachi, 1994).

The timing of many of the events comprising fertilization in sea urchins is given in Table 1.1. Studies using the gametes of other invertebrates and vertebrates indicate that a comparable sequence of events is

Table 1.1 Timing of fertilization events in the sea urchin, *Lytechinus pictus* (incubated at 16–18°C); timing of these events depends markedly upon species and temperature (See Whitaker and Steinhardt, 1982)

Membrane potential	
Ca^{2+}–Na^+ activation potential	Before 3 s
Na^+ activation potential	3–120 s
K^+ conductance increase	500–3000 s
Intracellular calcium release	40–120 s
Cortical granule reaction	40–100 s
NAD kinase activation	40–120 s
Reduced nicotinamide nucleotides increase	40–900 s
Acid efflux	1–5 min
Intracellular pH increase	1–5 min
Oxygen consumption increase	1–3 min
Protein synthesis initiation	5 min onwards
Amino acid transport activation	15 min onwards
DNA synthesis initiation	20–40 min

initiated in both groups of animals; sperm–egg attachment leads to the 'turning on' of new synthetic activities and developmental programs (Metz and Monroy, 1985).

Since metabolic and morphological changes of echinoderms, mollusks, ascidians, amphibians and mammalian gametes during fertilization have been well characterized experimentally, studies employing these organisms provide the major thrust of what is discussed herein. The two principal events of fertilization considered are (1) the initial interaction and activation of the sperm and egg and (2) the concluding events, involving pronuclear development and association that eventually lead to cleavage.

The spermatozoon 2

2.1 SPERM MATURATION

Sperm are highly differentiated by the time they leave the testis. In many species, particularly mammals and invertebrates in which sperm are transferred from the male to the female and held by the female for a period of time (Clark and Griffin, 1988; Wikramanayake et al., 1992), they may not have the ability to move progressively or to interact with and fertilize eggs. Sperm gain this ability while in accessory portions of the male reproductive tract, e.g. the epididymis, and/or in portions of the female reproductive tract, e.g. the seminal receptacle. This situation is in contrast to that seen in many invertebrates (e.g. sea urchins, starfish and clams) and lower vertebrates (e.g. fish and amphibians) where testicular sperm are fully capable of fertilization. Unlike many invertebrates in which spawned sperm are fairly homogeneous structurally and functionally and possess the ability to fertilize, in mammals sperm populations tend to be somewhat heterogeneous and require additional periods of maturation in order to fertilize. These changes are described below and are achieved in subpopulations at different rates (Ward and Kopf, 1993; Yanagimachi, 1994).

The location within the epididymis where mammalian sperm acquire the ability to fertilize varies according to the species. In man, the functional relation between sperm and successive regions of the epididymis is very flexible. Alternatively, the epididymis may not have an absolutely essential role in maturation and acquisition of fertilization ability as the attainment of these properties may also be partly a matter of time.

During epididymal maturation, the spermatozoon acquires the capacity for progressive movements (Hamilton, 1977; Hoskins et al., 1978). Sperm taken directly from rodent testes are either nonmotile or weakly motile. Sperm from the cauda epididymis, however, display active, progressive movement. This change does not appear to depend upon the available energy supply. Cyclic nucleotide phosphodiesterase inhibitors markedly increase respiration (three- to fourfold) and motility of cauda epididymal sperm. The observed stimulating effects on motility are mediated by an increase in intracellular cAMP (Garbers *et al.*, 1973a, b). The effect of cAMP on mammalian sperm motility may be direct or mediated by protein phosphorylation. Mammalian sperm have a cAMP-dependent protein kinase and phosphorylation of a 55 000 Da molecular weight protein is correlated with sperm motility, suggesting that the effect of cAMP is mediated via protein phosphorylation (Garbers *et al.*, 1973a, b). Major proteins of the sperm tail, e.g. tubulin and dynein, are among polypeptides phosphorylated in association with cAMP-enhanced motility. Although the change in sperm motility involves components of the axonemal complex, it is believed to be regulated by alterations associated with the plasma membrane (Yanagimachi, 1994). Such a postulation is based on a large number of studies demonstrating that during epididymal transit the sperm plasma membrane undergoes significant reorganization in its molecular architecture and some changes are brought about by direct interaction with epididymal secretory proteins (Brown *et al.*, 1983).

Contributions to the development of sperm motility include:

1. a transfer of glycerol-3-phosphorylcholine (Infante and Huszagh, 1985) and forward motility protein (Hoskins *et al.*, 1989; Acott and Hoskins, 1981; Acott *et al.*, 1983) from the epididymal fluid to the sperm;
2. an alteration in the cAMP-modulated protein kinase system (Wooten *et al.*, 1987);
3. the development of a mechanism to maintain low internal calcium (Yanagimachi, 1994);
4. changes in plasma membrane components.

The distribution of lectin binding sites on the sperm surface may increase, decrease or remain unchanged, depending upon the species, functional domain and protocol employed (Nicolson and Yanagimachi, 1979). Surface labeling studies using ^{3}H-sodium borohydrate or lactoperoxidase catalyzed iodination have been used to demonstrate differences in membrane proteins from caput and cauda sperm (Olson and Danzo, 1981).

Additional changes occur with respect to membrane lipids (Wolf and Vogelmayr, 1984; Nikolopoulou *et al.*, 1985; Parks and Hammerstedt, 1985; Schlegel *et al.*, 1986), the distribution of intramembranous proteins (Olson, 1980; Reger *et al.*, 1985; Suzuki, 1990) and the transfer of cholesterol to the sperm plasma membrane (Seki *et al.*, 1992: Suzuki, 1990). Immunochemical probes have shown a loss and/or modification of glycoproteins during epididymal maturation of mouse and boar sperm (Feuchter *et al.*, 1981). The sperm plasma membrane possesses integral and surface-absorbed proteins when leaving the testes. Some of the intrinsic proteins undergo changes in location; others are altered, masked or replaced during epididymal passage (Bellvé and O'Brien, 1983; O'Rand, 1985; Dacheux *et al.*, 1989; Phelps *et al.*, 1990). There is also an active glycosylation (Hamilton *et al.*, 1986) of the sperm surface by galactosyltransferase and sialyltransferase in the epididymal fluid (Bernal *et al.*, 1980; Hamilton, 1980; Tulsiani *et al.*, 1993). In addition to modifications of intrinsic plasma membrane proteins, glycoproteins are secreted by cells of the male accessory reproductive organs that become associated with the sperm's surface. The association of these components with the spermatozoon may account for surface changes such as charge density and adhesiveness.

In addition to changes in the sperm surface, as the mammalian spermatozoon passes through the epididymis it undergoes changes which ultimately give rise to its mature form. The cytoplasmic droplet migrates caudally along the middle piece and is eventually lost. The acrosomes of some mammalian sperm, including guinea pig and marsupials, undergo considerable morphological changes during epididymal transit (Fawcett and Hollenberg, 1963; Hoffer *et al.*, 1981; Bedford and Hoskins, 1990). Extensive crosslinking of nuclear protamines, as well as proteins of the outer dense fibers and fibrous sheath, also take place in the epididymis (Calvin and Bedford, 1971; Bedford and Calvin, 1974; Pellicciari *et al.*, 1983; Bedford and Hoskins, 1990; Kosower *et al.*, 1992).

2.2 SPERM DEPOSITION AND STORAGE

In many mammals semen is deposited within the vagina (e.g. cows, sheep and primates) while in others (e.g. pig, horse and dog) sperm deposition is in the uterus. Most of the sperm are eliminated (Drobnis and Overstreet, 1992) and only a few ever reach the site of fertilization so that in some species sperm–egg ratios in the ampulla can be 1:1 or less (Zamboni, 1972; Cummins and Yanagimachi, 1982; Shalgi and Phillips, 1988). The female reproductive tract prevents many abnormal sperm from reaching the ampulla. The tubal–uterine junction appears to be the major barrier. In the hamster less than 0.001% of the 10^8–10^9 sperm enter each oviduct from the uterus.

Nonmammals have mechanisms of transferring sperm to individuals at coitus similar to that of mammals (Austin, 1965). Coition with introitus is employed in reptiles, some amphibians, some teleosts, elasmobranchs, most insects, some annelids and some gastropods. Coition without introitus is seen in birds, and involves the ejaculation of semen from the genital papilla of the male into the cloaca of the female. Most frogs and toads employ amplexus, i.e. the male grasps the female and sperm and eggs are shed into the water. In spiders, crustaceans and cephalopods, a limb or a palp is modified for sperm or spermatophore transfer from the male to the female.

In spawners (e.g. fish and many marine invertebrates), which release their gametes into an aqueous environment to fertilize, several types of adult behavior are employed, such as:

1. synchronization of spawning;
2. aggregation of adults;
3. the use of spawning pheromones (Miller, 1989);
4. the movement of adults into shallow water;
5. an increase in sperm numbers released (Babcock *et al.*, 1994).

The number of gametes released is dependent upon the number of animals participating and the intensity of spawning, but the probability of fertilization varies in direct proportion to these factors.

2.3 SPERM ACTIVATION

2.3.1 INTERACTION WITH EGG COATS

Behavior and morphological changes of the sperm, including its activation and attraction to the egg, are brought about by cellular and noncellular components associated with the egg, which are often referred to as egg coats. The eggs of virtually all organisms are associated with one or more cellular or acellular layers that often play key roles in fertilization. In fact, in many species the earliest events of fertilization involve interactions between sperm and egg coats. Such interactions usually lead to functional and morphological changes in the sperm, which are collectively referred to as sperm activation. In turn, the coats of most species are modified to serve a variety of functions, such as a block to polyspermy and protection of the developing embryo.

Because of their great variety in form, composition and function, as well as the numerous ways investigators have elected to study the surface layers of eggs, there have been relatively few attempts to systematically organize the different types of coat associated with the eggs of animals (see Anderson, 1974). Hence, generalities are difficult and an apparent confusion of terms exists, which may have different meanings depending on the egg in question.

Eggs that are spawned (e.g. sea urchin and frog) often possess a jelly layer, a complex, carbohydrate-rich covering that is added to the egg within the ovary or within an accessory organ (e.g. the oviduct). In the sea urchin egg jelly consists of a three-dimensional interconnected network of a fucan sulfate polymer (Bonnell *et al.*, 1993). Bound to this matrix are biologically active components such as chemoattractant peptides and acrosome reaction-inducing glycopeptides (Keller and Vacquier, 1994; Bonnell *et al.*, 1994). *Xenopus* egg jelly appears to follow a similar structural pattern (Bonnell *et al.*, 1996). In addition, the plasma membrane of the eggs of many species of marine invertebrates (e.g. mollusks and echinoderms), as well as a number of other organisms, possesses a rather well developed glycocalyx which is referred to as a vitelline layer. In some species (e.g. sea urchins and starfish) the vitelline layer is greatly modified at fertilization and plays a significant role in polyspermy prevention. In other species (e.g. surf clams and mussels), the vitelline layer appears to undergo little, if any, change at insemination.

Some eggs, such as those of the sand dollar, possess pigment bodies, which are embedded within the surrounding jelly coat. Ascidian eggs are surrounded by a layer of test cells, a noncellular layer referred to as a vitelline coat and an outer layer of follicle cells (Koch *et al.*, 1993). All these layers must be traversed by the sperm in order for it to interact and fuse with the egg. Follicular cells are also present in association with insect eggs: they form an epithelium that lines the outer margin of an acellular layer, the chorion, that in turn surrounds the egg. At one pole of the egg, the chorion possesses an opening, the micropyle, which permits the passage of the sperm to the egg's surface (plasma membrane/oolemma). Some eggs, such as those of fish, are also surrounded by a relatively thick chorion possessing a micropyle. Eggs of eutherian mammals are surrounded by an acellular coat referred to as the zona pellucida. The zona pellucida, which consists of glycoprotein, lies immediately adjacent to the egg and in most eutherians is surrounded by cumulus cells and their matrix, hyaluronic acid (2500 kDa), which is conjugated to protein (Virji *et al.*, 1990). In some mammals (e.g. cows and sheep) the cumulus cells are shed either before or immediately after ovulation. Hence, in these cases the zona pellucida is the only coat the sperm must penetrate in order to reach the surface of the egg.

Although it was once held that sperm bring about dispersion of the cumulus, recent experiments indicate that normally very few sperm are present near the egg at fertilization. Sperm move through the cumulus and reach the zona without undergoing the acrosomal reaction (Talbot, 1985). It has been speculated that hyaluronidase on the sperm surface and not the enzyme present in the acrosome, facilitates cumulus penetration (Talbot, 1985). In a number of species, the presence of cumulus cells has been shown to improve *in vitro* fertilization by minimizing individual variations of male fertility (Yanagimachi, 1994). Consistent with this observation are results of experiments demonstrating that cumulus factors stimulate sperm motility and promote the acrosome reaction. Others are believed to prolong the fertilizable life of the egg and aid in zona penetration (see Yanagimachi, 1994).

2.3.2 MOTILITY AND BEHAVIORAL CHANGES: CHEMOKINESIS AND CHEMOTAXIS

As a rule, sperm are stored inactive until fertilization. Upon spawning or some time following ejaculation they become activated, which initially is manifested by a significant change in motility and respiration. Other changes may follow, which are also part of sperm activation, and these are described later in sections devoted to capacitation and the acrosome reaction. Stimulation of sperm motility and respiration, chemokinesis, may be brought about by suspending sperm in an aqueous environment – by dilution of seminal/testicular fluid – or as a result of the exposure of sperm to a chemical signal from the egg or its investments (Ward and Kopf, 1993). In most species this constitutes the first level of recognizable interaction between the sperm and egg. Chemotaxis is the directed movement of the sperm in response to a chemical concentration gradient to the egg and may be correlated with a change in wave form of the flagellum. In many cases it is difficult to distinguish chemokinesis from chemotaxis since a chemokinetic factor by definition is a substance that stimulates sperm motility. The increase in motility itself can bring the sperm into close proximity to the egg by an increase in the rate of random interactions. Although in theory no net increase in the accumulation of sperm in the vicinity of the egg might be expected with chemokinesis, since the movement toward and away from the egg would be the same, sperm are often trapped in extraneous layers surrounding the egg. Such phenomena give the false impression that sperm are attracted to the egg. A true test of chemotaxis is to demonstrate a change of direction of the sperm towards an increasing concentration of chemoattractant (Miller, 1985; Ward and Kopf, 1993).

Chemokinetic and chemotactic agents may be separate molecules, or both activities may reside in the same chemical species. These biochemical signals have been identified in virtually all major groups of organisms, including humans (see Ward and Kopf, 1993). Their advantage is that they increase the probability of a sperm reaching the egg. In a sense, they effectively increase the volume of the egg as a target for the sperm. Generally, sperm responses to chemokinetic and chemotactic factors are believed to be mediated by receptors as the responses of sperm are similar to those seen in somatic cells where receptor involvement has been more clearly defined. Responses of sperm include increases in cyclic nucleotides, intracellular ions, including Na^+ and Ca^{2+}, pH and net H^+ and K^+ effluxes. In this context, Vanderhaeghen *et al.* (1993) have demonstrated that members of the olfactory receptor gene family are preferentially expressed in dog testes with little or no expression in olfactory mucosa. Antibodies to one of the testes expressed proteins localized to late spermatids and mature sperm. Vanderhaeghen *et al.* (1993) speculated that such receptors may function as sensors for chemical signals in the regulation of sperm maturation, migration and/or fertilization.

In sea urchins, dormancy of sperm is partly maintained in the testes by a high CO_2 tension due to a low intracellular pH (Lee *et al.*, 1983; Trimmer and Vacquier, 1986). Dynein, the ATPase that powers the flagellum, is inactive below pH 7.3. Sperm activation upon spawning is initiated by a series of changes in ion content resulting in an increase in internal pH from 7.0 to 7.4. There is a sodium-dependent acid release coincident with an increase in sperm motility upon the dilution of semen into sea water. The activating factor may not be sodium *per se*, as incubation of sperm in solutions of high external pH (pH 9) or NH_4Cl, which presumably increase the intracellular pH, also activate motility in the absence of sodium. At pH 7.4, flagellar dynein hydrolyses ATP and the ADP produced stimulates mitochondrial respiration approximately 50-fold.

When sea urchin sperm are incubated in sea water at an acid pH they are immotile and have a low respiration rate. However, if egg jelly is added respiration and motility increase dramatically (Hansbrough and Garbers, 1981a, b; Hardy *et al.*, 1994). The active agent of sea urchin egg jelly has been shown to be a glycine-rich polypeptide of approximately 2000 molecular weight. The material from egg jelly of *Strongylocentrotus* (speract) and *Arbacia* (resact) stimulate respiration and motility in a species-specific fashion. They are potent at picomolar concentrations and in normal sea water (pH 8) they increase Na^+–H^+ exchange and intracellular levels of cyclic nucleotides. At acidic pH values they increase respiration rates by as much as tenfold; at alkaline pH only marginal changes in respiration are observed (Ward and Kopf, 1993). Since the pH of sea water and of egg jelly is alkaline, the physiological roles of resact and speract are unclear. It has been postulated that these factors may function to regulate respiration and/or motility in concert with other factors in egg jelly (Ward and Kopf, 1993).

In the presence of speract at pH 6.6, sperm undergo a sodium influx and a sodium-dependent proton efflux (Hansbrough and Garbers, 1981b). Proton release appears to be responsible for motility initiation and the increase in respiration, since both can be induced in sea water at pH 6.6 by monensin, a sodium-proton ionophore. If sodium is not present, the increase in respiration fails to occur. Coincident with an increase in respiration and motility, speract also brings about an increase in sperm cAMP and cGMP (Hansbrough and Garbers, 1981a, b). Observations demonstrating that similar changes are elicited by bromo-cGMP and that analogs of cAMP are without effect suggest that cGMP may mediate or be closely associated with increases in sperm respiration and motility induced by speract. Chemical cross-linking studies have identified a single sperm membrane protein of 75–77 kDa as a specific speract receptor (Dangott et al., 1989). The receptor may be associated with the membrane form of guanylate cyclase in sperm (Bentley et al., 1988).

A model relating changes in waveform and swimming paths produced by speract is shown in Fig. 2.1 (Cook et al., 1994).

Activation of cGMP synthesis opens K^+ channels, thus hyperpolarizing the plasma membrane (V_m), preventing Ca^{2+} entry and perhaps promoting Na^+–Ca^{2+} exchange to lower internal Ca^{2+}. A negative threshold V_m activates Na^+–H^+ exchange to elevate internal pH, which coordinates the termination of cGMP synthesis and K^+ channel activity with the initiation of cAMP synthesis and activation of Ca^{2+} channels to transiently elevate internal Ca^{2+}. Return to basal internal Ca^{2+} presumably follows inactivation of Ca^{2+} channels and operation of internal Ca^{2+} homeostatic mechanisms. For sperm swimming up a chemotactic gradient, K^+ channels remain active and suppress Ca^{2+} channel activity. If the sperm diverges from this path and no longer senses an increase in chemoattractant concentration, K^+ channel activity decreases, the membrane depolarizes and internal Ca^{2+} increases. The resultant increase in flagellar asymmetry will result in a turn, which may lead to a more optimal path.

Resact from Arbacia has been shown to be a Ca^{2+}-dependent chemoattractant (Ward et al., 1985). In response to the binding of resact to isolated sperm membranes, guanylate cyclase is inactivated and the molecular form of the enzyme (Mr = 160 kDa) is modified such that its mobility coincides with an apparent molecular weight of 150 kDa. Coincident with the mobility change is a dephosphorylation of the enzyme (Bentley et al., 1986).

Aside from increases in motility and respiration, when sea urchin sperm are incubated in solubilized egg jelly they undergo an agglutination which has been referred to as swarming (Lopo, 1983; Fig. 2.2).

Similar phenomena have been described for the sperm of other organisms. The affected sperm aggregate into dense clusters 2–4 mm in diameter and 5–10 min later they spontaneously disperse. The dispersed sperm do not swarm again with the addition of fresh jelly. The swarming of sperm when exposed to egg jelly provided the basis for the fertilizin–antifertilizin theory of Lillie (1919). This 'isoagglutination' reaction was thought to represent a reversible complex formation comparable to the interaction of antibody with antigen. Reversibility

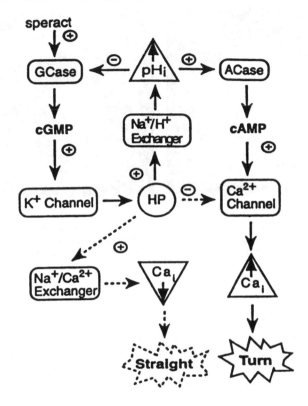

Fig. 2.1 Model for sperm chemotaxis in sea urchins. Activation of cGMP synthesis opens K^+ channels, thus hyperpolarizing the plasma membrane (V_m), preventing Ca^{2+} entry and perhaps promoting Na^+–Ca^{2+} exchange to lower Ca_i. A negative threshold V_m activates Na^+–H^+ exchange to elevate pH_i which coordinates the termination of cGMP synthesis and of K^+ channel activity with the initiation of cAMP synthesis and activation of Ca^{2+} channels to transiently elevate Ca_i; return to basal Ca_i presumably follows inactivation of Ca^{2+} channels and operation of Ca_i homeostatic mechanisms. Thus the left and right legs of this pathway, which respectively determine low and high Ca_i, recover rapidly from speract stimulus. It is proposed that the left, negative feedback loop forms the basis for gradient detection. Whereas swimming is more linear at low Ca_i and more curvilinear at high Ca_i, these mechanisms produce swimming responses that enhance fertilization. Reproduced with permission from Cook *et al.*, 1994.

was brought about by a modification of agglutinating substances of the egg jelly (fertilizin) by sperm from a multivalent to a univalent form (Metz, 1967). More recent investigations, however, indicate that this process is not a true agglutination as the sperm do not contact one another. Furthermore, the process is reversibly blocked by respiration inhibitors, indicating that it is dependent upon sperm motility (Collins,

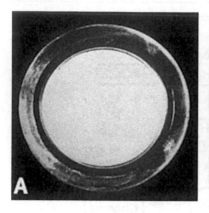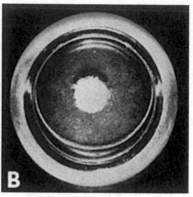

Fig. 2.2 Agglutination of *Megathuria* (mollusk) sperm induced by egg water. **(A)** Dense sperm suspension before addition of egg water. **(B)** aggregated sperm 10 min after the addition of egg water. Reproduced with permission from Tyler, 1940.

1976). The possible role of sperm swarming during fertilization has not been clearly defined; in fact, it does not appear to be obligatory since successful fertilization can occur in its absence. Interpretation of its possible significance at fertilization is paradoxical. It may function to keep excess sperm from reaching the egg and, ultimately, act to reduce the concentration of sperm in the immediate vicinity of the ovum, thereby preventing polyspermy. Conversely, it may attract sufficient numbers of sperm to the vicinity of the egg, thereby promoting successful fertilization.

Sperm chemotaxis has been demonstrated in a relatively large number of metazoa (Miller, 1985). In hard and soft corals at least three different lipid components have been found that are involved in sperm activation and chemotaxis (Coll *et al.*, 1990). The capsule of the siphonophore *Muggiaea* has been shown to attract sperm (Sardet *et al.*, 1982; Fig. 2.3).

Sperm approaching capsule material reduce the curvature of their trajectories from slightly curved paths to small circles so that eventually a cloud of sperm forms around the attractant source. During this attractive phase, sperm flagellar beat frequency and velocity are not noticeably modified. The active substance isolated from capsules has an apparent molecular weight of about 67 000. Calcium is required for sperm to respond to the attractant. This calcium dependency can be mimicked by treating sperm with the calcium ionophore A-23187, suggesting that the attractant induces a calcium flux, thereby increasing the asymmetry of the flagellar beat pattern.

Fig. 2.3 Accumulation of sperm at the animal pole of a *Muggiaea* egg. See Carre and Sardet, 1981.

In most viviparous teleosts, sperm become highly motile on contact with water. In herring, sperm are motionless in sea and brackish water where spawning occurs (Yanagimachi *et al.*, 1993). Initiation of sperm motility is dependent on a sperm motility inducing factor, which has been shown to be localized to the area of the micropyle (Yanagimachi *et al.*, 1992). Two different molecules from herring eggs have been identified that stimulate sperm movement: one a 105 kDa glycoprotein (Pillai *et al.*, 1993) and the other an 11 kDa protein (Morisawa *et al.*, 1992).

Evidence has been presented for a sperm attractant in human follicular fluid which enhances motility and hyperactivation. The factor is believed to be nonhydrophobic and less than 10 kDa (Ralt *et al.*, 1991, 1994). There is little evidence to suggest that it is a protein.

What turns off the chemotactic stimulus in any of the species examined to date has not been determined. Sperm chemotaxis disappears following fertilization or artificial activation, possibly as a result of cortical granule exocytosis, although the involvement of other egg changes has not been eliminated.

2.4 CAPACITATION

Mammalian sperm require an additional phase of maturation, occurring within the female reproductive tract, to prepare them for the acrosomal reaction and fertilization (Yanagimachi, 1981, 1994; Moore and Bedford, 1983). This is referred to as capacitation (Chang, 1951; Austin, 1951).

Many of the changes that occur in sperm during capacitation are concerned with plasma membrane alterations, including rearrangements of intramembranous particles, removal of sperm surface components and a decrease in net negative charge. Despite years of rather intensive study by many laboratories, its molecular basis is not fully understood.

Whether or not the sperm of invertebrates undergo capacitation is not clear. Some reasons for this lack of clarity may be related to the relatively short period it takes for the gametes of many invertebrate species, particularly spawners, to interact and fertilize. Hence, the period of capacitation, if it exists, may be extremely brief. Another is that relatively few studies have looked for this process in invertebrates. If it does occur as a general phenomenon in invertebrates, it is most likely to be found in groups in which the male transfers sperm directly to the female, i.e. in non-spawners. In the mating penaeoidean shrimp, *Sicyonia ingentis*, sperm is transferred from the male to the female and stored in a seminal receptacle, an organ not directly connected with the female reproductive tract. During storage sperm undergo a process that is required in order that they may obtain the capacity to undergo the acrosome reaction and fertilize (Wikramanayake *et al.*, 1992). Seminal fluid of the·sea urchin *Arbacia* reportedly prevents the rapid metabolic decline of sperm. If sperm are added to egg jelly in the presence of seminal fluid, respiration and viability are prolonged. It has been speculated that this may be a result of a surface modification such that the activity of the spermatozoon is altered (Shapiro and Eddy, 1980). This finding is similar to capacitation in mammals and indicates that this process may not be apparent in spawning animals due to the rapid dissociation of seminal components from the spermatozoon.

The site of capacitation in mammals appears to vary from one species to another, although successive regions of the female reproductive tract differ in their potential (Yanagimachi, 1981, 1994; Moore and Bedford, 1983). In species where semen is deposited in the uterus, capacitation occurs primarily in the oviduct. In species where sperm is deposited in the vagina, it usually starts there and continues in higher regions of the female reproductive tract. The oviduct is more effective than the uterus; however, the fact that capacitation usually occurs more rapidly when sperm are allowed to pass along the entire tract suggests a synergism. Crossfertilization experiments have demonstrated that the tract of one species is capable of capacitating sperm of a different species. Different types of *in vitro* fertilization procedure in humans indicate the flexibility of parameters affecting capacitation (Yanagimachi, 1994). Successful fertilization and pregnancy following direct intraperitoneal insemination or gamete intrafallopian transfer, where sperm are placed within the peritoneal cavity or both gametes are placed into the ampulla of the oviduct, respectively, indicate that human sperm do not need to move through specific portions of the female reproductive tract in order to become capacitated.

The sperm of mice, rats, guinea pigs, hamsters, rabbits, humans and a number of other mammalian species can also be capacitated in defined media (Yanagimachi, 1981, 1994), again verifying the fact that capacitation is possible without any contribution from the female reproductive tract. What is the role then of the female reproductive tract in capacitation? Since the mechanism of capacitation is generally believed to be a cell-surface phenomenon, the ability of some sperm to capacitate *in vitro* may mean that components of the sperm surface are eluted or are modified in artificial media. Enzymes in the female reproductive tract, blood cells and cumulus cells have been implicated as active participants in the modification of the sperm surface during capacitation in many species.

Since the lifespan of capacitated sperm is limited, it would seem expedient that completion of the process be synchronized temporarily and spatially with ovulation. The processes of capacitation may have evolved in response to the development of investments (cumulus cells, extracellular matrix and zona pellucida) that the mammalian sperm must penetrate (Fig. 2.4).

2.4.1 INTRACELLULAR AND SURFACE CHANGES

Generally, during capacitation major changes occur in intracellular ions, metabolism, adenylate cyclase and in organelles such as the nucleus,

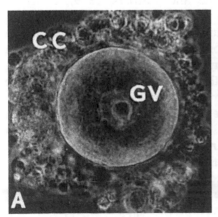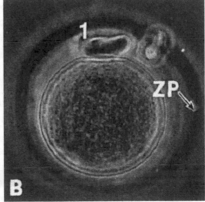

Fig. 2.4 (A) Mouse oocyte surrounded by cumulus cells (CC). Because of the density of cumulus cells the zona pellucida is not readily apparent. **(B)** Mature mouse ovum treated with hyaluronidase to disperse cumulus cells making the zona pellucida (ZP) apparent. GV = germinal vesicle; 1 = first polar body.

plasma membrane and acrosome (Clegg, 1983; Yanagimachi, 1994). Whether changes in Na^+ and K^+ occur during *in vivo* capacitation has not been clearly established. The status of intracellular free Ca^{2+} in capacitated sperm is controversial (see Yanagimachi, 1994). Much of the intracellular Ca^{2+} of sperm is stored or sequestered in the mitochondria and with calcium binding proteins. Whether or not and how these sites of Ca^{2+} accumulation function in events subsequent to capacitation have yet to be determined.

Calcium is present in the secretions of the male and female reproductive tracts and it has been shown that bovine sperm from the cauda epididymis are able to rapidly accumulate exogenous calcium (Babcock *et al.*, 1979). Calcium uptake by sperm is prevented or delayed through the action of a protein in seminal fluid that apparently acts upon the sperm plasma membrane. The loss of such a protein may allow capacitating sperm to accumulate calcium. Calcium-ATPase activity has been shown to be associated with membranes of sperm from a number of different species and membrane vesicles derived from sperm actively accumulate calcium (Gordon *et al.*, 1978; Bradley and Forrester, 1980). Perhaps this calcium transport system is modulated by seminal plasma proteins.

Calcium involvement in the activation of adenylate cyclase during capacitation has been reviewed (Clegg, 1983; see also Tash and Means, 1982). Adenylate cyclase activity reportedly increases under conditions that yield capacitated sperm. When guinea pig sperm are added to medium containing calcium, they undergo a 30-fold increase in cAMP within 30 s; when added to medium lacking calcium only a threefold increase in cAMP is observed (Hyne and Garbers, 1979). D-600, a calcium transport antagonist, blocks the calcium-dependent increase in cAMP. Sperm capacitated in calcium-free medium demonstrate an increase in cAMP 1 min after exposure to calcium-containing media and a maximum number of acrosome reactions within 10 min. Phosphodiesterase inhibitors decrease the time required to obtain the acrosome reaction in calcium-containing media. These results indicate that the mammalian sperm acrosome reaction is associated with both a primary transport of calcium and a calcium-dependent increase in cAMP. Because cAMP analogues do not induce an acrosome reaction in the absence of calcium, the increase in sperm cAMP, induced by additions of calcium, possibly reflects one of a number of calcium-dependent events associated with the acrosome reaction.

Extensive changes in sperm structure do not appear to occur during capacitation. Human sperm nuclei undergo a loss of Zn^{2+} and an increase in stability (Le Lannou *et al.*, 1985). A conversion of proacrosin to acrosin in the boar sperm acrosome by glycosaminoglycans in the female reproductive tract has been reported (Wincek *et al.*, 1979). Modifications in the sperm plasma membrane as a result of capacitation have been demonstrated by a variety of techniques. In fact, the removal or alteration of material associated with the surface of the sperm is believed to be an important part of capacitation (Yanagimachi, 1994). Capacitated sperm show a change in lectin binding in the acrosomal region and do not bind specific antibodies derived from uncapacitated sperm surface antigens (Koehler, 1978). Both the changes in antibody and lectin binding may be affected by a masking and/or removal of surface components. Since capacitation can be reversed by exposure to seminal plasma, it is possible that interactions of the seminal plasma and the sperm surface are key features of this process. The removal of 'coating' or decapacitation factors, either by high ionic strength media or by glycosidase digestion, suggests that absorbed components, such as glycoconjugates, on the sperm surface are released coincident with capacitation.

Surface galactosyltransferases on uncapacitated mouse sperm are preferentially loaded with poly-N-acetyllactosamine substrates (Shur and Hall, 1982a). With capacitation in calcium-containing medium these substrates are released from the sperm surface, thereby exposing galactosyltransferase for zona pellucida binding (Fig. 2.5).

Galactosylation of endogenous polylactosaminyl substrate is (1) reduced when sperm are incubated in calcium-free medium or treated with antiserum that reacts to galactosyltransferase substrate, and (2) increased by galactosylation of exogenous N-acetylglucosamine and binding to the zona pellucida. When added back to an *in vitro* fertilization assay, glycosides function as 'decapacitation factors', inhibiting sperm–egg binding by competing for sperm surface galactosyltransferase.

Investigations employing freeze-fracture replication have shown that areas of the sperm plasma membrane associated with the acrosome are cleared of intramembranous particles during capacitation (Friend, 1980). The binding of filipin, an agent that indicates the presence of β-hydroxysteroid, also decreases in the plasma membrane of the acrosomal region during this period. Both the changes in intramembranous particle distribution and sterol content of the sperm plasma membrane are believed to be in preparation for the acrosome reaction and render the sperm plasma membrane and the acrosome membrane fusigenic. The mechanisms for these membrane alterations have not been demonstrated. They may represent removal of components along the outer surface and/or within the membrane that restrict the mobility of intramembranous particles and, hence, change the fluidity of the membrane. Dramatic redistribution of surface antigen, resulting from a migration of molecules originally present on the posterior tail, occurs during capacitation of guinea pig sperm (Myles and Primakoff, 1984).

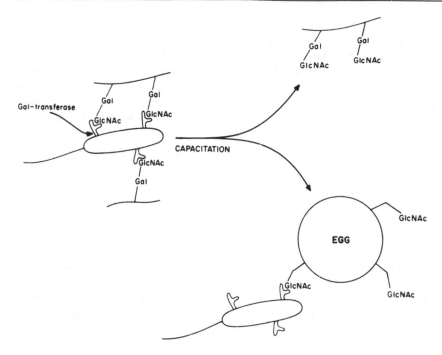

Fig. 2.5 Diagram illustrating the release of poly-N-acetyllactosamine glycosides from the sperm surface during capacitation. Glycoconjugate release could be facilitated by either dilution in the oviduct, increased ionic strength, glycosidase digestion or UDP-Gal-mediated catalysis of the galactosyltransferase reaction. Only the terminal disaccharides of the 'decapacitation factors' are illustrated. Reproduced with permission from Shur and Hall, 1982a.

This rearrangement of pre-existing surface molecules may act to regulate sperm functions during fertilization.

In addition to changes in membrane structure, capacitation in murine and bovine sperm results in hyperpolarization of membrane potential (V_m). Zeng *et al.* (1995) suggest that such V_m alterations regulate the activation state of sperm. How V_m alterations may define the molecular physiology of sperm capacitation has not been determined. Internal pH (pH$_i$) controls a wide range of cellular processes and provides a means of integrating diverse metabolic activities. In sperm alkaline shifts in pH$_i$ are associated with the initiation and modulation of flagellar motility, capacitation and the acrosome reaction (Yanagimachi, 1994). Capacitation of bovine and murine sperm are associated with elevation of pH$_i$ that are mediated by transmembrane acid/base transport mechanisms requiring functional Na$^+$-, Cl$^-$- and HCO$_3^-$-dependent acid-efflux pathways (see Zeng *et al.*, 1996).

Although many changes occur during the period of capacitation, which one or ones are essential components of this process remains an enigma. A major problem in establishing which of the many processes is critical lies in the fact that under the best of conditions not all sperm survive and capacitate. Additionally, there has been considerable confusion as to what actually constitutes this process during the 40+ years since its discovery. For example, the literature is replete with studies in which investigators claim that certain agents block sperm capacitation because *in vitro* fertilization did not occur in the presence of the agent. Although successful fertilization *in vitro* implies that the fertilizing sperm underwent capacitation, unsuccessful fertilization does not necessarily mean that the sperm failed to capacitate since there are a multitude of factors that influence the success of fertilization. Similar erroneous conclusions have been made based on the use of the acrosome reaction as a indicator of capacitation.

2.4.2 HYPERACTIVATION

Sperm of some mammalian species undergo frantic movement as a consequence of capacitation and before the acrosome reaction. For example, hamster sperm are weakly motile in the vas deferens. When suspended in capacitation medium they undergo a brief period of movement followed by a relatively quiescent period. After about 2 hours, the sperm become very active, move vigorously and form figure-8s by a whiplash-like beating of the flagellum (Fig. 2.6).

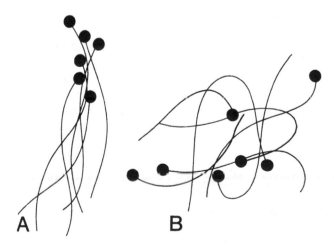

Fig. 2.6 Shapes of typical flagellar movements of epididymal **(A)** and hyperactivated **(B)** hamster spermatozoa. See Yanagimachi, 1981.

This vigorous activity, consisting of nonprogressive movements interspersed with episodes of linear movements, has been termed 'hyperactivation' (Yanagimachi, 1981). Accompanying such a change in activity is an increase in sperm metabolism, i.e. glycolysis and oxygen consumption. Whether hyperactivation is a general phenomenon, common to all mammals, and whether or not it is part of the activation process of nonmammalian sperm have not been determined. Nevertheless, activated motility is also seen in rabbit sperm and is characterized *in vitro* and *in vivo* by episodes of nonprogressive swimming with large-amplitude whiplash-like flagellar undulations, alternating with low-amplitude swimming. Although the physiological significance of this change in activity is unclear, the increase in activity may represent a means whereby sperm can exert strong thrusting movements during transit through the viscous fluids of the oviduct and layers surrounding the egg: the cumulus and the zona pellucida.

Alterations in the distribution of intramembranous particles occur within the plasma membrane of the sperm middle piece at the time of hyperactivation and may be related to this change in sperm motility (Koehler and Gaddum-Rosse, 1975; Friend *et al.*, 1977).

Epididymal and ejaculated hamster sperm treated with Triton X-100 and incubated in buffer containing ATP are able to exhibit hyperactivation, suggesting that the flagellum is prevented by some mechanism from affecting the activated type of movement. The mechanism of motility initiation in quiescent sperm is not clear. Hence, it should not be surprising that the mechanism of hyperactivation is also poorly understood. Extracellular calcium is not only involved in the initiation and maintenance of sperm motility, but is also required for the initiation and maintenance of hyperactivation. Yanagimachi (1994) has proposed a working model of hyperactivation in which adenylate cyclase, cAMP protein kinase and protein phosphorylation are key players (Fig. 2.7).

Upon completion of capacitation, presumably with the alteration of the sperm surface, receptors are activated, stimulate G-protein, which in turn activates (1) Ca^{2+} channels, allowing a transient Ca^{2+} influx, and (2) Na^+–H^+ channels, allowing a rise in intracellular pH. The Ca^{2+} that enters the sperm then stimulates adenylate cyclase to initiate a cAMP protein kinase cascade resulting in phosphorylation of axonemal proteins and ultimately sliding and bending of the axoneme.

2.5 ACROSOME REACTION

The eggs of many teleosts possess a small hole or micropyle in the thick covering surrounding the ovum, which provides a portal for sperm entry (Fig. 2.8).

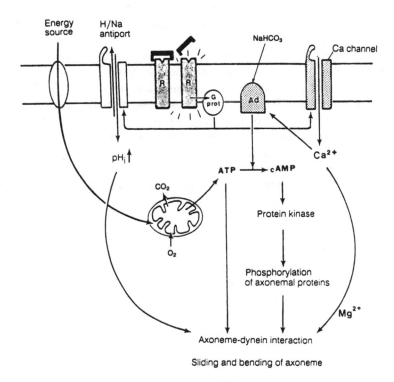

Fig. 2.7 Hypothetical model of sperm hyperactivation. Reproduced with permission from Yanagimachi, 1994.

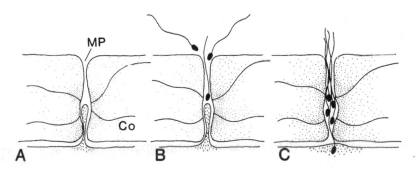

Fig. 2.8 Entry of sperm into a sturgeon egg. **(A)** The micropyle (MP) traverses the chorion (Co) and contains a projection of egg cytoplasm. **(B)** Sperm traversing the micropyle contact and fuse with the egg. **(C)** Incorporation of a spermatozoon. Adapted from Austin, 1965.

The presence of a channel through the layer(s) surrounding the teleost egg may be correlated with the absence of an acrosome in the sperm. However, this relation appears to be more complex, as some organisms, such as insects, possess eggs with micropyles and sperm with acrosomes. The need for an acrosome in these cases is not entirely clear. This organelle may be merely residual or it may facilitate the passage of the sperm through substances present in the micropyle.

Although the necessity of the acrosome reaction as an essential prerequisite for gamete fusion has been questioned by some investigators, extensive studies of fertilization in different species have shown that, in forms possessing acrosomes, only acrosome-reacted sperm fuse with the ovum. Among many invertebrates and vertebrates it is generally found that the acrosome undergoes a change when the spermatozoon comes into contact with the ovum (Dan, 1967; Colwin and Colwin, 1967). In mammals, during the acrosome reaction, the plasma membrane overlying the acrosome and the outer acrosomal membrane fuse at multiple sites and produce an array of vesicles (Barros et al., 1967). The vesicles that form have been shown to be mosaics consisting of membrane derived from both the plasma and the outer acrosomal membranes (Russell et al., 1979). At the level of the equatorial segment, the plasma membrane and the acrosomal membrane fuse to maintain a continuous membrane that delimits the contents of the spermatozoon (Fig. 2.9).

In the sperm of echinoderms (starfish and sea urchins; Usui and Takahashi, 1986) and the arthropod Limulus (Tilney, 1980), the acrosomal vesicle membrane and overlying plasma membrane lack intramembranous particles (IMP). Similar IMP-clear areas have been shown in the plasma membrane overlying the acrosome of guinea pig sperm, where the IMP-deficient patches are induced during capacitation (Yanagimachi and Suzuki, 1985). Although it has not been conclusively demonstrated, it is suspected that (1) the IMP-free areas represent protein deficient (lipid-rich) regions of the sperm plasma membrane and acrosome membrane that are susceptible for bilayer fusion and (2) the IMPs that surround the region of opposed plasma and acrosome membrane destabilize lipids within these membranes to facilitate their fusion.

As a result of the acrosomal reaction, vesicles derived from the fusion of the plasma and acrosomal membranes and the contents of the acrosome are released to the surrounding environment (Fig. 2.9). The release of acrosomal contents in this manner is akin to exocytosis in secretory cells. The substances that are released in this case (or carried on the sperm surface) are lysins, which dissolve a hole in the layer(s) surrounding the egg through which the sperm swims (Hoshi, 1985). In addition, the apex and much of the lateral aspect of the sperm head is

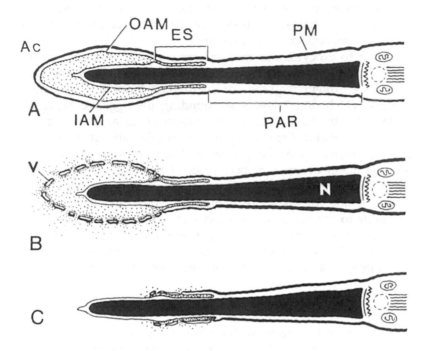

Fig. 2.9 Diagram of the acrosomal reaction in the hamster. **(A)** Before, **(B)** during the vesiculation of the outer acrosomal (OAM) and plasma membranes (PM) and release of acrosomal contents and **(C)** after the acrosomal reaction. Ac = acrosome; PAR = postacrosomal region; N ▬ nucleus; ES = equatorial segment; IAM = inner acrosomal membrane; V = vesicles formed as a result of multiple fusions between the plasmalemma and the outer acrosomal membrane. Lines depicting the plasma and the outer acrosomal membranes have different thicknesses to illustrate their fates subsequent to the acrosome reaction.

delimited by membrane derived from the inner acrosomal membrane. Hence, the acrosomal membrane becomes continuous with the sperm plasma membrane, and in invertebrates and nonmammalian vertebrates it is this portion of the delimiting membrane of the spermatozoon that fuses with the egg plasma membrane.

The equatorial segment of mammalian sperm has features that may prevent it from participating in the acrosome reaction (Moore and Bedford, 1983). The inner and outer acrosomal membranes in this region of the sperm head have an unusual pentalaminar structure while material located between the membranes is organized in a septate fashion. The plasma membrane associated with the rostral end of the acrosome is rich in anionic lipids. This is consistent with a high membrane fluidity in this area which would be conducive for membrane fusion. In contrast, membrane of the postacrosomal cap, which does not participate in fusion events of the acrosomal reaction, has a lower concentration of anionic lipids.

In addition to exocytosis of the acrosomal vesicle, other events occur as a part of the acrosome reaction. For example, in many species the exposure of a protein required for sperm–egg binding, the release of lytic agents which digest layers surrounding the egg, the rapid polymerization of actin leading to the formation of an acrosomal process, changes in the metabolism of cyclic nucleotide, and membrane and cytoplasmic protein phosphorylation and dephosphorylation are just some of the major events that take place at the time of or as a consequence of acrosomal vesicle exocytosis. Notable exceptions to this generality include the sperm of mammals, which undergo sperm–egg binding prior to the acrosome reaction – in fact, events associated with sperm–egg binding appear to be anticipatory to acrosomal vesicle exocytosis (see below). In addition, mammalian sperm do not form an acrosomal process as observed in many invertebrate species.

2.5.1 ACROSOMAL PROCESS

In many invertebrates, fusion of the plasma and acrosomal membranes is restricted to the apex of the acrosome and, consequently, few, if any, vesicles are formed as in the case of mammalian species (Dan, 1967; Figs 2.10–2.12).

Concomitantly, in species such as sea urchins, starfish and sea cucumbers, monomeric actin, which in the unreacted spermatozoon is

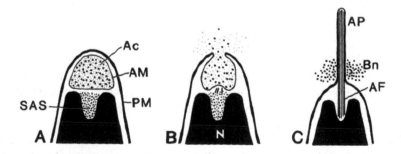

Fig. 2.10 Diagram of the acrosome reaction in the sea urchin *Arbacia*. **(A)** Intact acrosome (Ac). **(B)** Exocytosis of acrosomal contents via fusion between the plasma membrane (PM) and the acrosomal membrane (AM). **(C)** Formation of the acrosomal process (AP) containing polymerized actin (AF) and surrounded by bindin (Bn). SAS = subacrosomal space containing unpolymerized actin; N = nucleus. Lines depicting the plasma and the outer acrosomal membranes have different thicknesses to illustrate their fates subsequent to the acrosome reaction.

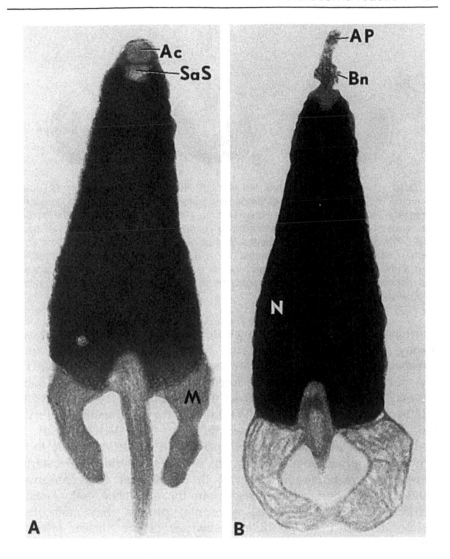

Fig. 2.11 Electron micrographs of *Arbacia* sperm that are intact **(A)** and have undergone the acrosome reaction **(B)**. Ac = acrosome; N = nucleus; M = mitochondria; AP = acrosomal process; Bn = bindin; SaS = subacrosomal space containing unpolymerized actin.

confined to the subacrosomal or periacrosomal region, polymerizes into a core of filaments and an acrosomal process is formed (Tilney, 1975a; Figs 2.10–2.12). Polymerization of actin filaments is polarized and in the sea cucumber, *Thyone*, the actomere located at the base of the

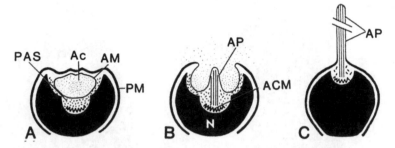

Fig. 2.12 Diagram of the acrosome reaction in the starfish *Asterias*. **(A)** Intact acrosome (Ac). **(B)** Exocytosis of acrosomal contents via fusion of the acrosomal membrane (AM) and the sperm plasmalemma (PM) and formation of the acrosomal process (AP) by the polymerization of actin within the periacrosomal space (PAS). The actomere (ACM) nucleates the polymerization of actin. **(C)** Acrosome reaction completed with the formation of a long acrosomal process containing actin filaments; N = nucleus. Lines depicting the plasma and the outer acrosomal membranes have different thicknesses to illustrate their fates subsequent to the acrosome reaction.

acrosomal process, acts as a nucleating site for actin polymerization (Tilney, 1978). Monomeres diffuse to the tip of the lengthening acrosomal process where they are assembled on elongating filaments. This is believed to provide the force for elongation of the acrosomal process (Tilney, 1985). The length of the acrosomal process can be quite variable: 11 µm in starfish, up to 90 µm in the sea cucumber (*Thyone*) and approximately 1 µm in sea urchins. The derivation of the membrane that accommodates the increase in surface area due to the formation of an acrosomal process has not been determined. In starfish sperm new membrane is believed to be derived from precursors situated in the pericentriolar region; in the horseshoe crab, *Limulus*, membrane for the elongating acrosomal process may come from vesicles derived from the outer membrane of the nuclear envelope (Tilney, 1985).

In other invertebrates such as the pelecypods *Mytilus* and *Spisula*, the acrosome surrounds a process containing actin filaments that is borne with the acrosome reaction (Fig. 2.13).

In these cases, there is no massive polymerization of actin to form an apical projection. In *Limulus* the acrosomal process is a 60 µm long actin bundle comprising about 100 hexagonally packed actin filaments wound into a supercoil at the base of the nucleus (Bullitt *et al.*, 1988). Upon stimulation, the supercoil is released and the bundle rapidly uncoils, extends through the nucleus and out of the anterior end of the sperm (Tilney, 1975b; Bullitt *et al.*, 1988).

The site of acrosome reaction in invertebrates is variable and in some

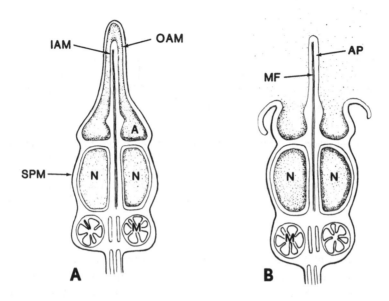

Fig. 2.13 Acrosome reaction in the mollusk *Mytilus*. **(A)** Sperm with intact acrosome (A). **(B)** Exocytosis of acrosomal contents and exposure of the acrosomal process (AP). Unlike *Arbacia* and *Asterias* sperm, in which the acrosomal process is formed via the polymerization of actin, in *Mytilus* actin filaments (MF) are present in the unreacted spermatozoon. IAM and OAM = inner and outer acrosomal membranes; SPM = sperm plasma membrane; N = nuclei; M = mitochondria.

cases controversial (Dale and Monroy, 1981). In starfish and the horseshoe crab the acrosome reaction occurs at the outer margin of a well-defined jelly layer; a long acrosomal process is formed that contacts and fuses with the egg plasma membrane (Fig. 2.14).

Penetration of the jelly and vitelline layers in these instances is a result of mechanical forces generated by filament extension and lytic products formerly within the acrosome. In sea urchins the acrosome reaction normally takes place when the sperm contacts the vitelline layer (Vacquier, 1979; Aketa and Ohta, 1977). It is important to note in this case that the tip of the acrosomal process is the sperm structure that first makes contact with the surface of the egg and is involved in gamete membrane fusion, i.e. fusion of the egg plasma membrane and membrane delimiting the spermatozoon. As will be described below, this situation differs from that of mammalian sperm, in which contact and fusion with the egg plasma membrane are localized to a different region of the sperm surface.

When the acrosome reaction in the normal course of mammalian fertilization occurs was at one time a controversial issue (see Moore

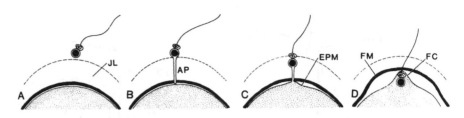

Fig. 2.14 Interaction of starfish (*Asterias*) sperm and egg. **(A)** Approach of the spermatozoon. **(B)** Acrosome reaction induced by contact with the jelly layer (JL). An acrosomal process (AP) is formed that projects to the surface of the egg. **(C)** Fusion of the acrosomal process and the egg plasma membrane (EPM). **(D)** Incorporation of the sperm nucleus within a fertilization cone (FC). FM = fertilization membrane.

and Bedford, 1983). It has been demonstrated that sperm do not initiate a true acrosome reaction *in vivo* until they come in contact with the zona pellucida. Mammalian sperm, however, are able to undergo an acrosome reaction in the absence of a zona pellucida. Such an event is referred to as a spontaneous acrosome reaction and is usually looked upon as being an unphysiological event (Yanagimachi, 1994). The acrosomes of moribund sperm may undergo changes that appear to resemble a true acrosome reaction. Such changes in dead sperm are referred to as false acrosome reactions.

2.5.2 REGULATION OF THE ACROSOME REACTION: INVERTEBRATES

Egg water, i.e. aqueous solutions in which eggs are incubated to allow their jelly layers to dissolve, and alkaline pH initiate the acrosome reaction of sperm from many marine invertebrates (Dan, 1967). In the starfish *Asterias amurensis* three components have been shown to be involved in the induction of the acrosome reaction (Hoshi *et al.*, 1990):

1. a very large, sulfated glycoprotein containing a novel saccharide structure and referred to as ARIS (acrosome-reaction-including substance);
2. sulfated steroid saponins, Co-ARIS;
3. sperm activating peptides.

Experimentally, ARIS is capable of inducing the acrosome reaction in alkaline or high calcium sea water; however, both ARIS and Co-ARIS are required in normal sea water. Although not obligatory for induction of the acrosome reaction, sperm-activating peptides facilitate the reaction when induced by ARIS and Co-ARIS.

In sea urchins a fucose-sulfated glycoconjugate ($\approx 10^6$ kDa) of the egg jelly coat is responsible for initiating the acrosomal reaction (SeGall and Lennarz, 1979; Kopf and Garbers, 1980). Although a sperm surface glycoprotein of 210 kDa is believed to be a receptor for jelly-coat-derived ligands, it has not been established that the fucose-sulfated glycoconjugate binds this glycoprotein (see Kopf and Ward, 1993). The response of the sperm to fucose-sulfated glycoconjugate includes Na^+ and Ca^{2+} uptake and efflux of H^+ and K^+ (Ward and Kopf, 1993). Calcium uptake occurs via two mechanisms at the time of the acrosome reaction (Schackmann et al., 1978; Schackmann and Shapiro, 1981). The first takes place very early, is sensitive to D-600 (an agent that blocks calcium channels) and is associated with a change (X) that allows the acrosome reaction to proceed if other requirements are met (Fig. 2.15).

The second is sensitive to respiratory inhibitors and is insensitive to D-600, suggesting that it is mitochondria-derived. The latter follows extension of the acrosomal process and appears to have no role in the acrosomal reaction, although it accounts for much of the calcium that is taken up by reacted sperm. In addition to the Na^+-dependent increase in internal pH (0.4 units) due to the activation of Na^+–H^+ exchange that occurs when sperm are diluted in sea water, there is a further alkalization of sperm when they are exposed to egg jelly. The additional increase in internal pH is dependent on external Na^+, induces the polymerization of actin and is a prerequisite for the acrosome reaction.

There appears to be a species specificity with respect to the ability of the fucose-sulfated glycoconjugate to stimulate the acrosome reaction. Notably, calcium uptake in sperm does not occur when the fucose sulfate component fails to induce an acrosome reaction. Mixing reactive and unreactive jelly layer components does not inhibit the uptake of calcium in affected sperm and is consistent with a receptor type of interaction. These observations correlate with those of Summers and Hylander (1975) who showed that heterologous crosses between different sea urchin species induce the acrosomal reaction, although there is no sperm–egg binding or fusion. These results indicate that species specificity of fertilization is not limited to the induction of the

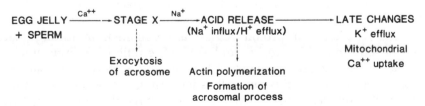

Fig. 2.15 Ionic changes associated with the acrosome reaction. See Schackmann and Shapiro, 1981.

acrosome reaction exclusively, but also resides with events that normally follow the acrosome reaction.

The fucose-sulfated glycoconjugate of sea urchin egg jelly induces a calcium-dependent activation of adenylate cyclase, and an increase in cAMP which results in the activation of protein kinase A activity (Garbers and Hardman, 1976). The acrosome reaction and increase in cAMP are tightly correlated and temporally consistent with an involvement of cyclic nucleotides in this modification of the acrosome. The initial step in the acrosome reaction appears to involve an increase in calcium permeability, suggesting that calcium–calmodulin may mediate this reaction (Garbers and Kopf, 1980). Although increases in cyclic AMP precede the acrosome reaction, elevation of this second messenger is not the sole mediator of the acrosomal reaction since cyclic AMP and its membrane-permeable analogs do not induce this event (Garbers and Kopf, 1980).

Sea urchin sperm do not undergo an acrosome reaction at a pH below 7.5 and at pH values greater than 8.8 the reaction is triggered spontaneously in the absence of egg jelly. These observations are consistent with experiments demonstrating a proton efflux during the acrosome reaction (Schackmann and Shapiro, 1981; Fig. 2.15). Unlike the acid release which is related to the burst in respiration when sperm are mixed with sea water, this increase in internal pH is not sensitive to respiratory inhibitors.

Although calcium is required for the exocytosis of the acrosome, it does not appear to be involved in the formation of the acrosomal process. Using the ionophores A-23187 and X-536SA, Tilney et al. (1978) induced actin polymerization and the formation of the acrosomal process in *Thyone* (sea cucumber) and found an acid release in isotonic sodium chloride or potassium chloride at pH 8. At pH 6.5 there is no acid release and no polymerization of actin to form an acrosomal process. The correlation between acid release and actin polymerization suggests that the increase in intracellular pH permits actin to be released from a 'bound' form (i.e. a complex consisting of profilin and nonfilamentous actin) and to assemble into filaments. Coincident with the acrosome reaction and the proton efflux, sea urchin sperm incubated with egg jelly accumulate sodium (Schackmann et al., 1978; Schackmann and Shapiro, 1981). When sodium uptake is prevented, the acrosome reaction and proton release are inhibited, indicating that fluxes of sodium and hydrogen are linked. (The ratio of sodium uptake to proton release is approximately 1:1, indicating a coordinate change.) Inhibitors of potassium flux, as well as increasing potassium concentrations, inhibit the acrosome reaction in sea urchins, demonstrating that the acrosome reaction is also accompanied by an efflux of potassium (Lopo, 1983). Measurements by Darszon et al. (1988) demonstrate that egg jelly induces a fast, transient hyperpolarization of sperm and the acrosome reaction. The hyperpolarization may be mediated by the opening of K^+ channels and is believed to be necessary to activate the Na^+-H^+ exchanger and to produce the alkalization necessary for the acrosome reaction.

Coincident with the acrosome reaction in starfish sperm are significant alterations in nuclear structure, including the disappearance of the anterior fossa containing the acrosome vesicle and periacrosomal material, as well as other nuclear regions, primarily the posterior pole (Dan, 1967; Usui and Takahashi, 1986; Longo et al., 1995). Amano et al. (1992) observed the initiation of degradative changes in nuclear histones coincident with the acrosome reaction in starfish sperm. Modifications in nuclear histones involving phosphorylation of sperm histones have been described for sea urchin sperm treated with egg jelly to induce acrosome reaction (Vacquier et al., 1989; Porter and Vacquier, 1986). These structural and biochemical changes induced by egg jelly prior to sperm–egg fusion suggest that the sperm nucleus is activated prior to its entrance into the egg cytoplasm, before any detectable morphological transformation into a male pronucleus.

2.5.3 REGULATION OF THE ACROSOME REACTION: MAMMALS

Aspects dealing with the zona pellucida as an inducer of the acrosome reaction are considered here; the role of the zona in sperm–egg binding is discussed in Chapter 3. As described above, there has been considerable controversy over the site and inducer(s) of the physiological relevant acrosome reaction in mammals and attention has been given to substances and sites within the female reproductive tract, the ovarian follicle and the egg itself. Consideration of this myriad of details is beyond the scope of this book; for more specific information the reader is referred to Yanagimachi (1994). The generally accepted view, however, is that after traversing the cumulus, the spermatozoon binds to the zona pellucida and is then induced to undergo the acrosome reaction (Florman and Storey, 1982; Storey et al., 1984). An alternative view states that progesterone trapped in or produced by cumulus cells can also initiate the acrosome reaction (Osman et al., 1989). Recent experiments by Roldan et al. (1994) demonstrate a primary role for progesterone in this reaction (see also Shi and Roldan, 1995).

The zona pellucida is an acellular coat that surrounds the eggs of mammals and is considered to be part of the extracellular matrix through which the sperm must pass in order to fertilize the egg (Ward and Kopf, 1993). It is a product of the egg (Wassarman, 1988) and mediates a number of reactions in addition to acrosomal exocytosis: sperm–egg recognition, binding of acrosome-intact sperm to the egg, sperm activation and an egg-induced block to polyspermy (Florman *et al.*, 1984). In the mammalian species studied thus far, the zona pellucida consists of two to four glycoproteins which demonstrate considerable charge heterogeneity due to the degree of glycosylation and/or sulfation of individual polypeptides. In the mouse, the zona pellucida is composed of three major sulfated glycoproteins referred to as ZP1, ZP2, and ZP3 (Wassarman, 1988; Fig. 2.16).

ZP1 (200 kDa) is a dimer connected by intermolecular disulfate bonds and maintains the three dimensional structure of the zona pellucida by crosslinking filaments composed of repeating heterodimers of ZP2/ZP3 (Fig. 2.17).

ZP2 (120 kDa) is believed to mediate the binding of the acrosome-reacted sperm to the zona pellucida (Mortillo and Wassarman, 1991). ZP3 (83 kDa) has two functions: (1) sperm binding and (2) induction of the acrosome reaction (Bleil and Wassarman, 1983). The O-linked carbohydrate of ZP3 and not the polypeptide itself appears to be responsible for the sperm-binding activity of ZP3 (Florman and Wassarman, 1985; Bleil and Wassarman, 1988). In contrast, the acrosomal-reaction-inducing activity of ZP3 is believed to be confined by both the carbohydrate and protein portions of the molecule (Wassarman *et al.*, 1985). Following fertilization and as a consequence of proteases released from the egg as a result of the cortical granule reaction, ZP3 is modified to ZP3f and, concomitantly, there is a loss of both of its biological activ-

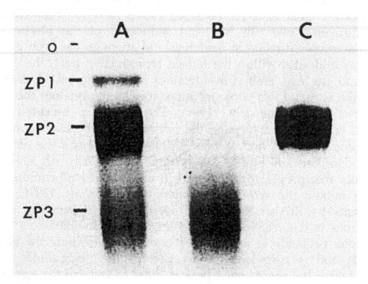

Fig. 2.16 SDS-PAGE analysis of purified mouse zona pellucida glycoproteins. **(A)** The three zona glycoproteins (ZP1, ZP2 and ZP3). **(B)** Purified ZP3. **(C)** Purified ZP2. Reproduced with permission from Bleil and Wassarman, 1986.

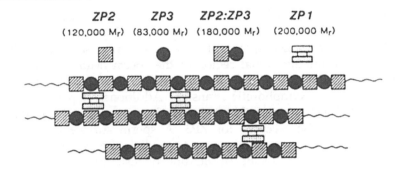

Fig. 2.17 Representation of the arrangement of glycoproteins ZP1, ZP2 and ZP3 in the mouse zona pellucida. Reproduced with permission from Wassarman, 1988.

ities, i.e. sperm binding and induction of acrosome reaction (Bleil and Wassarman, 1983).

In addition to mouse eggs, the zonae pellucidae of numerous other mammals have been characterized biochemically, including man (Chamberlin and Dean, 1990), and chemical homologs to ZP3 have been identified (see Ward and Kopf, 1993; Wassarman, 1993). To what extent similar biological activities are associated with chemical homologs of mouse ZP1, ZP2 and ZP3 glycoproteins in other mammals has not been clearly established.

With respect to the induction of the acrosome reaction in mammals, the zona pellucida component, ZP3, is viewed as the ligand (Bleil and Wassarman, 1980a, b; Florman and Wassarman, 1985) which binds to a sperm plasma membrane receptor, initiating a cascade of signaling events culminating in exocytosis of the acrosome. The ligand-bound receptor is believed to activate effectors such as heterotrimeric G protein, tyrosine kinase and/or phosphatases; these components activate other second messengers which carry out the sperm's response to zona binding. Although there is considerable evidence that ZP3 is the ligand involved in acrosomal exocytosis, the sperm receptor and subsequent signaling mechanisms are not well understood. A number of sperm surface molecules have been implicated in ZP3-induced acrosomal exocytosis. The egg binding protein (putative receptor) in mouse sperm that recognizes ZP3 is associated exclusively with the plasma membrane overlying the sperm head and is present in 10^4–10^5 copies (Mortillo and Wassarman, 1991).

To identify the putative receptor for ZP3 a variety of approaches have been tried, each yielding a different candidate molecule (Ward and Kopf, 1993). Crosslinking experiments by Bleil and Wassarman (1990) have identified a 56 kDa protein of acrosome-intact sperm.

Immunofluorescence studies (Leyton and Saling, 1989a), employing an antiphosphotyrosine antibody that recognizes polypeptides of 52, 75 and 95 kDa from mouse sperm plasma membranes, show a positive reactivity along the acrosomal region of the sperm head. Interestingly, labeled ZP3 binds to a 95 kDa sperm protein (p95) in nitrocellulose blots following electrophoretic transfer. Additional experiments by Leyton *et al.* (1992) suggest that p95, which possesses tyrosine kinase activity, may be the receptor for ZP3. A sperm surface β-galactosyl transferase activity has been proposed to mediate the interaction of the sperm and the zona pellucida by binding oligosaccharide residues on ZP3 (Miller and Shur, 1994). The membrane-bound sperm surface galactosyl transferase (GalTase) can galactosylate ZP3 but not ZP1 and ZP2 (Miller *et al.*, 1992). In contrast, milk-soluble GalTase is capable of galactosylation of all three zona glycoproteins, indicating that the lack of galactosylation of ZP1 and ZP2 by sperm GalTase is not due to an absence of terminal N-acetylglucosamine, and suggesting a specificity of the sperm-surface GalTase. Other sperm-surface molecules (e.g. PH-20 and mannosidase) have been identified that may be involved in adhesion to the zona pellucida (see Ward and Kopf, 1993) and the list of candidates may be confusing as there is no overwhelming evidence to support one over the other. Further efforts here will, no doubt, help to clarify this issue. Additionally, it should be noted that with the different approaches taken to identify putative receptors, it is possible that the relationship between the zona pellucida and a specific molecule on the sperm surface may be much more elaborate than first envisioned. That is, ZP3 may aggregate receptors on the sperm surface (Leyton and Saling, 1989b) and interact with a number of proteins that form a functional ZP3 receptor (Ward and Kopf, 1993). Of the molecules thus far identified as a putative receptor for ZP3 only one, p95, has been shown to contain a signaling domain that becomes activated upon binding ZP3.

Possible mechanisms involving a receptor ligand type of interaction leading to the acrosome reaction have been proposed by a number of different investigators (see Ward and Kopf, 1993; Yanagimachi, 1994). One model, which is one of the more comprehensive, hypothetical pathways for zona-induced acrosome reaction, is shown in Fig. 2.18.

It proposes that a receptor (Rzp), activated by ZP3, stimulates G protein (Endo *et al.*, 1987, 1988), which in turn acts on phospholipase C (PLC) in the sperm plasma membrane (Kopf and Gerton, 1991; Ward and Kopf, 1993; Yanagimachi, 1994). Phospholipase C cleaves phosphatidyl inositol diphosphate (PIP_2) into diacylglycerol (DAG) and inositol triphosphate (IP_3). Inositol trisphosphate increases internal Ca^{2+} by releasing Ca^{2+} from intracellular stores. Diacylglycerol activates Ca^{2+}-dependent protein kinase C (PK-C), which phosphorylates proteins. A

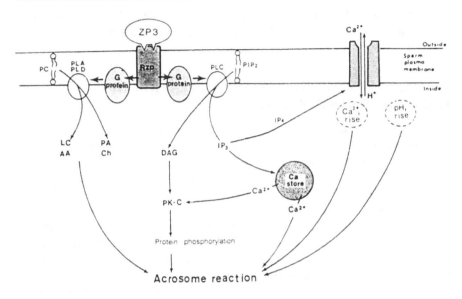

Fig. 2.18 Hypothetical pathways in the zona-induced acrosome reaction. See text for explanation. Reproduced with permission from Yanagimachi, 1994.

portion of the inositol trisphosphate pool is converted into inositol tetraphosphate (IP$_4$), which effects the opening of voltage-dependent Ca^{2+} channels, thereby promoting a massive influx of extracellular Ca^{2+} (Florman *et al.*, 1992; Florman, 1994). It is also postulated that some of this Ca^{2+}, as well as the formation of phosphatidyl choline (PC), arachidonic acid (AA) and phosphatidic acid (PA), which are derived from the activities of phospholipases A$_2$ (PLA) and D$_1$ (PLD)$_1$ respectively, act on membrane phospholipids to facilitate membrane fusion. G-proteins also activate adenylate cyclase which stimulates cAMP production. Some of the cAMP produced is believed to promote a Na$^+$ influx and a H$^+$ efflux (via a Na$^+$–H$^+$ antiporter) and brings about a rise in intracellular pH. Whether or not and how other agents that induce sperm to undergo the acrosome reaction, e.g. progesterone, interact in this model has not been determined. Furthermore, how the two signaling pathways, phosphotyrosine kinase and G protein, may be linked to induce acrosome exocytosis has yet to be determined (Ward and Kopf, 1993).

Recent investigations have localized a number of elements of the phosphoinositide system in mammalian sperm (Glassner *et al.*, 1991; Walensky and Snyder, 1995). G protein, phospholipase C and IP$_3$ receptor have been identified along the acrosomal region. The presence of these components, as well as the ability of thapsigargin (a sequiterpene lactone inhibitor of microsomal Ca^{2+}-ATPase) to induce the acrosome reaction, implicate IP_3-gated calcium release in the acrosome reaction.

2.6 SPERM LYSINS AND SPERM PENETRATION OF PERIOVULAR COMPONENTS

The acrosomes from sperm of a number of different animals have been shown to contain lysins which are released upon the acrosome reaction. Because of the presence of lysins within the acrosomal vesicle, this organelle has been looked upon as a lysosome or a zymogen granule. One function of sperm lysins is to produce a hole in the investments surrounding the egg, such as the jelly coat, vitelline layer and zona pellucida, through which the sperm is able to pass and gain access to the egg surface for gamete membrane fusion. Sperm-derived lysins have also been shown to be involved in sperm–egg binding in invertebrates (Hoshi *et al.*, 1994) and in mammals (see below).

According to Hoshi (1985), the following criteria have to be met in order that a sperm-borne agent be considered a lysin.

1. It is available in its active form at the time of lysis.
2. It is able to digest, disperse or affect the egg investment in question by itself or in combination with another agent.
3. If its activity is inhibited, sperm fail to fertilize.

The dissolution of egg investments by lysin is either enzymatic or stoichiometric (nonenzymatic). The sperm of many mollusks have been shown to contain lysins whose mechanism of action is stoichiometric. Experiments by Vacquier and co-workers (Vacquier and Lee, 1993; Shaw *et al.*, 1994) have demonstrated that sperm of the abalone contain proteins that cause the tightly intertwined 13 nm filaments comprising the vitelline envelope of the egg to lose cohesion to one another so that they 'unravel' and splay apart. As a consequence of this action, a 3 μm diameter hole is made through which the sperm can swim. How the abalone sperm lysin affects this change in the vitelline layer has not been clearly demonstrated.

In the sperm of many animals the acrosomes have been shown to contain a variety of hydrolytic enzymes (Hoshi, 1985). The acrosomal enzymes of mammalian sperm have been intensively studied and Table 2.1 lists some that are present in this group.

Hyaluronidase is localized within the mammalian sperm acrosome and is released during the acrosome reaction (Morton, 1975; Yanagimachi, 1994). Penetration of acrosome-intact sperm through the cumulus requires vigorous motility and is believed to be facilitated by hydrolysis of the cumulus extracellular matrix by sperm plasma membrane hyalur-

Table 2.1 Enzymes from mammalian sperm reported to be of acrosomal origin (Reproduced with permission from Yanagimachi, 1994)

First reported before 1980	First reported after 1980
Hyaluronidase	N-acetylexosaminidase
Acrosin	Galactosidase
Proacrosin	Glucuronidase
Acid proteinase	L-fucosidase
Esterase	Phospholipase C
Neuraminidase	Cathepsin D
Phosphatase	Cathepsin L
Phospholipase A	Ornithin decarboxylase
N-acetylglucosaminidase	Calpain II
Arylsulfatase	Metalloendoprotease
Arylamidase	Caproyl esterase
Collagenase	Peptidyl peptidase

onidase activity (Perreault *et al.*, 1979; Yanagimachi, 1994; Thaler and Cardullo, 1995). Recent evidence indicates that PH-20, a bifunctional protein that is involved in binding of acrosome reacted sperm to the zona (see Chapter 3; Ramarao *et al.*, 1994) and cumulus penetration via its hyaluronidase activity, may exist in two forms (Cherr *et al.*, 1996). The major soluble hyaluronidase is a molecule of 53 kDa which is acid-active. It may be generated as a result of enzymatic activities which could alter the neutral-active, 64 kDa membrane-bound form of PH-20.

Localization of acrosin to the acrosome by histochemical procedures includes the use of proteinate, fluorescently-labeled anti-acrosin antibody and horseradish-peroxidase-labeled antibody to acrosin (Stambaugh *et al.*, 1975; Garner *et al.*, 1975; Morton, 1975). The proteolytic activity of mammalian acrosin has been demonstrated by sperm smears on gelatin films (Gaddum-Rosse and Blandau, 1972). Acrosin is present in the mammalian sperm acrosome as a zymogen (proacrosin) and is autocatalytically activated (Meizel and Mukerji, 1975). Both proacrosin and acrosin have been implicated in sperm–zona binding (Yanagimachi, 1994). However, the absolute requirement for acrosin in fertilization processes is in question as a result of gene knockout experiments in which homozygous null mice for the acrosin gene were shown to be fertile (Baba *et al.*, 1994).

Lytic enzymes having trypsin and chymotrypsin-like properties have also been demonstrated in invertebrate sperm (Green and Summers, 1980; Hoshi, 1985). Penetration of the vitelline layer is a requirement in order for sea urchin sperm to fuse with the egg plasma membrane. This may occur mechanically by the action of the sperm flagellum, with the acrosomal process acting as a spear. In addition, portions of the vitelline layer may be digested by lysin present in the acrosome and externalized by the acrosome reaction. Fertilization decreases in the presence of the protease inhibitor tosyl-phenylalanine-chloromethyl-ketone (TPCK). If the vitelline layer is removed, TPCK is not inhibitory, suggesting that normally the vitelline layer is digested by acrosome-derived enzymes. Further evidence that hydrolases participate in the penetration of the vitelline layer comes from investigations with other invertebrates. Ascidian ova are enveloped by a complex of follicle cells, chorion and test cells that presumably presents a formidable barrier to successful gamete fusion. The entire surface of the *Ciona* chorion is apparently available for sperm binding and, despite lacking a typical acrosome, ascidian sperm are able to penetrate these barriers (DeSantis *et al.*, 1980). Trypsin and chymotrypsin inhibitors block the fertilization of ascidian eggs (Hoshi *et al.*, 1981). However, if the chorion is removed, the egg can be fertilized in the presence of inhibitors.

Penetration of the zona pellucida by the acrosome-reacted sperm requires 4 min in hamsters and about 20 min in mice (Yanagimachi, 1981). The movement of the sperm through the zona pellucida appears to be facilitated by a bobbing movement of the head. In this case the perforatorium, a small projection at the apex of the sperm head, may help to cut a slit in the zona. Lytic agents from the acrosome are also believed to participate in penetration, primarily because of their presence in the acrosome and observations demonstrating that trypsin inhibitors block fertilization. Despite evidence that proteinase inhibitors block fertilization there are doubts as to the specificity of this reaction as relatively high concentrations of inhibitors are required. Furthermore, it is uncertain whether the acrosome reaction can take place in the presence of the inhibitors. Hence, failure of inhibitor-treated sperm to penetrate the zona pellucida and fertilize may be due to a blockage of the acrosome reaction. Whether or not acrosomal enzymes play a role in zona penetration has not been clearly established (see Yanagimachi, 1994). If they are relevant, evidence is required demonstrating that they are present at the time of and at sufficient concentrations to achieve hydrolysis and penetration.

Sperm–egg binding   3

From a historical perspective, numerous experiments employing sea urchins have suggested the presence of sperm receptors in the egg vitelline layer. Sea urchin eggs treated with proteolytic enzymes, such as trypsin, show a decrease in fertility due to a failure in sperm–egg binding (Schmell *et al.*, 1977). Unfertilized eggs incubated in cortical granule protease also undergo a decrease in their ability to bind sperm, presumably from an alteration in receptors (Vacquier *et al.*, 1973; Carroll and Epel, 1975). Large glycoproteins isolated from vitelline layers of sea urchin eggs show a species-specific inhibition of fertilization (Aketa, 1973; Schmell *et al.*, 1977; Glabe and Vacquier, 1978; Glabe and Lennarz, 1981). Protein isolated from the surface of sea urchin (*Hemicentrotus*) eggs binds homologous sperm and antibody directed to it blocks fertilization (Aketa and Tsuzuki, 1968; Aketa and Onitake, 1969). Membrane protein preparations from one species of sea urchin (*Arbacia*) eggs inhibit fertilization by sperm of the same species but have no effect on the ability of sperm from a different species (*Strongylocentrotus*) to fertilize homologous eggs. In the presence of soluble components derived from the egg, acrosome-reacted sperm exhibit a species-specific binding to each other, suggesting that specific receptors are present on the ovum surface that facilitate gamete binding (Kinsey *et al.*, 1980). Sperm bind only to the external surface of isolated vitelline layers and also exhibit saturation kinetics consistent with the notion of a sperm receptor on the ovum surface (Vacquier and Payne, 1973).

3.1 BINDIN

The acrosomes of sea urchin sperm contain a 30 500 molecular weight protein called bindin (Vacquier and Moy, 1977; Vacquier, 1980; Lopez *et al.*, 1993; Hofmann and Glabe, 1994). A similar acrosomal component has also been identified in sperm from other invertebrates, e.g. the marine worm *Urechis* (Gould *et al.*, 1986). Bindin from acrosome-reacted sea urchin sperm coats the acrosomal process and has a species-preferential affinity for the vitelline layer of unfertilized eggs. How bindin is organized on the apex of the fertilizing sperm and how it is presented to the egg have not been determined. Evidence that bindin does, in fact, mediate gamete attachment comes from investigations indicating that (1) antibodies specific to this protein localize to the site of sperm–egg attachment (Moy and Vacquier, 1979) and (2) isolated bindin agglutinates unfertilized eggs; the protein is found in the region of contact between eggs. This agglutination is blocked by a glycopeptide fraction from the egg surface, suggesting that bindin interacts specifically with molecules (receptors) within the vitelline layer. Bindin has been shown to bind to sulfated fucan polysaccharides, components of the jelly layer and vitelline layer of sea urchin eggs (Hofmann and Glabe, 1994).

Bindin exhibits membrane fusogenic properties (Glabe, 1985a, b), i.e. it has a hydrophobic segment that displays similarities to viral fusion proteins (Glabe and Clark, 1991) and hence may play a role in gamete membrane fusion (Hofmann and Glabe, 1994). The amino acid sequence of bindin has been determined and recent efforts have been devoted to an analysis of divergent amino- and carboxyl-terminal flanking regions where species-specific egg adhesion is believed to exist (see Hofmann and Glabe, 1994). Interestingly, although application of bindin to eggs induces their adhesion to one another, this molecule does not trigger activation (Glabe and Vacquier, 1978). However, issues of the 'presentation' of bindin to the egg have complicated the interpretation of this negative result (Foltz, 1994). As will be discussed later, an acrosome-derived peptide from sperm of *Urechis* stimulates *Urechis* eggs to activate (Gould and Stephano, 1989, 1991).

Generation of proteolytic fragments of the extracellular domain of a putative sperm receptor has been achieved with sea urchin eggs (see Foltz and Lennarz, 1993; Ohlendieck *et al.*, 1993). The fragment is a glycoprotein of approximately 70 kDa that binds sperm and inhibits their ability to fertilize. It has also been demonstrated that the 70 kDa fragment interacts with purified bindin particles in a species-preferential fashion (Foltz and Lennarz, 1990; Foltz, 1994) and hence represents a putative bindin receptor. Antibodies raised to the 70 kDa fragment recognize a single protein of approximately 350 kDa on immunoblots of egg surface preparations (Foltz and Lennarz, 1992) and have been used to screen an expression library made from immature sea urchin ovary (Foltz *et al.*, 1993). This egg receptor to sperm has been shown to be a novel cell recognition molecule with at least one transmembrane domain and sites for N- and O-glycosylation (Fig. 3.1).

Analysis of its primary sequence revealed that its external domain

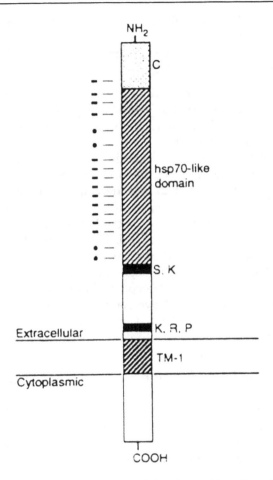

Fig. 3.1 Diagram of the sea urchin sperm receptor oriented in the egg plasma membrane. Only a single transmembrane domain (TM-1) is shown. The NH$_2$-terminal, Cys-rich domain (C), the hsp70-like domain and the two charged sequences – the domain rich in Ser and Lys (S and K) and the domain rich in Pro, Arg and Lys (P, R and K) are denoted. Putative N-linked (black circles) and O-linked (black bars) glycosylation sites are indicated. Reproduced with permission from Foltz *et al.*, 1993.

shares similarity to hsp70 proteins (Foltz *et al.*, 1993; Foltz, 1994). The extracellular domain exhibits a species-specific primary structure while the cytoplasmic domain is conserved among all of the echinoderms examined thus far; it has no homology with known proteins and contains five conserved tyrosine residues.

A number of functions have been postulated for the sperm receptor (Foltz and Lennarz, 1993; Foltz, 1994):

1. cell adhesion, thereby facilitating sperm–egg membrane contact and possibly fusion;
2. signal transduction from the egg surface upon sperm binding;
3. organization of the cytoskeleton.

Supporting evidence for each of these postulated functions exists (Foltz and Lennarz, 1993), but conclusive proof that this novel cell recognition molecule is actually involved in one or more of these processes awaits further testing.

A number of basic questions regarding sperm receptor structure and function have been raised (Foltz, 1994). Based on sequence analysis, the sperm receptor is determined to be a transmembrane protein. How the protein is situated in the plasma membrane relative to components of the vitelline layer, i.e. in a manner such that its sperm binding portion is accessible, has not been determined. It is claimed that a majority of the sperm receptors are located at microvilli. Such an organization may have an effect on sites of sperm–egg membrane fusion. Additionally, the question of how the term 'receptor' should be applied in cases of sperm–egg interaction is a controversial issue. According to Foltz (1994) receptors may be loosely defined as membrane glycoproteins that recognize molecules that come into contact with the cell surface. Three types have been distinguished:

1. transmembrane receptors that bind a ligand and trigger a biochemical response within the cell;
2. molecules that are involved in the binding and internalization of a ligand via receptor-mediated endocytosis;
3. molecules that bind molecules of other cells or of the extracellular matrix, i.e. cell adhesion molecules.

In this context, the sperm may be viewed as having receptors for egg coat proteins. In fact, such a situation exists with respect to mammals, for the sperm protein that binds ZP3 is described as the ZP3 receptor. In a sense, this problem is somewhat artificial and is really a matter of perspective – in other words looking at the situation from the sperm's or the egg's point of view.

3.2 ZONA BINDING

Investigations of sperm–egg binding in mammals have concentrated, for the most part, on the interaction of the sperm and the zona pellucida which may assume several forms. That which is easily disrupted by pipetting, occurs at 2° or 37°C, and is not species-specific, is generally referred to as 'attachment' (Hartmann et al., 1972). After a short period a tenacious union occurs – binding – which is species-specific, calcium- and temperature-dependent (Gwatkin, 1977) and ultimately leads to the acrosome reaction (Wassarman, 1990).

Investigation of sperm–egg binding in mammals is an emerging one that is centered on (1) events involving sperm–zona binding in the mouse and (2) the importance of zona carbohydrates via two different approaches. Interestingly, these two approaches, which are outlined below, have begun to converge. Binding of mouse sperm to the zona pellucida is dependent upon O-linked oligosaccharides (Bleil and Wassarman, 1980a, b; Wassarman 1990, 1993). In fact, there is evidence suggesting that a particular class of ZP3 O-linked oligosaccharides of approximately 3900 Da is responsible for the glycoprotein's sperm receptor activity. As previously described in connection with induction of the acrosome reaction (Chapter 2), there are several candidates under consideration which interact with ZP3:

1. a 56 kDa protein (p56);
2. a 95 kDa protein (p95) that is a tyrosine-kinase substrate;
3. the enzyme galactosyltransferase (GalTase; Shur and Hall, 1982a, b; Miller and Shur, 1994).

None of these putative receptors have been definitely characterized as having both ZP3-binding and signal-transducing components.

Although p56 binds tightly to ZP3, its role in sperm–zona binding and in signal transduction has not been elucidated. Tyrosine phosphorylation of p95 increases in response to incubation of mouse sperm with ZP3 suggesting, as described in Chapter 2, that it may be involved in signaling the acrosome reaction upon zona binding (Leyton and Saling, 1989a). A 95 kDa phosphotyrosine-containing protein has been identified in human sperm that undergoes tyrosine phosphorylation during capacitation (Burks et al., 1995) and is believed to mediate sperm–zona interactions. GalTase is found on the surface of cells, where it functions in adhesion by binding to its specific glycoside ligand on adjacent cells or in the extracellular matrix. Since the extracellular space lacks a sugar nucleotide donor it is believed that GalTase forms a stable complex with terminal N-acetylglucosamine residues on specific oligosaccharides. This is supported by experiments in which the addition of UDP Gal dissociates cell surface GalTase from immunobolized substrates containing terminal N-acetylglucosamine residues (Miller and Shur, 1994). GalTase is found on the sperm of several different mammals on the dorsal, anterior aspect of the sperm head. Hence, it is in the right spot to function in sperm–zona binding (Fig. 3.2).

Antibodies to the enzyme and the GalTase substrate modifier, α-lactalbumin, inhibit sperm–zona binding. Alternatively, if the GalTase binding site on the zona is enzymatically blocked or removed, sperm binding is lost. Sperm-binding oligosaccharides on ZP3 have been

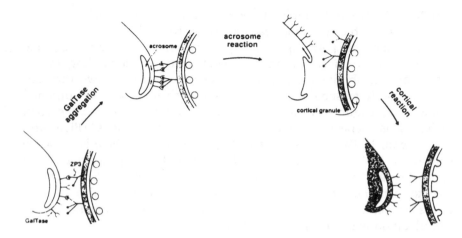

Fig. 3.2 Model of mouse fertilization involving a GalTase mechanism for sperm–egg binding and the acrosome reaction. Sperm GalTase binds to ZP3 in the zona pellucida. ZP3 binds 2-3 GalTase molecules, possibly aggregating GalTase and signaling sperm to undergo the acrosome reaction via a G-protein. During the acrosome reaction, GalTase moves laterally on the sperm head. N-acetylglucosaminidase is released and removes the GalTase-binding site (terminal N-actyl-glucosamine residues) enabling sperm to proceed through the zona. After penetration of the zona, sperm bind and fuse to the egg membrane, activating the egg and inducing the release of cortical granules. Cortical granules also contain N-acetylglucosaminidase, which removes the GalTase binding sites from ZP3, preventing additional sperm from binding to the zona. See Miller *et al.*, 1992.

shown to be ligands for sperm GalTase (Miller *et al.*, 1992). When intact sperm are incubated with solubilized zona glycoproteins and UDP-[^{3}H]Gal, zona glycoproteins that bind sperm GalTase, specifically ZP3, are labeled by the addition of [^{3}H]-galactose. How GalTase binding activity is linked to signal transduction mechanisms that occur subsequent to sperm–zona binding is unclear. Recent studies (Gong *et al.*, 1995), however, have suggested that sperm surface GalTase activates a G protein.

The question of how the acrosome-reacted sperm remains bound to and then moves through the zona pellucida has been examined (Wassarman, 1993; Miller and Shur, 1994). Binding of the acrosome-reacted sperm is believed to be maintained by proteins associated with the inner acrosomal membrane and ZP2. The sperm protein involved in this binding has been postulated to be proacrosin and/or acrosin and evidence suggesting that binding occurs either at the lectin-like or the structural domain of the molecule has been presented (Wassarman, 1993). An alternative suggestion (Miller and Shur, 1994), is based on the potential impedance of zona penetration by GalTase, which would bind exposed N-actylglucosamine residues of zona glycoprotein. It has been suggested that the high levels of β-hexosaminidases released with the acrosome reaction would provide an efficient means of zona penetration. This may occur by β-hexosaminidase removal or masking zona N-acetylglucosamine residues that could potentially interact with GalTase, although other possibilities have also been considered but are untested (Miller and Shur, 1994).

Using monoclonal antibodies that inhibit function, Myles and Primakoff (see Ramarao et al., 1994) have demonstrated that a protein referred to as PH-20 and originally identified and characterized in guinea pig sperm functions in sperm–zona binding. (The involvement of this bifunctional molecule in penetration of the cumulus has been discussed in Chapter 2.) PH-20 consists of two disulfide-linked fragments of 41–48 kDa and 27 kDa. When sperm are induced to undergo the acrosome reaction, PH-20 migrates from the posterior head plasma membrane into the inner acrosomal membrane. In this connection, PH-20 has been shown to be attached to the plasma membrane by a glycosyl phosphatidylinositol anchor. Removal of PH-20 by treatment of sperm with phosphatidylinositol-specific phospholipase C reduces the ability of acrosome-reacted sperm to bind the zona pellucida. Antibodies to PH-20 also inhibit binding of acrosome-reacted sperm to the zona pellucida, suggesting a specific role for this protein in sperm–zona binding. The potential role of PH-20 in signal transduction is not clear since glycosyl-phosphatidylinositol-linked proteins have no intracellular domain and are generally believed to be incapable of transducing a signal directly.

In addition to considerable remodeling of the sperm plasma membrane and the outer acrosomal membrane, the acrosome reaction brings about the rearrangement of a number of sperm surface molecules (Myles et al., 1987; Lopez and Shur, 1987). During spermatogenesis and epididymal transit, proteins on the sperm surface become localized to specific domains, which may be altered following the acrosome reaction. Examples of such proteins include PH-20 of guinea pig sperm which has been described above and GalTase of mouse sperm. Surprisingly, GalTase is not lost from the sperm after the acrosome reaction and undergoes a change in its distribution from the dorsal surface of the anterior sperm head overlying the intact acrosome to the lateral surface of the sperm coincident with the acrosomal reaction. It has been postulated that redistribution of GalTase to the lateral aspect of the sperm facilitates sperm–zona binding after the acrosome reaction as well as before (Lopez and Shur, 1987). Results of these and similar studies indicate that some proteins within the sperm plasma membrane, which are capable of diffusion, are maintained in specific domains. Myles *et al.* (1987) suggest that barriers to membrane protein diffusion exist at the equatorial region, the posterior ring and anulus and are responsible for maintaining a localized distribution of surface proteins. The migration of surface proteins is postulated to result from an alteration of these barriers, i.e. a change in protein structure to facilitate diffusion or active transport.

The role of zona glycoproteins in sperm–egg binding as observed for mammals may have functional correlates for gamete interactions in other groups of animals. For example, recent observations suggest that a 220 kDa glycoprotein from the vitelline coat of *Unio* (a mollusk) eggs interacts with sperm and that its O-linked oligosaccharide component is responsible for gamete binding (Focarelli and Rosati, 1995).

Gamete fusion and sperm incorporation 4

The site of gamete fusion and sperm incorporation in many eggs is restricted to a specific region of the ovum. Eggs with micropyles, such as fish and insects, are highly specialized in this respect as a portal in the layers investing the egg permits access of the sperm to a specific site on the ovum (Hart and Donovan, 1983; Hart et al., 1992). The egg plasma membrane beneath the micropyle is the region where the sperm normally fuses with the egg, although experiments with dechorionated eggs of some species show that sperm can fuse at any site along the egg surface. In a number of animals (e.g. cnidarians, hydrozoans, siphonophores and amphibians) the sperm fuses with the egg near the site of polar body formation (Freeman and Miller, 1982). In the amphibian *Discoglossus* sperm enter at a depressed region on the animal pole called the animal dimple (Campanella, 1975). These examples of the localization of the site of gamete interaction and fusion may be due to the production of a substance that attracts or traps sperm at the site of fertilization, or the presence of a differentiated surface that limits sperm–egg fusion to one site along the ovum surface. In echinoderms sperm appear to be able to fuse and enter the egg anywhere along the surface of the egg (Schroeder, 1980). The same appears to be the case for molluskan eggs, although one seldom, if ever, observes sperm–egg fusion at the site where the meiotic spindle is located (Longo, 1983). The fact that sperm appear to enter both at the animal and vegetal poles in some urodele eggs may be related to the polyspermy that occurs in this species. In the frog *Rana pipiens* sperm penetrate the jelly and vitelline layers and apparently approach the plasma membrane anywhere along the perimeter of the egg. Sperm entry, however, occurs only in the animal half of the egg (Elinson, 1980). Sperm incorporation in mammals, particularly the mouse and hamster, occurs along the region containing microvilli, away from the microvillus-free area in which the meiotic apparatus is located (Nicosia et al., 1977; Yanagimachi, 1981; Fig. 4.1).

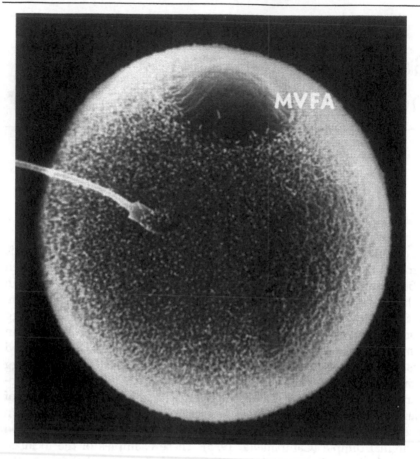

Fig. 4.1 Scanning electron micrograph of a hamster spermatozoon fusing with an egg. Zona-free hamster eggs were inseminated *in vitro* with acrosome-reacted sperm and fixed about 10 min later. The site of gamete fusion is restricted to the area of the egg projected into microvilli. The meiotic spindle is located in the elevated microvillus-free area (MVFA). Reproduced with permission from Yanagimachi, 1981.

Rabbit sperm after passing through the zona pellucida appear to move vigorously in the perivitelline space before attaching to the plasma membrane; 20 min later the entire sperm is located within the ovum. Hamster sperm attach to the egg plasma membrane apparently without wandering within the perivitelline space. In the meantime the beating sperm tail vibrates and rotates the egg within the perivitelline space. Finally, when the entire sperm is in the perivitelline space its vigorous movements cease.

Participation of microvilli in gamete fusion has been observed with transmission and scanning electron microscopy in hamsters (Yanagimachi and Noda, 1970a; Shalgi and Phillips, 1980a, b). Association of the acrosomal process and its fusion with the egg surface is believed to be mediated at microvilli in sea urchins (Vacquier, 1980; Figs 4.2, 4.3).

Gamete membrane fusion in the lamellibranches *Mytilus* and *Spisula* is believed to occur in the same manner. In the annelid *Chaetopterus* the morphology of the egg and the manner in which the gametes interact demonstrate rather vividly that, at least in this species, sperm–egg fusion is mediated at the tip of the egg's microvilli. Initial contact of sperm and egg involves an unreacted acrosome and the apices of microvilli that penetrate the vitelline layer. Apparently contact of the acrosome with the microvilli initiates the acrosome reaction; subsequently, the egg plasma membrane delimiting the apex of a microvillus fuses with the former inner acrosomal membrane (Anderson and Eckberg, 1983).

Relatively little is known about processes accompanying actual fusion of the gamete membranes. A number of models of membrane fusion have been proposed; all are derived as best estimates to explain experimental findings and none has been tested and verified (Zimmerberg, 1988; Monck and Fernandez, 1992). Generally, the process represents a

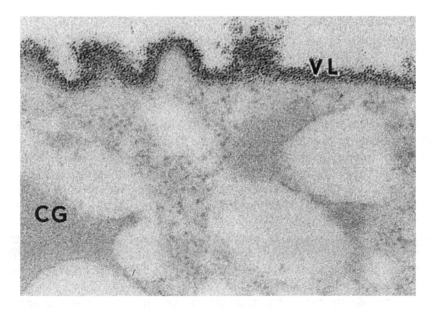

Fig. 4.2 Sea urchin (*Arbacia*) egg incubated with cationized ferritin to demonstrate the vitelline layer (VL). CG = cortical granule.

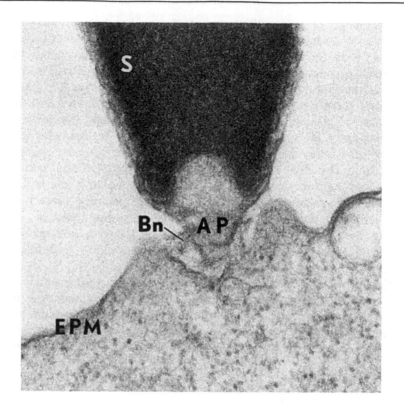

Fig. 4.3 Acrosome-reacted sea urchin (*Arbacia*) sperm in contact with the egg surface. AP = acrosomal process; Bn = bindin; EPM = egg plasma membrane; S = sperm nucleus.

merging of two membrane domains: first the outer leaflets and then the inner leaflets of the two lipid bilayers. This results in formation of a pore that allows for the confluence of the protoplasmic contents of both cells, the egg and sperm. Changes are believed to occur at the acrosomal reaction that render the sperm fusigenic with the ovum (Colwin and Colwin, 1967) and fusion is likely to be similar at the molecular level to that postulated for somatic cells. At the site of fusion, the development of membranous changes that accompany the acrosome reaction in guinea pig sperm may occur, where intramembranous-particle-free areas that are believed to be fusigenic emerge within the plasma membrane (Friend, 1980).

The precise moment of gamete fusion, i.e. the time at which sperm and egg plasmalemmae become continuous and the cytoplasms of both gametes are able to merge, is important in defining possible mechanisms and the chronology of early events of fertilization and their possible causal relations. Correlative ultrastructural and electrophysiological studies of sperm–egg interactions in the sea urchin *Lytechinus* indicate that egg activation, i.e. the initiation of the activation or fertilization potential, precedes sperm and egg plasma membrane fusion by 5–6 s (Hinkley *et al.*, 1986; Longo *et al.*, 1986). This temporal relationship has a bearing on the question of how sperm activation of the egg is achieved and is discussed below.

Fusion of mammalian sperm and egg is in striking contrast to gamete fusion in the marine invertebrates (Colwin and Colwin, 1967; Yanagimachi, 1988). In many of the invertebrates studied thus far, gamete fusion occurs between the egg plasma membrane and the tip of the acrosomal process, which is delimited by membrane derived from the acrosomal vesicle (Fig. 4.3). In mammals segments of plasma membrane over the equatorial segment (Bedford and Cooper, 1978) are the first to fuse with the egg plasma membrane (Yanagimachi and Noda, 1970b; Yanagimachi, 1994; Fig. 4.4).

4.1 REGULATION OF GAMETE FUSION

The presence of calcium is reportedly essential for sperm–egg fusion in mammals (Yanagimachi, 1981). In sea urchins calcium dependency for gamete fusion is controversial. According to Sano and Kanatani (1980), activated sperm added to eggs in the presence of the calcium chelator EGTA are unable to fertilize. However, other investigators have shown that sea urchin eggs can be fertilized in calcium-free media ($< 10^{-8}$ mol/l; Chambers, 1980; Schmidt *et al.*, 1982). The fusion of sperm and eggs in the absence of calcium is surprising in light of experiments showing that this ion is involved in membrane fusion processes of other cells and, in particular, the acrosome reaction. In fact, one possibility for the inability to fertilize in Ca^{2+}-free media may be due to blockage of the acrosome reaction. Sperm–egg fusion may be an exception; alternatively calcium may be involved but not essential. Magnesium ions have been shown to be indispensable for fusion between the sperm acrosomal membrane and the egg plasma membrane (Mohri *et al.*, 1994).

Based on the use of specific substrates and inhibitors, metalloendoproteases have been shown to be involved in gamete fusion of sea urchins and ascidians (Roe *et al.*, 1989; De Santis *et al.*, 1992). How such enzymes may function to promote fusion is currently unknown, although it has been postulated that proteolysis may lead to the establishment of a molecular organization, either by generating fusogenic sites or by removing charge or steric restraints to membrane apposition.

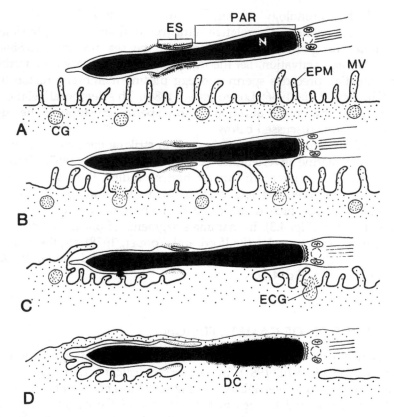

Fig. 4.4 Diagrammatic representation of a hamster sperm fusing with the egg plasma membrane (EPM). **(A)** Interaction of the sperm with the egg microvilli (MV). **(B)** Fusion of egg microvilli with, first, the equatorial segment (ES), followed by the postacrosomal region (PAR) of the sperm plasma membrane. **(C)** and **(D)** Subsequent stages of sperm incorporation. CG = cortical granules; ECG = exocytosing cortical granules; N = sperm nucleus; DC = dispersing chromatin. Lines depicting the plasma membranes of the sperm and egg and cortical granules have different thicknesses to illustrate their fates subsequent to gamete fusion. See Yanagimachi and Noda, 1970b.

Investigations with PH-30, a glycoprotein (≈ 60 kDa) referred to as fertilin and localized to the posterior region of the guinea pig sperm head, have demonstrated that it may play a role in gamete fusion (Myles, 1993; Ramarao *et al.*, 1994). PH-30 is composed of two subunits.

1. The α-subunit is a putative fusion peptide having an amino acid composition similar to that of viral fusion proteins.
2. The β-subunit, a binding domain related to integrin ligands (disintegrin; Fig. 4.5), may be used for recognition and binding of the egg plasma membrane.

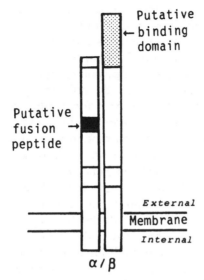

Putative
← binding
domain

Putative
fusion →
peptide

External

Membrane

Internal

α / β

Fig. 4.5 Model of guinea pig sperm PH-30. See Blobel *et al.*, 1992.

It has been postulated that the α-subunit may be involved in sperm–egg plasma membrane fusion (Blobel *et al.*, 1992). The recognition sequence for integrin binding proteins contains the conserved tripeptide RGD (arginine, glycine, aspartic acid) and peptides containing the RGD sequence block sperm–egg fusion (Bronson and Fusi, 1990). Interestingly, the precursor to the α-subunit has a metalloprotease sequence motif that may be functionally active during sperm maturation (Ramarao *et al.*, 1994). As pointed out by Yanagimachi (1994), the localization of PH-30 on the spermatozoon does not correlate with the site of gamete fusion. If, as indicated above, the plasma membrane over the equatorial segment fuses with the egg plasma membrane, why then is PH-30 located in the postacrosomal region and not in the equatorial segment? Mammalian eggs express a variety of integrins (Evans *et al.*, 1995) and recent evidence has been presented suggesting that integrin $\alpha_6\beta_1$ serves as a receptor for mouse sperm (Almeida *et al.*, 1995).

A major component of secreted glycoproteins, referred to as DE protein (37 kDa), is reportedly associated with the acrosomal region of the rat sperm head during epididymal transit (Rochwerger *et al.*, 1992) and migrates to the equatorial segment with the acrosome reaction. Complementary binding sites for DE protein have been identified on the egg plasma membrane. Because DE protein associates with the fusogenic region of acrosome-reacted sperm and the presence of complementary binding sites on the egg plasma membrane, Rochwerger *et al.* (1992) have proposed that DE protein has a role in sperm–egg fusion.

The competency of oocytes to fuse with acrosome reacted sperm is developed early during oogenesis. Studies with developing sea urchin oocytes (Longo, 1978b) have demonstrated that previtellogenic oocytes are capable of gamete fusion. In hamsters Zuccotti *et al.* (1991) have shown that oocytes are first competent to fuse with sperm when they are about 20 μm in diameter. How the oolemma gains the ability to fuse with the sperm is unclear. To search for candidate molecules involved with sperm–egg fusion many researchers have examined the retention of the ability of the egg plasma membrane to fuse with sperm after various enzyme treatments (for studies in sea urchins see Foltz and Lennarz, 1993; for studies in mammals see Ponce *et al.*, 1993; and Yanagimachi, 1994). Experiments with hamster eggs have shown that components of the egg plasma membrane required for fusion with sperm are resistant to proteolytic digestion (Ponce *et al.*, 1993). Only trypsin in calcium-free medium reduced the oolemma's fusibility significantly; pronase and α-chymotrypsin in calcium-free medium were without effect. From these findings and studies demonstrating an integrin-like molecule associated with human oocytes, it has been suggested that the oolemma contains fusion-mediating molecules similar to adherins (Ponce *et al.*, 1993).

The fusibility of the egg plasma membrane is reduced dramatically after fertilization and this ability is lost by the eight-cell stage in hamsters (Zuccotti *et al.*, 1991). Sea urchin embryos up to the eight-cell stage are capable of fusing with sperm. Whether or not later stage embryos are able to fuse has not been ascertained (Longo, 1989b).

A major requirement for sperm that possess acrosomes to fuse with eggs is to undergo the acrosome reaction. Requirements for sperm of species that lack acrosomes have not been clearly established. Strong sperm motility does not appear to be a requirement for gamete fusion in mammals (Yanagimachi, 1994). In some organisms, such as nematodes and crustaceans, sperm normally show limited movement yet are capable of gamete fusion. Sea urchin sperm in which the tails have been removed are capable of gamete fusion (Epel *et al.*, 1977). In this connection, the incidence of fertilization in mammalian and human eggs during *in vitro* fertilization can be increased by cutting a hole in the zona pellucida prior to insemination or by a subzonal injection of sperm. Such methods may be useful when sperm cannot penetrate the zona pellucida normally or when the zona is too rigid for sperm penetration.

4.2 INCORPORATION OF THE SPERM PLASMA MEMBRANE

That all of the sperm plasma membrane is incorporated into the egg plasma membrane at fertilization is more or less assumed in many instances, although experimental evidence has not verified this unequivocally. Electron microscopic studies of sperm incorporation in some invertebrates and mammals have demonstrated membranous elements, such as vesicles, at the site of gamete fusion that appear to be derived from the fused sperm and/or egg plasmalemmae (Colwin and Colwin, 1967; Bedford and Cooper, 1978). If true, this would indicate that portions of the egg and sperm plasma membranes are lost at the time of gamete fusion.

Investigations examining the integration of the sperm and egg plasma membranes at fertilization, where one of the gametes has been labeled, have been carried out in both invertebrates and mammals (Yanagimachi et al., 1973; Gabel et al., 1979; Longo, 1982, 1989b, c). Prior to sperm–egg fusion in hamsters the sperm plasma membrane of the postacrosomal region does not bind colloidal iron hydroxide. Once gamete fusion has been initiated, however, the former sperm plasma membrane is able to bind this marker. The rapid increase in colloidal iron hydroxide binding to the incorporating sperm head is believed to be a result of intermixing of sperm–egg membrane components comparable to the intermingling of antigenic determinant after fusion of somatic cells. Such results are not unexpected, for intrinsic glycoprotein and glycolipids, intercalated in a fluid bilayer, are freely diffusible in the plane of the membrane. These observations, however, do not exclude the possibility that colloidal iron hydroxide binding receptors are enzymatically added to sperm plasma membrane oligosaccharides after fusion or that colloidal iron-hydroxide-binding membrane components are inserted into the sperm plasma membrane following fertilization.

Similar experiments have been carried out with the surf clam, Spisula, in which concanavalin A binding to the egg, but not the sperm plasma membrane, has been demonstrated by the horseradish-peroxidase–diaminobenzidine reaction (HRP-DAB; Longo, 1982). Because of this dichotomy in lectin binding, changes in the affinity of the sperm plasmalemma following its fusion and integration with components of the egg plasma membrane can be followed (Fig. 4.6).

The plasma membranes of fertilized Spisula eggs react with concanavalin A-HRP-DAB and are associated uniformly with enzymatic precipitate except at sites of sperm incorporation by 1 min postinsemination. These portions of unstained plasma membrane are derived from the sperm and associated with the apex of the fertilization cone. From 2–4 min postinsemination HRP-DAB reaction product gradually becomes associated with all the membrane delimiting the fertilization cone (Fig. 4.7).

By 4 min postinsemination no difference in staining of plasma membranes derived from the egg or the sperm is detected. These observations are consistent with the movement of concanavalin A binding sites from the egg plasmalemma into the sperm plasma membrane.

Shapiro and co-workers claim that not all components of the sperm

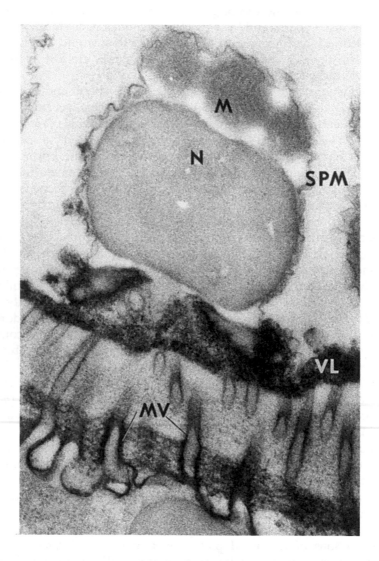

Fig. 4.6 Surf clam (*Spisula*) sperm attached to the surface of an egg. The gametes were treated with the lectin concanavalin A, followed by horseradish peroxidase and then reacted with H_2O_2 and diaminobenzidine. Dense reaction material is present along the surface of the egg microvilli (MV) and the vitelline layer (VL). The sperm plasma membrane (SPM) lacks reaction product due to the absence of concanavalin A binding. Because of this dichotomy in concanavalin A binding integration of the sperm and egg plasma membranes subsequent to gamete fusion may be followed. N = sperm nucleus; M = mitochondria. See Longo, 1982.

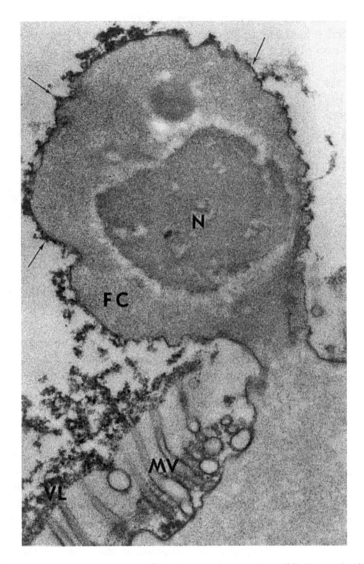

Fig. 4.7 Incorporated surf clam (*Spisula*) sperm nucleus (N) located within a fertilization cone (FC). The membrane at the arrows is derived from the spermatozoon and possesses concanavalin A receptors possibly because of their lateral movement within the plane of the egg plasma membrane into the sperm plasma membrane. MV = microvilli; VL = vitelline layer. See Longo, 1982.

and egg plasma membranes intermix following gamete fusion, but sperm surface components in fertilized sea urchin and mouse eggs remain as a discrete patch that may function in the determination of embryonic polarity (Gabel *et al.*, 1979). Further experiments have indicated that labeled sperm plasma membrane components are also internalized after fertilization (Gundersen *et al.*, 1982). However, localization of two plasma membrane proteins that are present in sperm and not in eggs, followed by immunofluorescence electron microscopy in fertilized sea urchin oocytes and eggs, indicate a very different situation (Longo, 1989a, b; see also Gaunt, 1983). Analysis of samples from a synchronous population of inseminated eggs demonstrated labeling of the egg plasma membrane starting at the site of gamete fusion; with time the area of label progressed to the opposite pole of the egg, indicating the incorporation and diffusion of sperm-specific plasma membrane components through the egg plasma membrane. The diffusion coefficient of the sperm plasma membrane polypeptides was calculated to be 6.7×10^{-10}–2.4×10^{-9} cm^2/s, a range that is comparable to that found for somatic cells.

4.3 FERTILIZATION CONE

At the site of gamete fusion a protuberance forms that has been referred to as the fertilization or incorporation cone (Fig. 4.8).

Formation of this structure consists of a movement of egg cytoplasm into the region surrounding the sperm nucleus, mitochondria and axonemal complex, resulting in a protrusion at the site of sperm entry (Longo, 1973). As the fertilization cone increases in size, incorporated sperm components move through it into the ovum's cortex. In many organisms the migration of the sperm nucleus into the egg cortex is accompanied by its rotation of 180° (Fig. 4.9). As a result of this turning movement, the sperm nucleus usually becomes located lateral to the site of entry with its apex directed to the egg plasma membrane.

Characteristically, fertilization cones are filled with numerous bundles of actin filaments that in sea urchin zygotes show a polarity when reacted with heavy meromyosin or the S1 fragment of myosin (Tilney and Jaffe, 1980; see also Maro *et al.*, 1984; Hart *et al.*, 1992). With S1 decoration the arrowhead complexes that form on actin filaments in the fertilization cone of sea urchins are directed to the center of the egg. The microfilaments are believed to polymerize *in situ* from cortical, monomeric actin. Whether or not and how actin in the fertilization cone might function to effect sperm incorporation is not entirely clear. The actin filaments of the sperm acrosomal processes are also polarized, with the heavy meromyosin or S1-actin arrowheads pointing to the sperm nucleus. Consequently, egg myosin could not bridge sliding actin filaments of both the fertilization cone and the acrosomal process to bring about sperm nucleus incorporation; both sets of actin filaments are polarized in the wrong direction when compared to the orientation of myosin and actin of a sarcomere. It is possible that actin filaments

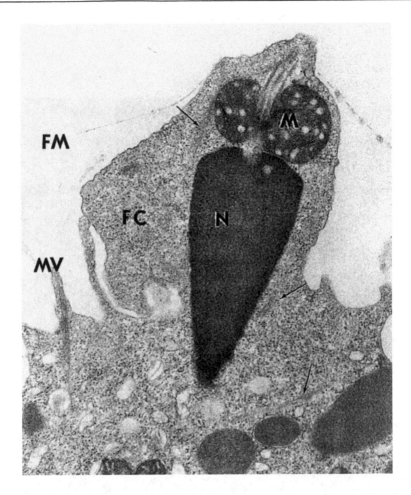

Fig. 4.8 Incorporated sea urchin (*Arbacia*) sperm within a fertilization cone (FC). The filamentous structures shown at the arrows contain actin. FM = fertilization membrane; M = sperm mitochondrion; N = sperm nucleus; MV = microvillus containing microfilaments. See Longo, 1980.

present at the site of sperm entry might be primarily involved in the elevation and enlargement of the fertilization cone; movements of sperm components through the egg cortex may be dependent on other mechanisms.

Cytochalasin B, a drug that disrupts actin microfilaments, has been shown to inhibit surface activity of fertilized sea urchin eggs, such as microvillar elongation and fertilization cone formation (Longo, 1980; Fig. 4.10).

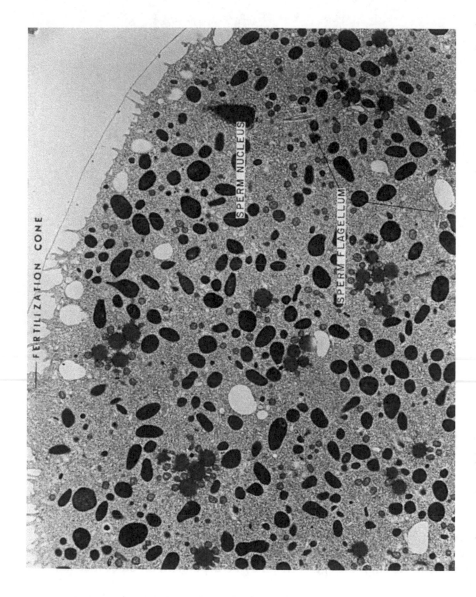

Fig. 4.9 Incorporated sea urchin (*Arbacia*) sperm nucleus having undergone rotation and located lateral to its site of entry, the fertilization cone. See Longo, 1973.

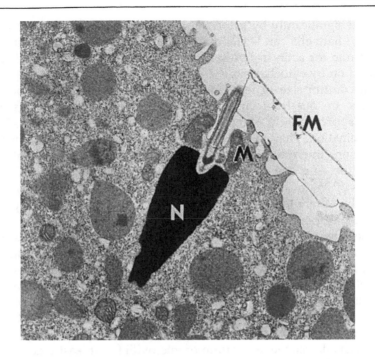

Fig. 4.10 Inseminated sea urchin (*Arbacia*) egg in which the formation of the fertilization cone has been inhibited with cytochalasin B. M = sperm mitochondrion; N = sperm nucleus; FM = fertilization membrane. See Longo, 1980.

Fertilized, cytochalasin-B-treated eggs undergo a cortical granule reaction, elevate a fertilization membrane and are metabolically activated. Treatment of eggs with cytochalasin B after sperm incorporation does not significantly affect migration or formation of the male pronucleus. These observations are consistent with the suggestion that at the site of gamete fusion there is a localized polymerization of actin that participates in the formation of the fertilization cone. In addition, experiments with cytochalasin B also indicate that treated sea urchin eggs can be activated by sperm but sperm fail to enter the egg (Gould-Somero *et al.*, 1977; Longo, 1978a). How the sperm is capable of activating the egg in this instance without entering it has not been determined. It is possible that the acrosomal process fuses with the egg plasma membrane but, since actin polymerization is impaired, the bridge linking the fused sperm and egg is weak and the sperm is removed from the egg surface by exocytosing cortical granules. Another possibility is that cytochalasin B inhibits fusion of the egg and sperm and that gamete contact/binding in this instance is sufficient for egg activation (Longo *et al.*, 1986). Interestingly, sperm incorporation is unaffected significantly by cytochalasin B in mouse or rat eggs (Maro *et al.*, 1984; Battaglia and Gaddam-Rosse, 1986), indicating either a different role for actin in fertilization or differences in the specificity of sites acted on by cytochalasin B.

The maximum size of fertilization cones varies depending upon the organisms in question; in mature sea urchin (*Arbacia*) eggs they measure approximately 6 μm in length by 4 μm in diameter, 5–7 min postinsemination. They then regress and are reabsorbed by 10 min post-insemination. Interestingly, in *Arbacia* the fertilization cones that form on immature eggs are much larger than those that develop on mature ova: sizes of 25 μm in length by 10 μm in diameter are not unusual. In some mollusks and starfish, fertilization cones tend to be small protrusions of the egg cytoplasm (Kyozuka and Osanai, 1994a, b).

At the site of sperm entry in anurans a microvillus-free bleb of cytoplasm forms, presumably functionally equivalent to a fertilization cone (Picheral, 1977). Eventually it disappears and is replaced by a small clump of elongate microvilli. The microvillus-free bleb may be pinched off leaving the microvilli. If this is the case then it is possible that plasma membrane components, as well as other sperm-derived structures, may be eliminated from the egg. The site of sperm entry reportedly remains detectable as a clump of microvilli for at least 2 hours.

In mammals (rats, rabbits and hamsters) following the fusion of the egg and sperm plasma membranes, tongues of cytoplasm surround the anterior portion of the sperm head and form a vesicle that is present for a time within the zygote (Yanagimachi and Noda, 1970a; Yanagimachi, 1994). At the site of gamete fusion a protrusion of cytoplasm forms which is homologous to fertilization cones seen in invertebrate eggs and is often referred to as an incorporation cone (Zamboni, 1971). In mouse eggs the protrusion is filled with cytoplasmic organelles found in other regions of the zygote, and along its plasmalemma there is a prominent layer of actin (Maro *et al.*, 1984). In this protrusion the early events of male pronuclear development take place. In fertilizing hamster eggs Webster and McGaughey (1990) demonstrate the presence of cortical actin filaments associated with the incorporating spermatozoon. They speculate that the actin filaments play a role in sperm penetration of the egg cortex.

Labeling experiments at the ultrastructural level of observation indicate that the plasma membrane of the fertilization cone is sufficiently fluid as to permit the free diffusion of different sperm and egg plasma components (Nishioka *et al.*, 1987; Longo, 1986, 1989). The diffusion of components within the plasma membrane delimiting the fertilization cone of inseminated sea urchin eggs has been described above (see Longo, 1989a, b). Additional labeling studies (Longo, 1986) demonstrate that:

1. more than 90% of the membrane that delimits the fertilization cone is derived from the egg plasma membrane in the form of microvilli; in other words, as the fertilization cone increases in size, microvilli retract and membrane components delimiting these projections move laterally within the plasma membrane and become associated with the fertilization cone;
2. this movement involves either a selective flux of components deficient in oligosaccharides or the modification of components possessing carbohydrates (glycolipids and glycoproteins); the net result is the formation of a fertilization cone that is distinguished by a delimiting membrane possessing significantly fewer carbohydrate residues than the remainder of the egg surface.

The significance of these observations lies in the fact that at the site of gamete fusion there is a dramatic rearrangement of surface molecules within the egg plasma membrane that gives rise to an asymmetrical membrane topography. The asymmetry is due to both the insertion of the sperm plasma membrane and to a redistribution of egg plasma membrane components. Hence, the egg plasma membrane is not static but is in flux. What role such changes may serve during and following insemination has not been established. Additional analyses of the egg surface involving fluxes of membrane components have been reviewed (Longo, 1988).

Reorganization of the plasma membrane has also been examined in fertilized human eggs (Dale *et al.*, 1995). In contrast to what is found in other organisms, studies examining the localization of major sugar groups and scanning electron microscopy indicate the egg plasma membrane is not topographically reorganized following fertilization.

Egg activation

5

5.1 MEIOTIC STAGES OF EGGS AT INSEMINATION

Essentially four types of eggs are recognized with respect to their state of meiosis at the time of insemination (Fig. 5.1).

Generally, fertilization follows ovulation and eggs are inseminated during or following their meiotic divisions. Several exceptions to this rule exist and include various species of marine annelids and free-living flatworms, in which the spermatozoon enters the egg prior to ovulation. Quite apart from these exceptions the classification that has been formulated is not particularly rigid with respect to animals comprising each class and the classes themselves. For example, in some cases, eggs are reportedly inseminated at the first or second anaphase of meiosis. Eggs of some vertebrates are inseminated at meiotic prophase, i.e. the germinal vesicle stage, rather than at the second metaphase of meiosis. In other animals, e.g. starfish, a number of classes may apply (II, III or IV; Fig. 5.1). Eggs included in class I are fertilized at meiotic prophase and are found in various nematodes, mollusks, annelids and crustaceans. The eggs of some mollusks, annelids and insects are inseminated at the first metaphase of meiosis and make up class II, while the eggs of most vertebrates are found in class III and fertilize at the second metaphase of meiosis. Eggs belonging to class IV are those of various echinoids and coelenterates and are inseminated at the completion of meiosis – the pronuclear stage.

In an examination of this classification two features may be noted. First, there is no apparent phylogenetic relationship between the different groups and the animals comprising them. Second, eggs belonging to classes I, II and III have not completed meiosis and are in a state of meiotic arrest at the time of insemination. Therefore, in these forms, the interaction of the sperm and egg triggers processes that initiate the resumption of meiosis.

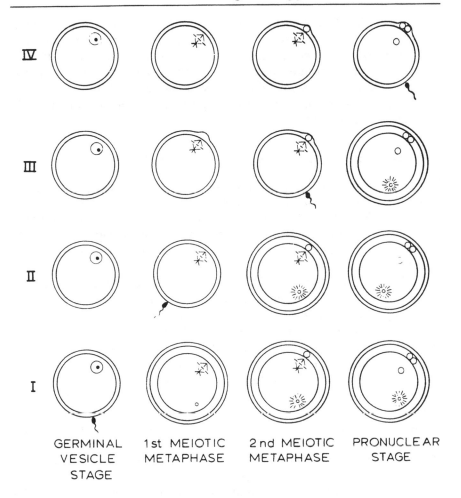

Fig. 5.1 Classification of eggs according to the meiotic stage at which they are inseminated.

In these instances fertilization also includes completion of meiotic maturation of the maternally derived chromosomes.

Changes in the ability to fertilize during the course of meiotic maturation have been investigated in both vertebrates and invertebrates (Usui and Yanagimachi, 1976; Longo, 1978b). Generally such studies indicate that sperm can penetrate oocytes at early stages of maturation and that male pronuclear development is usually retarded. The incidence of polyspermy is often high in these cases, indicating that mechanisms underlying the development of the block to polyspermy are established during oocyte maturation. For example, in sea urchins a close association of the cortical granules with the plasma membrane is important for their exocytosis (the cortical granule reaction) and the prevention of polyspermy. That immature sea urchin eggs do not undergo a cortical granule reaction and are polyspermic at low sperm concentrations is consistent with observations demonstrating that their cortical granules do not move to the plasma membrane until meiotic maturation is completed.

5.2 HOW DOES THE SPERM ACTIVATE AN EGG?

The question of how the sperm induces egg activation has puzzled biologists for well over 100 years. Contrasting ideas have been presented (Swann, 1993; Foltz, 1994; Nuccitelli, 1991) and the prevailing opinions fall essentially into two camps, one espousing the passage of a soluble activation factor from the sperm to the egg, and a second postulating a receptor-mediated process.

As discussed by Swann (1993), injection of cytosolic extracts of sea urchin sperm into eggs triggers the cortical granule reaction (Dale *et al.*, 1985; Dale, 1988), a clear sign of egg activation. These results are consistent with the idea that sperm contain a soluble activating factor that is 'injected' into the egg at or immediately following gamete membrane fusion. Identification of the factor has not been achieved. Several candidates have been proposed, including Ca^{2+} (Jaffe, 1983), inositol trisphosphate (Iwasa *et al.*, 1990) and an unidentified protein (Stice and Robl, 1990), but none has been conclusively proved to be the active substance (Whitaker *et al.*, 1989; see Swann, 1993). Whatever the activating substance might be, a central tenet of this hypothesis is the formation of a pore via gamete membrane fusion, which then allows the activating substance to move from the sperm into the egg.

The second hypothesis holds that sperm bind to a receptor within the egg plasma membrane that transduces a signal to a second messenger system (Nuccitelli, 1991; Foltz and Shilling, 1993). Evidence that a receptor is involved in egg activation comes from experiments employing (1) antibodies to egg or sperm surface components and (2) proteolytic enzymes that either stimulate egg activation or block sperm binding (see Foltz and Shilling, 1993). Studies by Gould and co-workers (Gould *et al.*, 1986; Gould and Stephano, 1991) have shown that a basic protein ($\approx$ 50% lysine and arginine) from acrosomal granules of the marine worm *Urechis* agglutinate and activate eggs. It was suspected that this reaction might be nonspecific, as poly-L-lysine was equally effective in activating *Urechis* eggs. However, an active peptide (Val-Ala-Lys-Lys-Pro-Lys) produced as a thermolytic peptide of the acrosomal protein induces the fertilization potential, completion of meiotic maturation and DNA replication. Experiments with *Xenopus* gametes indicate that an ATP receptor on the egg membrane is the target for ATP originating in sperm, suggesting that an ATP-induced increase in sodium permeability mediates the initial sperm-to-egg signal in the fertilization process (Kupitz and Atlas, 1993). Interestingly, a factor (< 1300 mol. wt) derived from an extract of sea urchin sperm induces Ca^{2+} transients and accompanying changes of membrane potential in fertilized denuded eggs but not unfertilized eggs (Osawa et al., 1994). One possible explanation offered for the failure of the sperm factor to activate unfertilized eggs is that it is unable to penetrate the vitelline layer and bind to its receptor on the egg plasma membrane.

Further evidence supporting the notion that egg activation is the result of a receptor-ligand-type interaction comes from experiments in which muscarinic cholinergic (Kline et al., 1988) or serotonin receptors (Kline et al., 1991) are expressed in Xenopus or starfish (Shilling et al., 1990) oocytes following injection of the appropriate mRNA. In these cases, experimentally injected oocytes activate when exposed to the appropriate ligand. Experiments with somatic cells have demonstrated that muscarinic and serotonin receptors span the plasma membrane and are coupled to G protein which in turn causes an increase in phospholipase C_β activity, resulting in inositol trisphosphate production and Ca^{2+} release (Berridge, 1993). Results with microinjected oocytes indicate that signaling pathways are in place in the oocyte's plasma membrane and implicate the involvement of G protein and phospholipase C isoforms in the activation process. Thus far, the putative sperm receptor that may couple the interacting sperm to the signaling transduction system and egg activation has not been identified.

The question of when sperm and egg plasma membrane fusion occurs during fertilization has been examined (Longo et al., 1986, 1994a; Hinkley et al., 1986) and may provide useful insights into the manner by which sperm activate eggs. Electrophysiological studies (McCulloh and Chambers, 1991, 1992) have shown that the abrupt increase in conductance of inseminated sea urchin eggs is localized to the site of sperm–egg contact. This change in conductance results from an establishment of electrical continuity between the gametes and represents the insertion of about 5 nS conductance in parallel across the egg plasma membrane. Clearly fusion of the gamete's plasma membranes would establish electrical continuity, as well as a pore uniting the cytoplasms of both the sperm and the egg. Such a pore may allow the passage of an activity factor. However, the question is, when do the egg and sperm plasma membranes fuse with respect to the increase in electrical conductance? Electron microscopic (Longo et al., 1986, 1994a) and dye passage (Hinkley et al., 1986) observations indicate that gamete membrane fusion does not occur until about 5 s after the conductance increase in fertilized sea urchin eggs. Results of both experiments are consistent with the idea that the sperm interacts with a receptor on the egg surface to initiate egg activation. Since gamete membrane fusion occurs about 5 s after the initiation of the activation potential (or current), the change in electrical activity of the egg may be the result of a nonfusion event, possibly a receptor–ligand type of interaction.

This interpretation makes the assumption that the fusing sperm and egg plasma membranes are sufficiently stable during fixation and subsequent processing and hence their specific morphological relationships to one another are maintained. To what degree this is true is unknown. Moreover, in studies by McCulloh and Chambers (1986a, b, 1991) capacitance of the egg plasma membrane at the site of gamete interaction was recorded as it would reveal the time when the sperm's plasma membrane capacitance would become electrically accessible from within the egg cytoplasm. Amazingly, it was determined that capacitance increased simultaneously with the conductance increase. The establishment of electrical continuity in this manner could result from the joining of the sperm and egg cytoplasm, i.e. as a consequence of gamete membrane fusion. Alternatively, electrical continuity might result from a transient increase in ionic permeability of apposed gamete membranes, preceding the fusion event. Such a prefusion event may be due to the formation of a gap-junction-like structure (Almers, 1990; Pollard et al., 1990) or a membrane modification, heretofore not characterized. (Articles and reviews by Longo et al. (1986, 1994a) and Whitaker et al. (1989) provide further critical analyses of these experiments.)

In summary, the question of how the sperm activates the egg has not been fully resolved. A major reason for this is that what happens during the brief period (approximately 5 s) from sperm–egg binding to egg activation is not well understood. Experiments to explore aspects of this period are often based on assumptions that may or may not be correct, and techniques to investigate what is occurring during this very critical phase of fertilization lack sufficient resolution (temporal, spatial, etc.) to accurately determine what is taking place. Ionic changes of the egg at fertilization, which are critical to egg activation, provide additional clues as to what the sperm might be doing to trigger egg activation and are discussed below.

5.3 DEPOLARIZATION OF THE EGG PLASMA MEMBRANE

The resting potential of sea urchin eggs as measured with microelectrodes falls into two general ranges: (1) –5 to –20 mV, which may be a consequence of current leakage, and (2) –60 to –80 mV (Shen, 1983). The resting potential has also been determined by tracer flux experiments and shown to be about –70 mV. The resting potentials of a number of invertebrate eggs are given in Table 5.1.

Table 5.1 Resting potentials for eggs of some marine invertebrates in sea water (Data compiled from Shen, 1983)

Organisms	Resting potential (mV)
Renella (coelenterate)	−70
Urechis (echiuroid)	−33
Spisula (mollusk)	−20
Hyanassa (mollusk)	−1 to −20
Dentalium (mollusk)	−70
Strongylocentrotus, Lytechinus (echinoid)	−60 to −80
Halocynthis, Ciona (tunicate)	0 to −20

Membrane potential changes at fertilization display characteristics which have been studied in eggs of echinoderms, polychetes, echiuroids and nemertean worms, mollusks, crustaceans, ascidians, fish, amphibians and mammals (see Nuccitelli *et al.*, 1989). In echinoids there is a transient depolarization (Whitaker and Steinhardt, 1982; Shen, 1983; Fig. 5.2), which is referred to as the activation potential (Chambers, 1989).

The activation potential consists of three phases (McCulloh and Chambers, 1991, 1992). Phase 1 corresponds in duration to the latent period (Whitaker *et al.*, 1989), the time between sperm–egg contact and the initiation of the calcium wave that causes egg activation. The attached sperm induces an increase in the conductance of the egg plasma membrane, which is localized to the site of sperm–egg interaction (McCulloh and Chambers, 1991). Phase 2 consists of a major increase in membrane conductance which is accompanied by the exocytosis of cortical granules. Unlike phase 1, phase 2 is global; it spreads from the site of sperm–egg interaction to the opposite pole of the egg (McCulloh and Chambers, 1991). The message responsible for evoking this wave-like opening of channels throughout the cytoplasm of the egg is believed to be an elevation of free calcium. Evidence in support of this suggestion includes the following:

1. Injection of calcium ionophore A-23187 stimulates a response resembling phase 2.
2. The calcium chelator, EGTA, eliminates the phase 2 increase in conductance.
3. A calcium wave has been shown to travel as a sharp zone across the egg with characteristics similar to that of the putative messenger (Hafner *et al.*, 1988).

The mechanism by which sperm cause these changes in conductance has not been elucidated. The return of membrane potential to its near resting state, i.e. its prefertilization level, constitutes phase 3.

Activation potential:

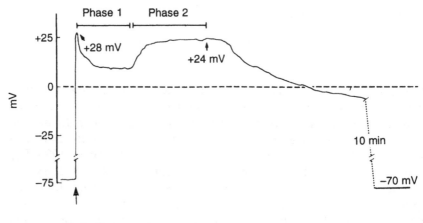

Activation current:

(*V*~m~ held constant by voltage clamp)

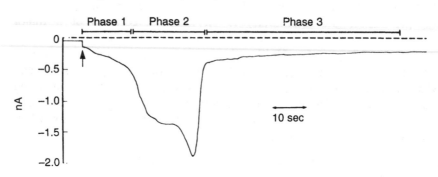

Fig. 5.2 Activation potential and activation current of monospermic sea urchin eggs. See text for details. Reproduced with permission from Chambers, 1989.

It is not clear what happens during the latent period but it is believed to represent the culmination of processes that trigger an autocatalytic calcium wave (Swann *et al.*, 1992). It has been postulated that the sperm (1) acts through a surface receptor coupled to G protein that stimulates inositol trisphosphate (IP_3) production and the Ca^{2+} increase or (2) introduces into the egg a soluble activation factor. If an aqueous pore forms between the sperm and egg during the latent period it is believed to be initially labile and reversible. These aspects have been discussed above (see section 5.2).

Depolarization of the sea urchin egg is related to the initiation of the rapid block to polyspermy postulated by earlier investigators of fertilization. Support for this contention comes from the following observations (Jaffe, 1976; Whitaker and Steinhardt, 1983).

1. Eggs with a fertilization potential greater than 0 mV are not polyspermic.
2. Ova with a more negative fertilization potential are polyspermic.
3. If current is applied to eggs to hold the membrane potential to +5 mV, fertilization is blocked; when the current is released, the membrane potential returns to normal and the eggs fertilize.
4. If current is applied to eggs so that the fertilization potential remains below −30 mV, the eggs become polyspermic.

These observations strongly indicate that an initial, rapid block to polyspermy coincides with an electrical depolarization of the membrane. Similar findings have also been made with eggs of starfish, echiuroid worms and amphibians.

Analyses of voltage-clamped sea urchin eggs (see Chambers, 1989) has provided new insights into our understanding of gamete interaction and egg activation (Figs 5.2, 5.3).

With this method the membrane potential is held invariant, set at a value chosen by the investigator, and changes in current are recorded. When the egg is clamped at a voltage greater than +17 mV, sperm attach to the egg; however, there is no electrophysiological response and the eggs do not activate (Jaffe, 1976). When eggs are clamped between +10 and −100 mV, sperm undergo phase 1 and, depending on the voltage clamp, three different responses may occur (Chambers and McCulloh, 1990; Fig. 5.3). Between +15 and −15 mV essentially every sperm that initiates an electrophysiological response causes a type I activation consisting of the three phases described above. As the clamp is made more negative fewer of the electrophysiological responses are associated with sperm entry and at values more negative than −75 mV sperm entry no longer occurs. An important observation is that each time sperm entry occurs a type I activation response is recorded. Type I responses are replaced by type II and III responses at membrane potentials more negative than −15 mV. Type II and III responses occur after sperm attachment, but sperm entry and cleavage do not occur. The type II response has a single phase and ends with a sharp cut off of current. The type III response is distinguished by a phase 1 current profile which cuts off and remains at the level of

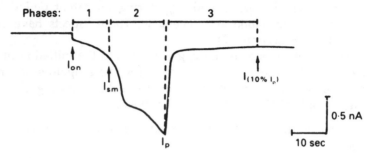

Type I activation — spermatozoon enters (V = −33 mV)

Phases: 1 2 3

I_{on} I_{sm} $I_{(10\% I_p)}$ I_p

0·5 nA

10 sec

Type II transient — no sperm entry (V = −58 mV)

I_{on} I_m

0·5 nA

10 sec

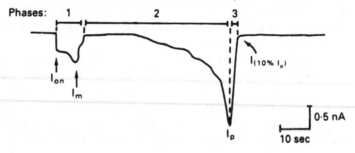

Type III, modified activation — no sperm entry (V = −45 mV)

Phases: 1 2 3

I_{on} I_m $I_{(10\% I_p)}$ I_p

0·5 nA

10 sec

Fig. 5.3 Three types of electrophysiological responses observed after insemination of voltage clamped sea urchin eggs. I_{on} = the initial onset of current; I_m = the maximum amplitude of current of Type II responses or of phase 1 of Type III responses; I_{sm} = the maximum amplitude of current during phase 1; I_p = amplitude of the major current peak during phase 2 of Types I and III responses; $I_{(10\% Ip)}$ = amplitude of current equivalents to 10% of I_p. See text for details. Reproduced with permission from Chambers and McCulloh, 1990.

current required to maintain the egg's membrane potential at the desired value before insemination. After a variable period (up to 40 s) phase 2 occurs, a wave of cortical granule exocytosis ensues, indicative of activation, and as the fertilization membrane elevates the sperm is carried off the egg's surface. These data demonstrate a voltage dependency of sperm entry which may be mediated by an effect of membrane potential on Ca^{2+} influx. Since the increase in membrane conductance during phase 1 is localized to the site of sperm attachment, an altered influx of Ca^{2+} at this site would be expected to affect the concentration of internal Ca^{2+} at the site of the entering sperm (Chambers and McCulloh, 1990).

As shown above, ionic currents associated with the fertilization potential and changes in intracellular ion activities are important for triggering development in sea urchins and also in eggs of other marine invertebrates (Chambers, 1976). External potassium, magnesium and calcium are reportedly not required for development, although divalent cations may be necessary for sperm–egg fusion (Schmidt et al., 1982). Sodium flux at fertilization is crucial since its removal immediately after insemination blocks cytoplasmic alkalinization and does not permit development (Chambers, 1976; see below).

Examination of cross-species fertilization indicates that the blocking voltage for sperm–egg fusion is determined by the sperm species. In studies where sea urchin eggs are inseminated with oyster sperm (Osanai et al., 1987), individual sperm neither produce a depolarization sufficient to block additional sperm entry nor stimulate egg activation. Simultaneous entries of multiple sperm, however, as reflected by the summation of consecutive depolarizations, induce cortical granule exocytosis and egg activation.

In Pallusia, an ascidian, sperm trigger a depolarization of the egg from -80 to $+30$ mV followed by a phase of membrane depolarization comprising two series of membrane oscillations (Goudeau et al., 1992). The oscillations coincide with transients in free Ca^{2+} and the resumption of meiosis (Speksnijder et al., 1990). Experiments by Speksnijder (1992) suggest that cortical components, most probably the endoplasmic reticulum in the cortical region of the myoplasm, are involved in the initiation of the repetitive calcium waves.

In amphibian eggs the potential changes caused by fertilization are due to a chloride flux, which in turn may be Ca^{2+} dependent (Schlichter, 1989). The maximum value is reached within 1–2 s and then decreases over a 20–30 min period. As in the case of sea urchins, if the membrane potential is increased, fertilization can be prevented (Cross and Elinson, 1980); eggs can be made polyspermic if the hyperpolarization that normally accompanies fertilization is prevented. The kinetics and duration of membrane potential changes in anurans have been related to the kinetics and duration of the cortical granule reaction. The membrane potential changes are rapid (1–2 s in duration) but transitory. The cortical granule reaction is much slower (39–117 s postinsemination) but permanent (Schmell et al., 1983). Both the membrane potential change and the cortical granule reaction at fertilization in anurans constitute blocks to polyspermy.

In mammalian eggs the fertilization potential consists of recurring hyperpolarizations coincident with calcium transients (see Miyazaki, 1989; Kline and Kline, 1992; Fissore and Robl, 1993). Fertilized hamster eggs are distinguished by repeated increases in calcium at intervals of 40–120 s where each transient lasts for 12–18 s. Apparently the Ca^{2+} increases throughout the egg cytoplasm. The first hyperpolarizing response occurs when the sperm flagellum stops beating, a sign that is believed to indicate that gamete membrane fusion has occurred (Miyazaki, 1989). Different cycles and amplitudes of Ca^{2+} transients have been noted in other fertilized mammalian eggs (Kline and Kline, 1992; Fissore and Robl, 1993; see Miyazaki et al., 1993 for a review). The first event of activation in the human oocyte is a slow outward current of 300 pA (fertilization current), which is Ca^{2+}-dependent and induces a gradual hyperpolarization of the plasma membrane (Gianaroli et al., 1994).

Direct experimental evidence for a block to polyspermy at the level of the plasma membrane in mammals is limited. The observation that many sperm can penetrate the zona pellucida of a rabbit egg, yet the ovum remains monospermic, is generally believed to indicate the presence of a block to polyspermy at the level of the plasma membrane. However, the mechanism by which this block is achieved has not been determined. Although membrane potential changes are found in hamster eggs that consist of recurring hyperpolarizations beginning with the approximate time of sperm entry, the relation of the hyperpolarization to ionic changes and the block to polyspermy has not been determined. A slow transient depolarization is recorded for fertilized rabbit eggs (McCulloh et al., 1983). The small amplitude of responses compared with the large variations of the resting potentials suggests that the depolarization may be sufficient to block polyspermy. Except for oscillations, the membrane potential of fertilized mouse eggs is constant up to 60 min postinsemination, suggesting that the polyspermy block that is established during this period is not electrically mediated (Jaffe et al., 1983). Polyspermy prevention in the mouse, and possibly in other mammals, occurs at several levels including restrictions in the number of sperm reaching the site of fertilization, blocks at the zona pellucida and at the plasma membrane, and elimination of supernumerary sperm (Yu and Wolf, 1981). The role of cortical granules exocytosis in the prevention of polyspermy is discussed below.

The molecular mechanism by which changes in membrane potential alter sperm attachment and fusion has not been determined. It is presumed that egg plasma membrane topography is altered and prevents further sperm interaction. In surf clam (Spisula) eggs there is no cortical granule reaction nor elevation of a fertilization membrane, yet a complete block to polyspermy is established within 15 s of insemination. It has been suggested that the change in intramembranous particle density of *Spisula* egg plasma membranes at insemination may be related to the development of a block to sperm incorporation (Longo, 1976a).

5.4 CALCIUM FLUXES

The above descriptions of egg membrane depolarization that occur at fertilization not only underline the importance of this change in polyspermy prevention and sperm entry but also emphasize the importance of calcium in egg activation. The first measurements of cellular calcium that drew attention to its potential role in fertilization and the initiation of the cell cycle were conducted by Mazia (1937). Total calcium content was unchanged, indicating that the concentration of free calcium increased at the expense of the bound form. Using calcium-45, fluxes of calcium following fertilization in sea urchins have been demonstrated (Azarnia and Chambers, 1976; Paul and Johnston, 1978). Shortly after Mazia's important observations Moser (1939) demonstrated that calcium was necessary for cortical granule exocytosis, a dramatic, visual demonstration of egg activation. Calcium influx does not appear to be a prerequisite for egg activation in sea urchins, since many parthenogenetic agents activate eggs in the absence of external calcium (Steinhardt *et al.*, 1974, 1977). These observations indicate that activation is accompanied by the mobilization of internal stores of bound calcium. In contrast to the situation in sea urchins, in the surf clam exogenous calcium is required for egg activation (Schuetz, 1975; Kadam *et al.*, 1990).

Using the calcium-dependent, luminescent protein aequorin or calcium-sensitive fluorochromes, a transient increase in calcium has been demonstrated in the eggs of a wide variety of animals including sea urchin, starfish, annelids, ascidians, fish, frog and mammals (Gilkey *et al.*, 1978; Moreau *et al.*, 1978; Eisen *et al.*, 1984; Speksnijder, 1992; Nuccitelli *et al.*, 1993; Miyazaki *et al.*, 1993; Eckberg and Miller, 1995). In sea urchins there is an initial, small Ca^{2+} release, due apparently to membrane depolarization. This is followed by a band of Ca^{2+} release that propagates to the opposite pole of the egg with a velocity of approximately 5 μm/s (Hafner *et al.*, 1988; Mohri and Hamaguchi, 1991; Stricker *et al.*, 1992, 1994). Confocal microscopic observations indicate that the calcium wave passes through the entire egg cytoplasm, although apparently faster around the cortex than through the center of the egg, and precedes exocytosis of cortical granules.

In bivalves Ca^{2+} fluxes differ from those seen in deuterostomes. At insemination there is a transient increase in intracellular Ca^{2+} followed by a period during which intracellular levels remain higher than the resting (unfertilized) level (Deguchi and Osanai, 1994a). During this period in some species there are oscillatory increases while in others there are no oscillations. In *Mytilus* internal Ca^{2+} increased uniformly over the whole of the oocyte.

In the shrimp, *Sicyonia*, resumption of meiosis and formation of the jelly layer and hatching envelope do not require the presence of sperm (Lindsay *et al.*, 1992). Oocytes activate when exposed to Mg^{2+} during spawning (see also Goudeau and Goudeau, 1989). External Mg^{2+} triggers a Ca^{2+}-propagated Ca^{2+} activation wave which is initiated along several fronts at the same time. How Mg^{2+} triggers the release of Ca^{2+} from internal stores within the egg is unknown. A second Ca^{2+} wave occurs which may be brought about by sperm–egg interaction, although conclusive proof of this is wanting (Lindsay *et al.*, 1992).

The possibility of egg activation being initiated by a receptor–ligand interaction of the gametes implies that some aspect of sperm–egg adhesion, mediated by acrosomal components, may trigger calcium release (Gould and Stephano, 1989). Attachment of bindin itself to sea urchin eggs, however, does not induce the cortical granule reaction although this negative result needs to be cautiously interpreted, as a variety of explanations are possible (Vacquier and Moy, 1977). Membrane depolarization at fertilization may trigger the initial calcium release; although depolarization by direct injection of current into sea urchin eggs does not bring about activation (Jaffe, 1976). Membrane fusion *per se* does not induce the cortical granule reaction, and presumably Ca^{2+} release, as demonstrated by the egg–egg fusion experiments of Bennett and Mazia (1981a, b). However, the fusion of mammalian oocytes results in the production of a cell capable of initiating embryonic development (see Nogués *et al.*, 1995). Whether or not activation in this case is dependent on a Ca^{2+} flux has not been determined.

The calcium influx associated with membrane depolarization suggests that it may take place via voltage-gated channels (see also Shen and Buck, 1993). Sea urchin eggs can be depolarized by elevated potassium, which opens calcium channels. Although the rate of calcium entry in this case is not equivalent to that observed at fertilization, the level achieved is the same as that of fertilized eggs, yet the ova do not activate (Schmidt *et al.*, 1982). This could mean that calcium influx alone is not adequate to activate the egg. Another possibility is that the egg in this situation sequesters entering calcium so that its free concentration never reaches a level sufficient for activation. The 'calcium bomb' hypothesis of egg activation (Jaffe, 1983) proposes that a small amount of calcium is released at fertilization, possibly from the fusing sperm itself; this in turn triggers additional calcium release from an internal store, the cortical endoplasmic reticulum, and a self-propagating calcium response. Because experiments injecting calcium into eggs have been generally unsuccessful in eliciting a propagated calcium wave (Swann and Whitaker, 1986) and numerous investigations demonstrating an involvement of various agonists (IP_3, cyclic ADP ribose) and receptors (G protein) in egg activation, this proposal has been extensively modified. It is apparent that the question of how the Ca^{2+} wave is initiated is unclear. Once initiated, however, the wave may be propagated by Ca^{2+} itself, i.e. by a Ca^{2+}-induced Ca^{2+} release or by a Ca^{2+} sensitized IP_3-induced Ca^{2+} release. Further aspects concerning the generation and propagation of the calcium wave in eggs and its role in cortical granule exocytosis are described in Chapter 6.

5.5 INTRACELLULAR pH CHANGE

In addition to the dramatic change in calcium that eggs undergo at fertilization there is an increase in intracellular pH. In sea urchin eggs the internal pH changes from 6.9 to 7.3 with a concomitant increase in acidity of the sea water; the egg remains alkaline for approximately 60 min (Johnson *et al.*, 1976; Shen and Steinhardt, 1978; Fig. 5.4).

There is evidence suggesting that the eggs of other organisms also undergo similar changes in internal pH during activation (Shen, 1983). Fertilization of starfish eggs (*Asterias*) includes the production of acid in the surrounding sea water which suggests that a rise in internal pH occurs with activation. Acid production is not universal among eggs of

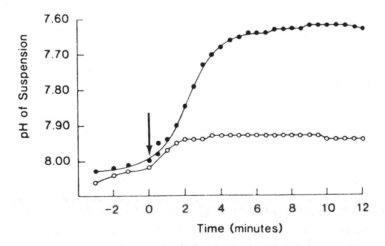

Fig. 5.4   pH of sea urchin (*Arbacia*) eggs suspended in sea water (filled circles) and sodium-free sea water (open circles) at various times after activation with calcium ionophore A-23187. Reproduced with permission from Carron and Longo, 1982.

invertebrates, e.g. it is not seen in fertilized *Mytilus* (lamellibranch) and *Ascidia* (tunicate) ova. Possible changes in intracellular pH in these species have not been extensively investigated. There have been few studies of pH changes in eggs of vertebrates (see Nuccitelli, 1991). Its role in amphibian eggs is unclear and its occurrence in fertilized mammalian eggs has not been investigated (Grandin and Charbonneau, 1989; House, 1994).

In sea urchins the increase in internal pH at fertilization requires sodium, as acid production is a result of an efflux of protons from eggs and is a function of the external sodium concentration (Johnson *et al.*, 1976). This also indicates that the sodium and proton fluxes are coupled. In addition, amiloride, which inhibits passive sodium flux, blocks acid production and eggs remain unactivated. The exchange of sodium and proton has a 1:1 stoichiometry that lasts for about 3 min and is not inhibited by sodium cyanide.

In addition to sodium–proton fluxes there is a sodium–potassium exchange, detectable within minutes of insemination, that causes the sodium content of the egg to fall below that of the unfertilized egg. This exchange is energy-dependent and reduced in sea water with a low potassium concentration (Girard *et al.*, 1982). That the ratio of sodium/potassium can be minimized by lowering the external potassium concentration suggests that sodium–potassium exchange is mediated by a sodium/potassium ATPase. Investigations with sea urchins (*Paracentrotus*) show that after fertilization the alkalinity of the cytoplasm is maintained by two mechanisms: a sodium–proton exchange with the same characteristics as in unfertilized eggs and an acid-extruding pump that is dependent on external sodium, is amiloride-sensitive and requires metabolic energy (Payan *et al.*, 1983).

Although acid production and cytoplasmic alkalinization are sodium-dependent and amiloride-sensitive, a number of observations indicate that they may not be directly coupled (Shen, 1983).

1. The role of acid production is linearly dependent on external sodium, while the rate of cytoplasmic alkalinization is independent of external sodium concentrations above a minimal concentration necessary for egg activation.
2. Eggs activated with NH_4Cl and then washed in fresh sea water release acid upon fertilization. Suspension of sea urchin eggs in NH_4Cl (5 mmol/l, pH 8.0) induces cytoplasmic alkalinization and stimulates their metabolism and ability to undergo chromosome replication and condensation without triggering the cortical granule reaction (Epel *et al.*, 1974). However, fertilization of NH_4Cl-pulsed eggs does not induce cytoplasmic alkalinization.
3. Acid production starts 30–60 s postinsemination and is completed

2–4 min later. The rise in intracellular pH begins 60–90 s postinsemination and is completed by 6–8 min postinsemination.
4. The amount of acid produced varies with the external pH; at pH 9 it
   is greater than at pH 7.

Observations from a number of different laboratories indicate that
phosphatidylinositol bisphosphate turnover occurs naturally during egg
activation, giving rise to IP_3 and diacylglycerol (DAG; see Nuccitelli,
1991). The role of IP_3 upon fertilization to release Ca^{2+} from intracellular stores is dealt with in Chapter 6; DAG in the fertilized egg has been
shown to stimulate the internal pH increase (Crossley et al., 1991). The
effect of DAG is 'indirect', i.e. via the stimulation of protein kinase C
(PKC) which in turn activates the Na^+–H^+ antiporter (Swann and
Whitaker, 1985; Shen and Burgart, 1986). Exposure of sea urchin eggs
to DAG stimulates PKC and the PKC-induced pH increase can be
stimulated via the Ca^{2+} calmodulin pathway (Shen, 1989). Shen (1989)
indicates that activation of the Na^+–H^+ antiporter can also occur by a
PKC-independent pathway.

The effects of increased intracellular pH in activated sea urchin eggs
are numerous and varied (Shen, 1983; Table 5.2).

Many of the pH-induced events are not interdependent; for instance,
the increased potassium conductance can be suppressed without
affecting protein synthesis. Activation of DNA synthesis can occur
without increased protein synthesis and protein synthesis can be
enhanced without DNA synthesis or chromosome condensation. The
fact that these pH-induced events are not mutually dependent on one
another suggests that fertilization induces a single pervasive change,

Table 5.2 Processes in sea urchin eggs affected by the
increase in intracellular pH at fertilization: ↑ = increased; – =
no effect; + = promoted (Data compiled from Shen, 1983)

Process	Effect*
Protein synthesis	↑
DNA synthesis	↑
Messenger RNA polyadenylation	↑
Glucose-6-phosphate dehydrogenase activity	↑
Pronuclear development	+
Chromatin condensation	+
Potassium conductance	↑
Glycogenolysis	↑
Thymidine uptake	↑
Cortical granule reaction	–

which then activates a series of independent fertilization responses (Epel, 1978). The mechanism by which the increase in pH induces such diverse responses is not known. One possibility is that cytoplasmic alkalinization affects specific enzymatic activities and protein–protein interactions, which lead to regulatory modifications such as phosphorylation. Activation of PKC and Ca^{2+} calmodulin kinase have been proposed as second messenger pathways coupling gamete fusion and Na^+–H^+ antiporter activity (see Shen and Buck, 1990).

In mollusks there is evidence to suggest that the regulatory effects of internal pH changes may be linked to specific stages of meiosis (see Deguchi and Osanai, 1994b). It has been proposed that an increase in internal pH is insufficient and may be unnecessary for germinal vesicle breakdown in *Spisula* and *Barnea*, which are normally fertilized at prophase I (Dubé, 1988; Dubé and Coutu, 1990). However, in *Patella* an increase in intracellular pH appears to be essential for germinal vesicle breakdown (Guerrier *et al.*, 1986). Experiments employing *Hiatella* oocytes treated with serotonin to reinitiate meiosis suggest that both an internal calcium increase and an internal pH rise are responsible for meiosis reinitiation from prophase I and that one can compensate for the loss of the other (Deguchi and Osani, 1995). Unlike prophase I oocytes, metaphase I oocytes did not demonstrate a significant internal pH increase following serotonin exposure or insemination.

In an effort to determine causal relationships between structural and

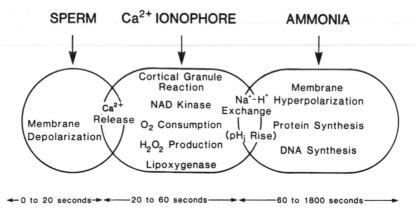

Fig. 5.5 Postactivation sequence of events by sperm, calcium ionophore or ammonia. The normal sequence, initiated by sperm, results in the initial membrane depolarization, calcium release and all subsequent changes. Calcium ionophore does not initiate the initial membrane depolarization, but triggers subsequent events. Incubation in ammonia or other weak bases induces the rise in intracellular pH. The earlier changes do not take place, but those changes following the intracellular pH rise are initiated. See Epel, 1980.

metabolic changes that are initiated by fertilization or artificial activation, Epel (1980) has called attention to the fact that the sequence of fertilization/activation can be divided into two temporarily distinct series of events (Fig. 5.5).

The first, which includes early processes, such as the cortical granule reaction, activation of NAD kinase, lipoxygenase and an increase in oxygen consumption, is triggered by the transient rise in calcium ions. The second, which includes relatively later processes, such as membrane hyperpolarization, an increase in protein synthesis and the initiation of DNA synthesis, arises from the earlier calcium increase and is due to the transient activation of a sodium–proton exchange with a resultant rise in pH.

Blocks to polyspermy and the cortical granule reaction

6

In most animals fertilization is normally monospermic, i.e. only one sperm enters the egg and the paternal and maternal chromosomes come together to constitute the embryonic genome. The entry of more than one sperm leads to abnormal development in many invertebrates and vertebrates and is referred to as pathological polyspermy. In some animals, particularly those with large eggs such as urodeles, reptiles and birds, multiple sperm entry is common. However, only one male pronucleus becomes associated with the female pronucleus to constitute the embryonic genome. This situation is known as physiological polyspermy. To prevent the lethal effects of polyspermy, specific mechanisms, or blocks to polyspermy, have evolved to allow only one sperm to enter the ovum or to participate in the development of the embryonic genome. Such mechanisms include the following:

1. a rapid block, characterized as transient and incomplete, which involves the depolarization of the egg plasma membrane (Chapter 5);
2. a slower block, usually involved with the exocytosis of cortical granules from the egg, which is complete and accomplished within minutes of insemination;
3. in some cases monospermy is ensured by specialized functions of the egg cytoplasm which include processes such as the degeneration of 'extra' incorporated sperm nuclei or the actual sloughing of cytoplasm containing 'extra' sperm nuclei from the egg.

The eggs of many animals incorporate more than one mechanism to prevent polyspermy. In this chapter, cortical granule structure, function and release are discussed.

6.1 CORTICAL GRANULES

The cortex of the sea urchin egg is lined with a layer of cortical granules about 1 μm in diameter (Fig. 6.1).

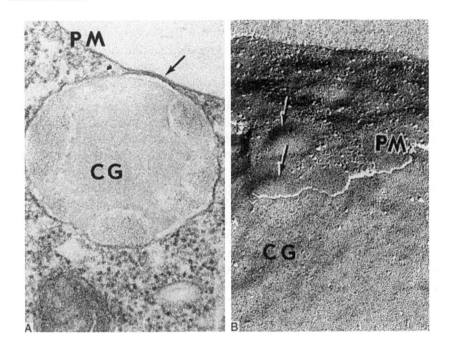

Fig. 6.1 (A) Cortical granule (CG) of a sea urchin (*Arbacia*) egg closely associated with the plasma membrane (PM). The arrow depicts a dome-shaped elevation where the two structures are closely associated. **(B)** Freeze-fracture replica of
a cortical granule (CG) from a sea urchin (*Arbacia*) egg and a portion of the P-
face of the plasma membrane (PM), showing dome-shaped elevations (arrows)
where the two structures are closely associated with one another. See Longo,
1981a.

In *Strongylocentrotus* there are about 18 000 of these organelles per
egg (Vacquier, 1981). They are manufactured by the Golgi complex and
become closely associated with the plasma membrane during oocyte
development (Anderson, 1968). Ultrastructurally, the cortical granules
of sea urchin eggs appear to constitute a homogeneous population of
organelles; however, the composition of those of *Strongylocentrotus* has
been shown to be heterogeneous (Anstrom *et al.*, 1988). Cortical
granules of sea urchins display variations in organization depending
upon the species. The content of *Arbacia* cortical granules is distinguished by a central scalloped mass, surrounded by some lenticular
material (Fig. 6.1(A)). In *Strongylocentrotus* there are two morphologically recognizable components:.

1. There is a spiral of electron-dense material which contains peroxidase activity (Katsura and Tominga, 1974), as well as a protein

having a LDL-receptor motif. This protein may be involved in specific protein–protein interactions that occur during cortical granule biogenesis or the cortical granule reaction (Wessel, 1995).
2. At one pole of the cortical granule is some amorphous material that consists of hyaline (Hylander and Summers, 1982a).

The complex organization of cortical granule internal components of other organisms has also been described, e.g. the mollusk, *Mytilus* (Humphreys, 1967; Fig. 6.2), and fish (Ginsburg, 1987).

The cortical granules in amphibian and mammalian eggs do not show unusually complex patterns and are filled with electron-dense granular material (Gulyas, 1980; Fig. 6.3).

At least two distinct morphological types are recognizable in mammalian oocytes, light and dense. Whether or not they represent differences in composition has not been determined.

Cortical granule content from sea urchins, amphibians and mammals has been examined directly by biochemical and cytochemical techniques and indirectly by analysis of the medium following their discharge (Schuel, 1978; Hoodbhoy and Tolbot, 1994). Serine protease and sulfated mucopolysaccharides appear to be universal components

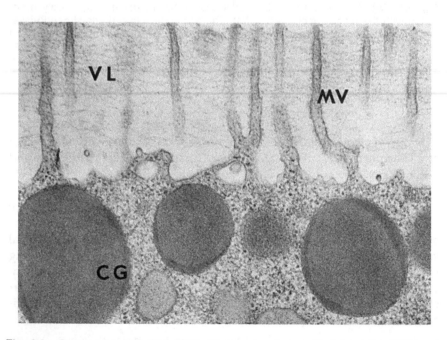

Fig. 6.2 Cortex of a surf clam (*Spisula*) egg showing cortical granules (CG) and microvilli (MV). VL = vitelline layer.

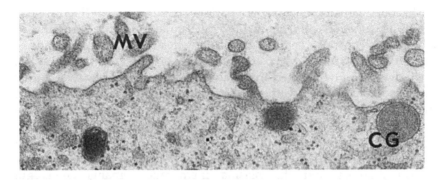

Fig. 6.3 Cortex of a hamster egg showing microvilli (MV) and cortical granules (CG).

of these structures. Peroxidase, β1-3-glucanase, hyaline protein, β-glucuronidase and other proteins are also present in the cortical granules of some organisms (Table 6.1).

The cortical granules of sea urchins possess a high concentration of calcium (Poenie and Epel, 1987; Gillot *et al.*, 1991); at present, there is no evidence indicating that cortical granule calcium participates in the increase of intracellular calcium at fertilization. A protein of 75 kDa, present within the cortical granules of mouse eggs and released upon

Table 6.1 Constituents of cortical granules; − = not present/not determined; + = present (See Schuel, 1978)

Component	Organism	Localization			
		Morpho-logical	*Bio-chemical*	*Cyto-chemical*	*Secretory*
Calcium	Sea urchin	−	+	−	−
Protease	Sea urchin	−	+	−	+
	Mammal	−	−	−	+
Peroxidase	Sea urchin	−	+	+	+
Sulfated	Sea urchin	−	+	+	+
mucopolysaccharide	Amphibian	−	−	+	−
	Mammal	−	−	+	+
β,3-glucanase	Sea urchin	−	+	−	+
Hyaline protein	Sea urchin	+	−	−	+
	Amphibian	+	−	+	+
Acid phosphatase	Sea urchin	−	−	+	−
	Mammal	−	−	+	−
β-glucuronidase	Amphibian, reptile and bird	−	−	+	−

fertilization/activation has been partially characterized (Pierce *et al.*, 1990); its function remains to be determined. Cortical granules of mammalian eggs also contain glycoconjugates having mannose and/or glucose residues and a trypsin-like protease (Cherr *et al.*, 1988). The function of the glycosylated material released from mammalian cortical granules is unknown, although it may participate in polyspermy prevention (Hoodbhoy and Talbot, 1994). Cortical granule proteinases modify ZP2 to $ZP2_f$, which may lead to zona hardening (see below).

Although the structure of the sea urchin egg cortex has been analyzed by a number of techniques, the nature of the association of cortical granules and the plasma membrane remains an enigma (Detering *et al.*, 1977). Analyses of the sea urchin egg cortex show the cytoplasmic region associated with the cortical granules and plasma membrane to be relatively unspecialized, lacking any apparent modification which might serve to attach the two structures. However, the connection of cortical granules to the oolemma is sufficiently strong to survive forces encountered during the isolation of plasma-membrane–cortical-granule complexes. The normal attachment of sea urchin eggs to the overlying plasma membrane can be disrupted by urethane and tertiary amines, suggesting that a special attachment exists between the two structures (Hylander and Summers, 1981). Two synaptosomal associated proteins, synaptobrevin and SNAP25, may be involved in cortical granule delivery, docking and fusion (Steinhardt *et al.*, 1994).

Modifications of sea urchin egg plasma membranes have been observed in areas occupied by cortical granules using freeze-fracture replication, scanning electron microscopy and filipin staining for 3-β hydroxysterol components (Longo, 1981a; Carron and Longo, 1983). The plasma membrane modifications seen with freeze-fracture replication are dome-shaped areas lacking intermembranous particles (Fig. 6.1). These may allow specific contacts between the plasma membrane and the cortical granules thereby facilitating bilayer fusion. Unique patterns of intramembranous particles and particle-free areas, within the plasma membrane and possibly induced by underlying structures, have been described in numerous cells having secretory activities. Distinct clearings of intramembranous particles have been observed in portions of the plasma membrane associated with secretory vesicles and are generally considered to represent areas depleted of membrane proteins in the fusion zone.

6.2 CORTICAL GRANULE REACTION: THE PROCESS

In sea urchins, which have been studied extensively regarding polyspermy blocks, the electrically mediated fast block to polyspermy is followed by the discharge of vesicles from the egg cortex – the cortical granule reaction. The cortical granule reaction is characteristic of eggs from many organisms, including echinoderms, fish, amphibians and mammals. In some animals, such as pelecypods and some annelids, cortical granules are present but do not undergo exocytosis nor change with fertilization, nor do extraneous layers develop around the egg. Nevertheless, a block to polyspermy takes place within seconds after fertilization or artificial activation in these organisms. The eggs of some organisms, e.g. the ascidian *Ciona*, reportedly do not have cortical granules. At insemination in *Ciona*, material reportedly contained in subcortical vesicles is released from the ovum (Rosati *et al.*, 1977). In *Ophiopholis* (an ophiuroid) the cortical granule population forms a layer five to six deep, and at insemination they move to and fuse with the egg surface (Holland, 1979). A similar massive exocytosis of cortical granules also occurs in the shrimp (*Penaeus*) and cnidarian (*Bunodosoma*) (Dewel and Clark, 1974; Clark *et al.*, 1980).

The plasma membrane of the sea urchin egg is reflected into relatively short microvilli that lack a core of actin microfilaments. The underlying cortical granules tend to be situated in areas that lack microvilli (Schroeder, 1979). Attached to the sea urchin oolemma is a glycocalyx, or vitelline layer (Anderson, 1968). It is this structure to which sperm bind via bindin and which, at the time of cortical granule exocytosis, becomes detached from the egg surface to form the fertilization membrane (Fig. 6.4).

The intervening space, i.e. the region between the elevated vitelline layer and the egg plasma membrane, filled with secretory materials derived from cortical granules, is the perivitelline space (Millonig, 1969; Chandler and Heuser, 1980). It is important to note that the elevation of the vitelline layer to form a fertilization membrane, as in the case of sea urchins, is not a feature common to the eggs of all animals that undergo cortical granule exocytosis at fertilization.

Studies show that the sperm initiates a wave of exocytosis that traverses the egg at about 10–20 µm/s (Jaffe, 1983). The kinetics of this wave are consistent with those of an autocatalytic process. There are observations, however, which are not entirely consistent with the notion that the cortical granule reaction is propagated by autocatalysis. For example, localized cortical granule discharges have been observed under a variety of experimental conditions (Chambers and Hinkley, 1979), suggesting that the release of a cortical granule does not automatically induce the discharge of neighboring granules. Furthermore, a wave of cortical granule discharge can pass through areas lacking cortical granules; in other words, transmission of the stimulus for exocytosis does not require the presence of cortical granules.

Exocytosis of cortical granules has been studied in echinoderms,

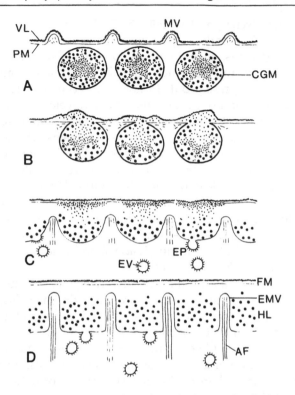

Fig. 6.4 Diagrammatic representation of cortical granule discharge, fertilization membrane formation, hyaline layer development, microvillar elongation and the initiation of endocytosis in sea urchin eggs. **(A)** Cortex of the egg depicting cortical granules, plasma membrane (PM), vitelline layer (VL) and short microvilli (MV). CGM = cortical granule membrane. **(B)** Cortical granule discharge and vitelline layer elevation. **(C)** and **(D)** A portion of the cortical granule contents has joined with the vitelline layer to form the fertilization membrane (FM). The remaining cortical granule material remains in the perivitelline space to become the hyaline layer (HL). Portions of the plasma membrane are involved in endocytosis as evident by the presence of endocytotic pits and vesicles (EP and EV). The surface of the fertilized egg is projected into elongate microvilli (EMV) containing a core of actin filaments (AF).

amphibians and mammals and appears to involve similar processes in these species (Figs 6.4, 6.5).

The potential roles of adhesion, dehydration and osmotic swelling of cortical granules at the time of cortical-granule-membrane–plasma-membrane fusions have been reviewed (Zimmerberg, 1988). Calcium may play a role in the electrostatic attraction of the cortical granules to the plasma membrane; activation of serine proteases and phospholipases may also be involved with the fusion of the cortical granule membrane and plasmalemma. Using transmission electron microscopy

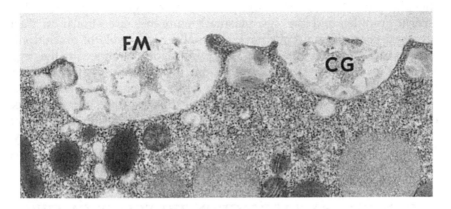

Fig. 6.5 Cortical granule discharge in the sea urchin *Arbacia*. CG = exocytosing cortical granule; FM = developing fertilization membrane.

Anderson (1968) and Millonig (1969) indicate that opening of a cortical granule may occur via multiple fusions between the cortical granule membrane and oolemma, thereby forming a series of vesicles, composed of membrane derived from both the cortical granule and oolemma, which are then released to the perivitelline space. Whether this scheme is correct is important in establishing qualitative and quantitative changes of the egg surface area at fertilization. Using freeze-fracture replication Chandler and Heuser (1979) were unable to find intermediate stages of cortical-granule-membrane–plasma-membrane fusion, suggesting that the fusion process is completed very rapidly. They indicated that a single pore is formed, which increases in size to allow dehiscence of the cortical granule contents. This suggests that virtually all the membrane delimiting a cortical granule is incorporated into the egg plasmalemma when the two structures fuse. The 'incorporation' of cortical granule membrane into the egg plasmalemma has also been demonstrated by changes in membrane capacitance (Jaffe *et al.*, 1978).

Phospholipase A_2, a calcium-dependent enzyme that cleaves the fatty acid at the second acyl position of phospholipids, may play a role in the cortical-granule reaction of sea urchin eggs. Unsaturated fatty acids such as arachidonic acid are found at the second acyl position and phospholipase A_2 activity leads to the formation of arachidonic acid and lysophosphoglyceride, which may mediate fusion of the cortical granules with the egg plasma membrane (Schuel, 1978). Several lines of evidence support the involvement of arachidonic acid and lysophosphoglyceride in the exocytosis of cortical granules (Ferguson and Shen, 1984). Phospholipase A_2 is present in the sea urchin egg and melittin, a phospholipase A_2 activator, triggers the cortical granule reaction. The cortical granules and the egg plasma membranes are similar in their fatty acid composition: both have unusually high levels of arachidonic acid. At fertilization the level of phosphatidylcholine in the egg decreases twofold; the level of arachidonic acid in phosphatidylcholine, phosphatidylinositol-phosphatidylserine and phosphatidylethanolamine also decreases. A transient activation of lipoxygenase converts free arachidonic acid to hydroxy fatty acid (HETE, hydroxyeicosatetraenoic acid), which may be important in regulating membrane permeability. The role of lipids, particularly phosphatidylinositol, in generating the calcium flux characteristic of egg activation is discussed below.

6.3 CORTICAL GRANULE REACTION: THE ROLE OF CALCIUM

The current paradigm regarding the mechanism of cortical granule discharge is that calcium functions as an essential intracellular messenger. This is based on the following observations in eggs from a variety of different organisms.

1. Isolated preparations of cortical granules (cortical granule lawns) can be made to exocytose by the addition of calcium (Vacquier, 1975).
2. Calcium ionophore, A-23187, triggers the cortical granule reaction in sea urchin eggs without external calcium (Steinhardt et al., 1974; Chambers et al., 1974).
3. Studies in the fish medaka and in sea urchins are consistent with a self-propagating calcium-induced calcium release that is initiated by the sperm and sweeps through the egg in a wave-like fashion (Gilkey et al., 1978; Eisen et al., 1984; Hafner et al., 1988; Stricker et al., 1992).
4. The cortical granule reaction is inhibited in sea urchin and Xenopus eggs microinjected with calcium chelators (Zucker and Steinhardt, 1978; Kline, 1988).

In addition, electrophysiological and histochemical studies (Kline and Stewart-Savage, 1994) demonstrate that cortical granule exocytosis in fertilized hamster eggs is closely coupled to periodic increases in intra-cellular calcium.

The increase in intracellular calcium that occurs at fertilization in sea urchins has been shown to be both necessary and sufficient for egg activation. There is considerable variation in sea urchins, but generally the level of free Ca^{2+} in the unfertilized egg is approximately 0.2 μmol/l and rises to 10–20 times this value at fertilization. Studies by Mohri and Hamaguchi (1991), however, indicate that more than 150 μmol/l Ca^{2+} is released at fertilization, but the increase is only to 2 μmol/l because of sequestration by components within the cytoplasm. Some

20–30% of the total Ca^{2+} is released from the egg at fertilization and this is believed to be from cortical granules (Gillot et al., 1990).

Any transient change in Ca^{2+} concentration in the cytosol has to be set against the buffering capacity of the cytoplasm (see Clapham, 1995) and reflects a balance between mechanisms that allow Ca^{2+} to move into the cytosol and mechanisms that sequester or bind Ca^{2+}. The ability of the cytoplasm to bind and sequester calcium is believed to be responsible for the localized activation induced by microinjection of Ca^{2+} into eggs (Swann and Whitaker, 1986) and the inability of such a procedure to propagate a wave of Ca^{2+} release in sea urchin eggs. This is in contrast to the Ca^{2+}-induced Ca^{2+} release that is observed in sea urchin egg homogenates (Galione et al., 1991). The injection of pure Ca^{2+} solutions rather than buffered solutions is problematic because eggs have a rather strong Ca^{2+} buffering capacity and vesicles may form around the injection pipette (Nuccitelli, 1991). Therefore, negative results when using unbuffered Ca^{2+} injection solutions are not conclusive. These and observations with somatic cells (Clapham, 1995) have stimulated researchers in this area to identify and characterize agents that regulate Ca^{2+} levels in eggs.

The threshold concentration of calcium necessary to induce the cortical granule reaction in vitro depends upon a number of factors including: the preparations of eggs, the ionic composition of the suspending medium, the concentration of ATP and the EGTA–calcium-binding constant used to compute the calcium concentration (Shen, 1983). A concentration of 9–18 μmol/l calcium results in cortical granule discharge from isolated cortical granule preparations (Steinhardt et al., 1977). Values of 1–3 μmol/l affect cortical granule discharge when ATP is supplied to the medium and the cortices are prepared in solutions closely resembling the ionic composition of the egg cytoplasm (Baker and Whitaker, 1978). With isolated cortical granule preparations from the sea urchin Hemicentrotus, two classes of calcium sensitivity in different media have been demonstrated (Sasaki, 1984). In Hemicentrotus there is also evidence for a dissociable protein that may regulate the calcium sensitivity of cortical granule discharge.

Although immature sea urchin oocytes possess cortical granules, they are unable to undergo cortical granule exocytosis when fertilized and do not develop a block to polyspermy (Longo, 1978b). Ultrastructural examination of developing sea urchin oocytes has revealed that cortical granules do not assume a close apposition with the plasma membrane until late oogenesis. The absence of a close association of the cortical granule membrane and the egg plasma membrane is believed to be one of the reasons for failure of cortical granule exocytosis. A similar situation has also been described for mammalian oocytes (Ducibella et al., 1988a, b): mouse germinal vesicle oocytes are not competent to establish a block to polyspermy and undergo cortical granule exocytosis. However, the basis for this lack of competency to undergo the cortical granule reaction may also be related to additional factors. For example, germinal vesicle oocytes can elevate increasingly higher levels of intracellular calcium as meiotic maturation progresses (Peres, 1990; Tombes et al., 1992; Fujiwara et al., 1993). Ducibella and Bruetow (1994) have shown that the capability of mouse oocytes to undergo a global sperm-induced cortical granule loss develops very near or at metaphase II. Insemination of earlier stages, e.g. metaphase I oocytes, is distinguished by a release of cortical granules localized to the site of sperm entry. There is no propagation of cortical granule exocytosis. The failure to undergo the cortical granule reaction in immature starfish oocytes is believed to be a result of a deficiency in the IP_3-induced Ca^{2+} release mechanisms, which is responsible for the spread of the wave of activation (Chiba et al., 1990). In hamster oocytes the IP_3-induced Ca^{2+} release mechanism develops in two phases during maturation (Fujiwara et al., 1993). Sensitivity to IP_3 increases gradually between the germinal vesicle stage and prometaphase of meiosis I. This is followed by the development of a nonlinear dose–response relation during meiosis II, and finally in the regenerative, propagating Ca^{2+} release of the mature egg.

6.3.1 ENDOPLASMIC RETICULUM AND CALCIUM

With respect to the role of calcium in egg activation and cortical granule exocytosis, two questions have been the impetus for research in this area for the past 10 years (Gillot et al., 1990). What is the source of intracellular calcium and how is the calcium wave initiated and propagated through the egg cytoplasm? Although the roles of other organelles in the initiation and propagation of the calcium wave at fertilization (e.g. mitochondria and cortical granules) have not been clearly defined, it is highly probable that the endoplasmic reticulum function is the major source of calcium for egg activation. Evidence to support this notion can be summarized as:

1. the organization of the egg's endoplasmic reticulum and its morphological similarity to the sarcoplasmic reticulum of skeletal muscle cells;
2. the presence of proteins involved with calcium sequestration and release;
3. the ability of microsomes derived from the endoplasmic reticulum to accumulate and release calcium (Oberdorf et al., 1986);
4. the direct visualization of calcium in cisternae of the endoplasmic reticulum.

Studies employing a variety of agonists and antagonists have demonstrated the release of calcium from different stores in sea urchin eggs (Zucker *et al.*, 1978). A-23187 and monoelectrolytic activators release calcium from an undefined source, as occurs during fertilization, whereas hypertonic activating treatments release calcium from what appears to be a different cellular compartment. The release of calcium appears to involve an all-or-nothing phenomenon and a repeat release is possible after about 40 min. This 40-min period is believed to represent the time required to recharge the internal store. Energy is required to fill the stores, since depletion of ATP blocks the cortical granule reaction and there is an ATP-dependent binding of calcium-45 in isolated cortices (Baker and Whitaker, 1978). Consistent with these observations are investigations demonstrating a calcium-dependent ATPase in sea urchin cortices (Mabuchi, 1973). Interestingly, agents that inhibit calcium release from the sarcoplasmic reticulum, such as sodium dantrolene, have no effect on cortical granule discharge in sea urchin eggs at fertilization.

Specialized regions of the egg endoplasmic reticulum that are associated with the cortical granules in amphibian (*Xenopus*), sea urchin and mouse ova have been identified morphologically and have been implicated as sites of Ca^{2+} storage and release (Gardiner and Grey, 1983; Fig. 6.6).

This suggestion is derived from observations demonstrating the following: Muscle contraction is mediated by calcium release from a specialized subsurface cisternum of endoplasmic reticulum, the sarcoplasmic reticulum. There is a striking morphological similarity of the plasma membrane–endoplasmic reticulum association is observed in *Xenopus*, mouse and sea urchin eggs to the transverse tubule and sarcoplasmic reticulum of muscle cells (Gardiner and Grey, 1983; Sardet, 1984; Luttmer and Longo, 1985). However, the significance of this morphological relationship, if any, with respect to the interaction and propagation of the Ca^{2+} wave in eggs has not been established. A temporal correlation in the development of the cortical endoplasmic reticulum and capacity of *Xenopus* eggs to propagate a wave of cortical granule exocytosis has been noted (Charbonneau and Grey, 1984; Campanella *et al.*, 1984). In the anuran *Discoglossus* egg the site of gamete interaction is distinguished by an invagination (dimple) that possesses an extensive arrangement of endoplasmic reticulum (Gualtieri *et al.*, 1992). At fertilization the organization of endoplasmic reticulum within the dimple undergoes a dramatic reorganization (Campanella *et al.*, 1988). As demonstrated in medaka and sea urchin (Gilkey *et al.*, 1978; Eisen *et al.*, 1984), the calcium wave is still present when cortical granules are centrifuged to one pole of the egg, indicating that the cortical granules are not the direct source of the calcium wave and another egg compartment is involved, i.e. the cortical endoplasmic reticulum.

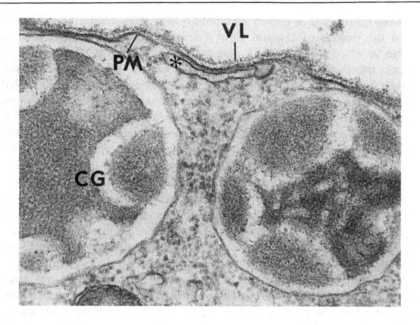

Fig. 6.6 Cortex of a sea urchin (*Arbacia*) egg showing cortical granules (CG) and a cisternum of endoplasmic reticulum (*) closely associated with the egg plasma membrane (PM). VL = vitelline layer. See Luttmer and Longo, 1985.

Electron microscopic studies (Luttmer and Longo, 1985) have shown that in sea urchins the endoplasmic reticulum in the egg cortex (cortical endoplasmic reticulum) forms a basket-like structure around individual cortical granules (see also Sardet, 1984; Fig. 6.7).

Using confocal microscopy of sea urchin eggs injected with the lipophilic dye dicarbocyanine (DiI), Terasaki and Jaffe (1991) have substantiated these ultrastructural observations and have shown that the endoplasmic reticulum of the egg consists of two interconnected forms (Fig. 6.8): (1) a tubular cortical network that appears as a honeycomb, particularly in isolated preparations of the cortex (Terasaki *et al.*, 1991; Fig. 6.9) and (2) a network of lamellar sheets located within the subcortex.

Interestingly, the endoplasmic reticulum undergoes a fragmentation at fertilization (Jaffe and Terasaki, 1993). The significance of this alteration may be related to the breakdown and reformation of the sperm nuclear envelope (see below) and/or to facilitate sperm entry/pronuclear migration. With respect to the latter possibility, the endoplasmic reticulum may be viewed as a meshwork that could impede the migration of large organelles; hence its breakdown would allow greater movement of the incorporated sperm nucleus and pronuclei.

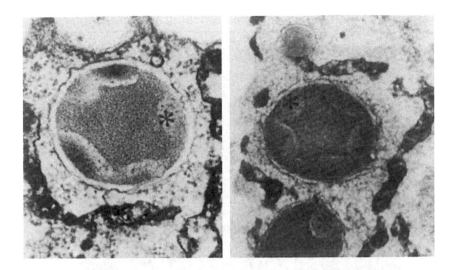

Fig. 6.7 Cortices of *Arbacia* eggs depicting cortical granules partially surrounded by cisternae of endoplasmic reticulum that has been impregnated with electron-dense material. Reproduced with permission from Luttmer and Longo, 1985.

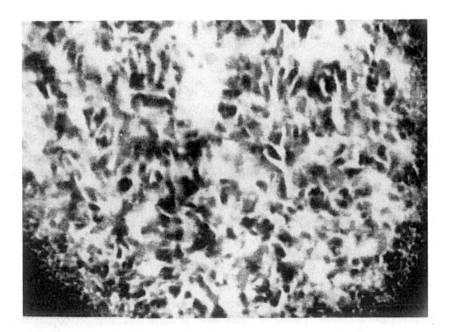

Fig. 6.8 Confocal image of a sea urchin egg injected with the lipophilic dye DiI. The dye was picked up by the endoplasmic reticulum and spread throughout this membranous system. Lamellar sheets are seen in the central (subcortical) region of the egg while a honeycomb network is present in the cortex. Reproduced with permission from Terasaki and Jaffe, 1991.

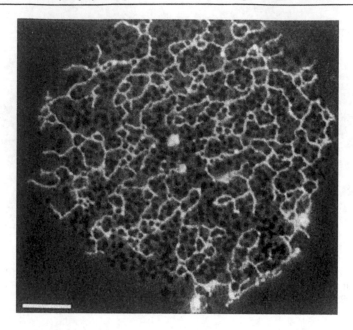

Fig. 6.9   DiI-labeled cortex of a sea urchin egg visualized simultaneously by phase-contrast and fluorescence microscopy. Cortical granules are often encircled by endoplasmic reticulum. Reproduced with permission from Terasaki *et al.*, 1991.

Immunochemical studies (McPherson *et al.*, 1992; Parys *et al.*, 1993) have demonstrated that not only do the cortical and subcortical endoplasmic reticulum of sea urchin eggs differ morphologically, but each is distinguished by virtue of their association with specific calcium channels. Ryanodine receptors have been found only in the cortical endoplasmic reticulum (Fig. 6.10), whereas IP_3 receptors are present in both the cortical and subcortical endoplasmic reticulum (Fig. 6.11).

The localization of these receptors to different compartments in the egg may be functionally significant with respect to the initiation and propagation of the calcium wave at fertilization (Sher and Buck, 1993; see below). The IP_3 receptor, but not the ryanodine receptor, has been localized in *Xenopus* eggs and oocytes (Parys *et al.*, 1993; Kume *et al.*, 1993) and occupies both the cortical and subcortical endoplasmic reticulum. In addition, the endoplasmic reticulum of sea urchin and frog eggs contains other proteins that are present in the sarcoplasmic reticulum of skeletal muscle cells (see Clapham, 1995). For example, calsequestrin and calreticulin (calcium buffering proteins) have been localized and shown to form a tubular network throughout sea urchin and *Xenopus* eggs (Oberdorf *et al.*, 1988; Henson *et al.*, 1989; Parys *et al.*, 1993).

Proteins that function in calcium sequestration and release have also

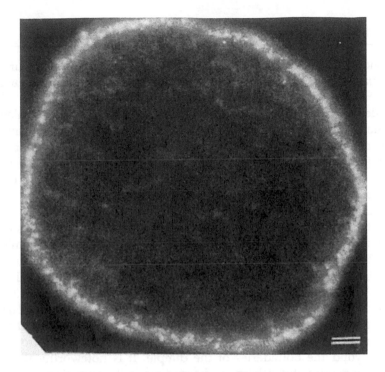

Fig. 6.10 Section of a sea urchin egg stained to demonstrate the cortical localization of the calcium release channel/ryanodine receptor. Reproduced with permission from McPherson *et al.*, 1992.

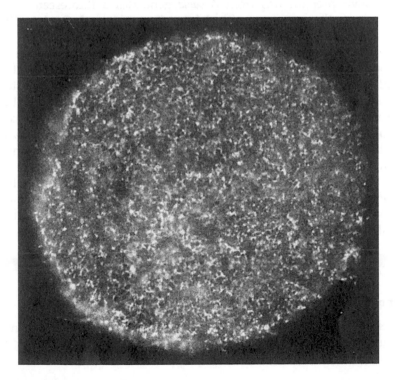

Fig. 6.11 Section of a sea urchin egg stained to demonstrate the cortical and subcortical localization of the IP₃ receptor.

been shown to be associated with microsomal preparations of sea urchin and *Xenopus* eggs. Oberdorf *et al.* (1988) demonstrated that sea urchin microsomal preparations are able to sequester calcium in an ATP-dependent manner and release calcium by additions of calcium ionophore, trifluoroperazine, calcium and IP$_3$. In addition to IP$_3$, microsomal preparations from sea urchin eggs have been shown to be sensitive to a variety of agents known to cause calcium release from isolated sarcoplasmic reticulum including: ryanodine, caffeine and cyclic ADP ribose (Clapper and Lee, 1985; Fujiwara *et al.*, 1990; Galione *et al.*, 1991; Lee, 1991).

Direct visualization of calcium in egg endoplasmic reticulum has been achieved by precipitation of calcium *in situ* followed by electron microscopy (Poenie and Epel, 1987; Gualtieri *et al.*, 1992) or by using fluo-3/AM ester and inspection by fluorescence microscopy (Terasaki and Sardet, 1991). In *Discoglossus* Gualtieri *et al.* (1992) found that 1–3 min after activation precipitated calcium appears on the smooth endoplasmic reticulum that fills the dimpled region where sperm–egg interaction occurs and that there is a decreasing gradient of precipitates from the center beyond the boundaries of the dimple. It is noteworthy that the gradient of precipitated calcium is directed along the animal/vegetal axis and this may portend an animal/vegetal gradient of releasable internal calcium. Fluo-3/AM ester loaded into the endoplasmic reticulum of cortical lawn preparations from *Arbacia* generates a fluorescent signal indicative of the presence of calcium (Terasaki and Sardet, 1991; Fig. 6.12).

When such preparations are incubated with IP$_3$ the fluorescent signal disappears consistent with a release of calcium from the cisternae of the endoplasmic reticulum. In summary, these observations demonstrate the presence of calcium binding and releasing proteins in a cellular compartment known to regulate intracellular calcium in somatic cells. The localization of the IP$_3$ and ryanodine releasing channels to the endoplasmic reticulum of eggs is significant with respect to the initiation and propagation of the calcium wave at fertilization (Shen and Buck, 1993). Despite these details, the presence of calcium channels in membrane delimiting cortical granules has not been fully explored and we have little idea as to the mechanism by which calcium brings about cortical granule fusion with the egg plasma membrane.

6.3.2 CALCIUM RELEASE MECHANISMS

(a) Inositol trisphosphate (IP$_3$)

The role of lipids, particularly phosphatidylinositol, in generating intracellular signals has been demonstrated in a number of different cell

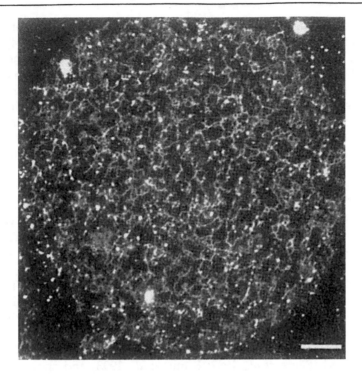

Fig. 6.12 Fluorescent signal from a cortex incubated in 10 µmol/l fluo-3/AM and 1 mmol/l ATP for 20 min. The reticular network is the endoplasmic reticulum, indicating that fluo-3/AM has been cleaved by esterases in the endoplasmic reticulum and that calcium concentrations in these conditions are sufficiently high to produce a signal from fluo-3. Reproduced with permission from Terasaki and Sardet, 1991.

types including eggs (Berridge and Irvine, 1984; Berridge, 1993). Phosphatidylinositol is hydrolyzed by phospholipase C to diacylglyceride (DAG) and IP_3 as part of a signal transduction mechanism for controlling a variety of cellular processes including secretion, metabolism, phototransduction and cell proliferation. DAG operates within the plane of the membrane to activate protein kinase C, whereas IP_3 is released into the cytoplasm to function as a second messenger for mobilizing intracellular calcium. Experiments to identify the inositol-sensitive release site indicate that it is located within the endoplasmic reticulum. Early studies examining the induction of cortical granule dehiscence with microinjections of IP_3 and the increase in polyphosphoinositide turnover in fertilized sea urchin eggs demonstrate that IP_3 and phospholipase C are involved in the autocatalytic cycle of calcium release at fertilization (Whitaker and Irvine, 1984; Turner et al., 1984).

It has been suggested that sperm induce an initial production of IP_3 to generate the calcium increase at fertilization (Stith et al., 1994). How the initial production of IP_3 is generated has not been established. The wave of Ca^{2+} release induced by IP_3 injection crosses Lytechinus eggs with a velocity of approximately 5μm/s (Swann and Whitaker, 1986). Experiments by Swann and Whitaker (1986) indicate that IP_3 diffuses much more readily than Ca^{2+} within the egg cytoplasm and that Ca^{2+}-stimulated production of IP_3 and IP_3-induced Ca^{2+} release from an internal store can account for the progressive release of Ca^{2+} at fertilization.

IP_3 initiates a Ca^{2+} wave in mollusks (Bloom et al., 1988), tunicates (Dale, 1988) and in Xenopus (Busa et al., 1985; Nuccitelli et al., 1988). Heparin blocks the action of injected IP_3 (Nuccitelli et al., 1988); caffeine and ryanodine are without effect in Xenopus. It is important to recognize that although heparin has been used extensively to discern mechanisms of Ca^{2+} wave initiation and propagation in eggs, it is a nonspecific competitive inhibitor of IP_3-induced Ca^{2+} release. There is evidence that it may affect ryanodine- and cyclic ADP ribose-induced Ca^{2+} release and inhibits the cortical granule reaction even when the Ca^{2+} increase is near normal (Shilling et al., 1994).

The IP_3 receptor of Xenopus eggs and oocytes has been purified and characterized (Parys et al., 1992; Kume et al., 1993) and shown to be a tetrameric complex with a monomer molecular mass of 256 kDa. The primary structure has been determined by sequence analysis of its cDNA (Kume et al., 1993); an IP_3 binding domain and putative calcium channel have been identified. Fluorescence observations demonstrate that IP_3 receptor reactivity is present in the perinuclear region, along the periphery of the germinal vesicle and throughout the animal and vegetal hemispheres of Xenopus oocytes and eggs (Parys et al., 1993, 1994; Kume et al., 1993). As in sea urchin eggs, reactivity to the IP_3 receptor in Xenopus eggs and oocytes is distributed within the endoplasmic reticulum.

Larabell and Nuccitelli (1992) proposed a model of Ca^{2+} wave propagation in Xenopus in which sperm–egg interaction brings about an activation of phospholipase C that in turn hydrolyses PIP_2 to IP_3 and DAG. IP_3 triggers release of Ca^{2+} and Ca^{2+} in turn triggers additional PIP_2 hydrolysis, presumably via activation of phospholipase C but a calcium-induced calcium release has not been ruled out. DAG activates protein kinase C, which helps to stimulate cortical granule exocytosis (Bement and Capco, 1989, 1990; see Bement, 1992) and this leads to a down-regulation of PIP_2 hydrolysis and phosphorylation of the IP_3 receptor. Consistent with this scheme are experiments demonstrating that antibodies to PIP_2 reduce cortical levels of Ca^{2+} in Xenopus eggs.

Recent confocal studies of immature *Xenopus* oocytes have demonstrated the modulation of IP_3-induced Ca^{2+} release by Ca^{2+} (Lechleiter et al., 1991; Lechleiter and Clapham, 1992). Spiral waves of Ca^{2+} release are induced by IP_3 and $GTP\gamma$ when injected into *Xenopus* oocytes (Fig. 6.13).

Wave propagation is contributed by IP_3-mediated Ca^{2+} release from internal stores but is believed to be modulated by cytoplasm concentration and diffusion of Ca^{2+}. Ca^{2+} acts then as a co-agonist with IP_3 to release Ca^{2+} from internal stores. At high concentration Ca^{2+} is inhibitory; once inhibition is removed and the store is replenished to a required set level the channel conducts Ca^{2+} into the cytoplasm. This scheme is suggestive of a sensitizing action of Ca^{2+} on IP_3 receptors,

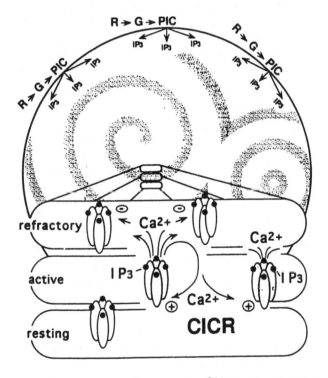

Fig. 6.13 Molecular model for regenerative Ca^{2+} signaling in *Xenopus* oocytes. Spiral and circular Ca^{2+} waves are generated at constant levels of IP_3, through the cyclical stimulation and inhibition of the IP_3 receptor channel by Ca^{2+}. Ca^{2+} release within the propagating active zone (gray bands) is due to Ca^{2+}-induced Ca^{2+} release (CICR) from IP_3 (black spheres of inset)-bound IP_3 receptors (shown as a tetramer). The refractory zone is due to subsequent inhibition of the IP_3 receptors by high Ca^{2+} levels. Ca^{2+} inhibition of the IP_3 receptor results in annihilation of colliding wavefronts. Reproduced with permission from Lechleiter and Clapham, 1992.

thereby serving a role in the regenerative process of Ca^{2+} release. Whether or not a similar scheme occurs in *Xenopus* eggs at fertilization has not been established. As pointed out by Larabell and Nuccitelli (1992) differences in Ca^{2+} wave characteristics of *Xenopus* oocytes and eggs may be related to differences shown to exist in the organization of the endoplasmic reticulum of oocytes *versus* eggs (Campanella *et al.*, 1984). Although at the present time there is no clear cut evidence for a calcium-induced calcium release mechanism (e.g. a ryanodine receptor) in *Xenopus* oocytes/eggs, experiments by Kobrinsky *et al.* (1995) have shown that cloned skeletal muscle ryanodine receptor expressed in *Xenopus* oocytes can substitute for the IP_3 receptor as the intracellular calcium release channel required for maturation.

In hamster and mouse eggs, IP_3 induces a propagated Ca^{2+} wave (Miyazaki, 1988; Miyazaki *et al.*, 1992, 1993; Kline and Kline, 1994). Observations that antibodies to the IP_3 receptor block IP_3-induced calcium release demonstrate the involvement of IP_3 and its receptor in the generation of the Ca^{2+} wave at egg activation in hamsters and mice (Miyazaki *et al.*, 1992; Xu *et al.*, 1994). IP_3-induced calcium release is looked upon as essential in the initiation, propagation and oscillation of sperm-induced Ca^{2+} waves (Miyazaki *et al.*, 1993). Experiments reviewed by Miyazaki *et al.* (1993) indicate that IP_3-induced Ca^{2+} release is sensitized by Ca^{2+}. This could be the basis for the generation of Ca^{2+} waves and oscillations that are observed in mammalian eggs and that have been postulated to effect such phenomena in *Xenopus* oocytes (Lechleiter and Clapham, 1992). Models for the mechanism of Ca^{2+} wave propagation in eggs have been proposed by Miyazaki *et al.* (1993) and are depicted in Fig. 6.14.

Experiments with both hamster and mouse oocytes demonstrate that IP_3-induced Ca^{2+} release develops with egg maturation (Fujiwara *et al.*, 1993; Mehlmann and Kline, 1994). In hamster oocytes IP_3-induced Ca^{2+} release develops in two phases (Fujiwara *et al.*, 1993). First, IP_3 sensitivity increases gradually between the germinal vesicle stage and prometaphase of meiosis. Second, a nonlinear dose–response relation develops during second meiosis to metaphase II, resulting in the ability of the egg to undergo a regenerative, propagating Ca^{2+} release the same as is induced by a threshold pulse of IP_3 in mature eggs. Development of the IP_3-induced Ca^{2+} release is thought to be a prerequisite factor for the acquisition of the ability of mammalian eggs to undergo normal fertilization (Fujiwara *et al.*, 1993). The size of the releasable Ca^{2+} store in immature mouse oocytes is similar in size to stores in mature eggs (Mehlmann and Kline, 1994), but the Ca^{2+} store in immature oocytes is less sensitive to IP_3. A similar relationship is also observed in starfish (Chiba *et al.*, 1990).

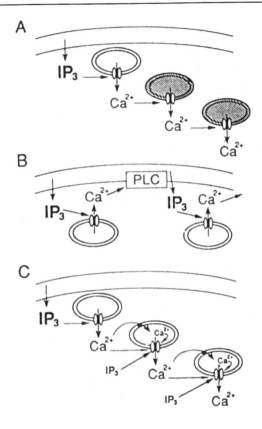

Fig. 6.14 Models of the mechanism of a Ca^{2+} wave starting from local IP_3-induced Ca^{2+} release. **(A)** A model based on CICR from stores that possess ryanodine receptors (striped stores). **(B)** A model based on a cycle between IP_3-induced Ca^{2+} release and Ca^{2+}-dependent production of $InsP_3$. **(C)** A model based on Ca^{2+}-sensitized IP_3-induced Ca^{2+} release. IP_3 = $InsP_3$; PLC = phospholipase C. Reproduced with permission from Miyazaki *et al.*, 1993.

(b) G protein- and tyrosine kinase-linked receptors

The production of IP_3 and DAG in cells is believed to be linked to GTP-binding proteins (via PLCβ1) or tyrosine kinase (via PLCγ1) (Berridge, 1993; Fig. 6.15).

Nuccitelli (1991) has presented essentially three hypotheses for egg activation, two of which embody G protein or tyrosine kinase. In other words, sperm–egg interaction stimulates:

1. G protein, which in turn triggers PIP_2 hydrolysis and the Ca^{2+} wave;
2. tyrosine kinase, and this in turn activates the egg via phosphorylation;

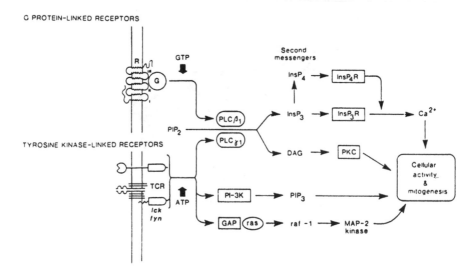

Fig. 6.15 Two major receptor-mediated pathways for stimulating the formation of inositol trisphosphate (InsP$_3$) and diacylglycerol (DAG). Many agonists bind to seven-membrane-spanning receptors (R), which use a GTP-binding protein (G) to activate phospholipase C-β1 (PLCβ1), whereas PLCγ1 is stimulated by the tyrosine kinase-linked receptors. The latter activate other effectors such as the phosphatidylinositol 3-OH kinase (PI-3K), which generates the putative lipid messenger phosphatidylinositol (3,4,5)-trisphosphate (PIP$_3$) and the GTPase-activating protein (GAP) that regulates *ras*. InsP$_3$R = InsP$_3$ receptor; PKC = protein kinase C. See Berridge, 1993.

3. the diffusion of an activator from the sperm to the egg.

The involvement of G proteins in the formation of IP$_3$ and DAG at fertilization has been indicated in the eggs of a wide variety of species (Nuccitelli, 1991; Shilling *et al.*, 1994; Moore *et al.*, 1994).

1. Injection of cholera toxin (a G protein activator) or GTPγS (a hydrolysis-resistant analog of GTP) brings about activation in sea urchin eggs (Turner *et al.*, 1986, 1987). Injection of GDPβS (a metabolic stable analog of GDP, which acts as a competitive inhibitor of GTP) blocks sperm-induced activation and, as originally reported, IP$_3$ could bypass this blockage (Turner *et al.*, 1986). Further experiments in this area, however, have shown that, depending on the timing of the injections, GDPβS can block the IP$_3$ response (Kline *et al.*, 1990) and, in addition, GDPβS does not block the wave of free Ca^{2+} even though the cortical granule reaction is prevented (Whitaker *et al.*, 1989). Therefore, the interaction of GDPβS and the egg (G protein?) in this case is not a simple one.

2. Mouse oocytes injected with an mRNA that codes for the human m_1 muscarine receptor (a member of the family of G-protein-coupled receptors) activate when exposed to acetylcholine (Williams et al., 1992; Moore et al., 1993).

Although G protein is present in eggs and its stimulation is capable of triggering activation, there is as yet no direct evidence that sperm-induced activation uses such a molecule (Moore et al., 1994). Furthermore, experiments by Rakow and Shen (1990) and Crossley et al. (1991), demonstrating that heparin reduces IP_3, and GTPγS-induced Ca^{2+} release but has no effect on sperm-induced Ca^{2+} release, are consistent with the suggestion that sperm activation involves a wave of Ca^{2+} release without relying exclusively on IP_3 production.

Evidence supporting the involvement of a tyrosine kinase receptor in the generation of IP_3 includes the following.

1. A $\approx$ 138 kDa protein similar to mammalian phospholipase Cγ is phosphorylated on tyrosine groups 15–60 s after fertilization (Peaucellier et al., 1988; see also Shilling et al., 1994).
2. Starfish and frog oocytes injected with mRNAs for receptors that are known to exert their action via tyrosine kinase activate when incubated with appropriate agonist (Shilling et al., 1994; Yim et al., 1994). Additionally, $p59^{fyn}$ kinase, a member of the src family of protein tyrosine kinases, is activated within minutes of fertilization in sea urchins (Kinsey, 1996).

These experiments indicate that eggs contain components necessary to respond to related receptors by signaling pathways including phospholipase Cγ. In sea urchin eggs protein tyrosine kinase inhibitors do not block early events of fertilization, such as elevation of the fertilization membrane, although later events including pronuclear migration, DNA synthesis, cleavage and gastrulation are inhibited (Moore and Kinsey, 1995; Kinsey, 1995). Although these results do not support the idea that tyrosine kinase plays an essential role in transducing the early responses of sea urchin eggs, such as activation of phosphoinositol turnover and the Ca^{2+} transient (Bement, 1992), they indicate that control mechanisms used in later development are established at fertilization and require the action of one or more protein tyrosine kinases (Kinsey, 1995). Whether sperm–egg interaction triggers G protein (utilizing PLCβ for the hydrolysis of PIP_2) or protein kinase (utilizing PLCγ for the hydrolysis of PIP_2) mediated pathways to generate IP_3 and DAG remains unresolved (Moore and Kinsey, 1995). It is possible that:

1. the sperm receptor may be a complex whose activation affects more than one signaling pathway at fertilization (Shilling et al., 1994);

2. protein tyrosine kinase may be part of a redundant system that can be bypassed or compensated for by activation of a G protein-coupled mechanism or another pathway not requiring protein tyrosine kinase activity;
3. sea urchin eggs may employ a mechanism of activation fundamentally different from other organisms.

The involvement of effectors downstream of G protein- and tyrosine-linked receptors in activation, e.g. protein kinase C, have been demonstrated in *Xenopus*, hamster and mouse (Nuccitelli, 1991; Bement, 1992; Gallicano *et al.*, 1993).

It is important to realize that of the variety of activation mechanisms available to cells few have been characterized to any extent in regards to egg activation. Hence it is possible that we are not even aware of all of the basic mechanisms regulating egg activation and how they interrelate to one another. For example, experiments by Whalley *et al.* (1992) provide data indicating that sea urchin eggs injected with cyclic GMP undergo an increase in Ca^{2+} and cortical granule exocytosis that are not blocked by heparin. Cyclic GMP does not stimulate the Na^+–H^+ antiporter when the Ca^{2+} transient is blocked by the Ca^{2+} buffer BAPTA. These observations suggest that cyclic GMP does not act through the phosphoinositide signaling pathway and the increased cyclic GMP that accompanies sperm activation is utilized to activate the egg. The latency ($\approx$ 20 s) that occurs when eggs are injected with cyclic GMP may represent the time required for activation of enzymes used for the generation of a more potent activating substance.

(c) Calcium release mechanisms: IP₃-independent

The presence of an intracellular IP_3-mediated Ca^{2+} store is well established and experiments employing different activating agents in combination with heparin treatment have indicated the presence of at least two Ca^{2+} stores in sea urchin eggs (Rakow and Shen, 1990; see also Zucker *et al.*, 1978). For example, heparin-treated eggs injected with IP_3 fail to undergo Ca^{2+} release, whereas a transient increase in Ca^{2+} is observed in heparin-treated eggs at fertilization. Considering the similarity of the egg's endoplasmic reticulum to the sarcoplasmic reticulum and that the major release channels in the sarcoplasmic reticulum required for excitation–contraction coupling include ryanodine and the IP_3 receptors, it was suspected that the egg might also possess an IP_3-independent Ca^{2+} store mediated, for example, by the ryanodine receptor. The ryanodine receptor is activated by calcium or by another second messenger, e.g. cADP ribose.

Ryanodine, a plant alkaloid known to effect sarcoplasmic reticulum

Ca^{2+} release, induces Ca^{2+} release in sea urchin eggs (Fujiwara et al., 1990; Sardet et al., 1992; Buck et al., 1992), in egg homogenates (Galione et al., 1991) and in cortical lawn preparations (McPherson et al., 1992). In order to achieve these results in sea urchin eggs rather high ryanodine concentrations, at least an order of magnitude higher than concentrations effective on permeabilized muscle fibers, need to be employed (Sardet et al., 1992). The meaning of this difference is not yet clear. A $\approx$ 380 kDa protein of sea urchin cortices is identified by immunoblot analysis with antibodies to the ryanodine receptor of muscle fibers (McPherson et al., 1992). The presence of a ryanodine sensitive Ca^{2+} release channel that is localized to the sea urchin egg cortex and the distribution of the IP$_3$ receptor throughout the entire endoplasmic reticulum of the egg demonstrate a spatial dichotomy that may serve different roles (see below). The natural ligand of the ryanodine receptor of sea urchin eggs has not been established unequivocally, although Galione et al. (1991) provide evidence that cyclic ADP ribose may trigger Ca^{2+} release from ryanodine receptors (Clapper et al., 1987; Lee et al., 1989).

Homogenates of sea urchin eggs display transients in intracellular free Ca^{2+} with the addition of Ca^{2+}, i.e. a Ca^{2+}-induced Ca^{2+} release. Ca^{2+} release is also induced by caffeine, ryanodine and cycle ADP ribose (Galione et al., 1991; Lee, 1991). Initial experiments with agonists and antagonists of IP$_3$ and ryanodine receptors revealed that both Ca^{2+}-mobilizing systems are activated during fertilization (Lee et al., 1993; Galione et al., 1993) and that blockage of either of the systems alone is not sufficient to prevent a sperm-induced cortical granule exocytosis. It has been postulated that the presence of these two systems may ensure the occurrence of a Ca^{2+} transient at fertilization as this signal and the increase in intracellular pH are directly responsible for initiation of the developmental program of the embryo (Lee et al., 1993). More recent experiments, however, indicate that IP$_3$-induced calcium release is the primary mechanism underlying the generation of the Ca^{2+} transient. When IP$_3$-induced calcium release is inhibited, other Ca^{2+}-releasing mechanisms are unable to replace the deficit in intracellular Ca^{2+} release (Mohri et al., 1995).

Using confocal microscopy of sea urchin eggs injected with various agonists and antagonists, Shen and Buck (1993) have demonstrated specific spatio-temporal patterns for IP$_3$-, cyclic-ADP-ribose- and ryanodine-regulated Ca^{2+} release mechanisms in the generation of the Ca^{2+} wave. Temporal differences in Ca^{2+} release and rise and fall times were observed with each agonist. Higher fluorescence intensity changes were present in the cortex versus the remainder of the cytoplasm with agonists to all three systems, which is consistent with morphological observations demonstrating the localization of the IP$_3$

receptor to the cortex and subcortex and the ryanodine receptor to only the egg cortex (McPherson et al., 1992; Parys et al., 1993; see also Stricker et al., 1992).

Although the importance of IP$_3$-sensitive Ca^{2+} stores has been demonstrated in mammalian eggs, recent experiments with mouse eggs have documented the presence of a ryanodine receptor that is responsive to either ryanodine or cADP ribose, promotes the conversion of ZP2 to ZP2$_f$, but does not induce the resumption of the cell cycle (Ayabe et al., 1995). As in sea urchin eggs, this Ca^{2+}-store is preferentially localized to the egg cortex. Conditions that block the release of Ca^{2+} from the ryanodine sensitive store had no effect on events of egg activation.

In summary, the precise mechanism(s) by which eggs regulate the Ca^{2+} release involved with activation has not yet been clearly defined. Two systems, IP$_3$-dependent and IP$_3$-independent, have been partially characterized; some eggs appear to possess both mechanisms while others have only the former. How each functions to initiate egg activation will be central to future investigations.

Calmodulin is associated with the sea urchin egg cortex and when isolated cortices are treated with antibody to calmodulin they lose their sensitivity to calcium (Whitaker and Steinhardt, 1982). These results and observations that calmodulin antagonists, trifluoroperazine and chlorpromazine, inhibit cortical granule exocytosis are consistent with the notion that calmodulin may also play a role in Ca^{2+}-mediated cortical granule discharge (Baker and Whitaker, 1979; Moy et al., 1983).

6.4 RESULTS OF CORTICAL GRANULE EXOCYTOSIS

Virtually all the cortical granules in eggs of sea urchins, amphibians and fish are discharged at fertilization. In mice a substantial number of cortical granules (about 25% of the population) are exocytosed before sperm−egg fusion, with considerable modification to the zona pellucida (see below); the remainder are dehisced at fertilization (Nicosia et al., 1979; Ducibella et al., 1990). In the annelid Sabellaria the cortical granule reaction is initiated when the eggs are spawned into sea water (Pasteels, 1965) prior to insemination. Modifications in the egg cortex involving microvillar elongation take place at fertilization in Sabellaria. In the echiuroid worm Urechis a subset of cortical granules is released at insemination; the remainder are discharged later with the elevation of the vitelline layer (Paul, 1975). In the starfish Pisaster a portion of the cortical granules are still present in the early gastrula, where they appear to contribute to the extracellular matrix of the developing embryo (Reimer and Crawford, 1995). In some mollusks, such as Mytilus, all of the cortical granules are retained until gastrulation, at which time they are discharged (Humphreys, 1967). For organisms whose eggs neither possess cortical granules nor undergo a cortical granule reaction, it is clear that cortical granule exocytosis is not required for fertilization or egg activation and development. Furthermore, the cortical granule reaction can be inhibited in sea urchins and later events, including those of fertilization and cleavage, are not impaired.

As a consequence of the cortical granule reaction there is the externalization of cortical granule contents, which have profound structural and physiological effects on the egg (Figs 6.4, 6.5). Such changes have been reviewed for sea urchins, frog, fish and mammals (Shapiro et al., 1989) and, in general, include the following processes, which remodel extracellular components surrounding the egg: (1) proteolysis, (2) assembly of proteins derived from the cortical granules and extracellular components and (3) crosslinking of perivitelline molecules. In sea urchins a portion of the cortical granule contents remains in the perivitelline space and, in the presence of calcium, polymerizes to form the hyaline layer (Citkowitz, 1971; Kane, 1973; Alliegro et al., 1992). Another protein stored in cortical granules and yolk granules, toposome, is deployed to the hyaline layer and plasma membrane following fertilization (Gratwohl et al., 1991). The hyaline layer prevents polyspermy, maintains blastomere adherence and participates in morphogenetic changes of the developing embryo. Not all of the hyaline layer material within the egg is released with the cortical granule reaction; a portion is retained in small vesicles (Hylander and Summers, 1982a). These vesicles may be involved in the regeneration of the hyaline layer during embryogenesis.

6.4.1 ASSEMBLY OF THE FERTILIZATION MEMBRANE

The material released from cortical granules in sea urchins also contains proteases that are believed to assist in the elevation of the vitelline layer to form the fertilization membrane and in the removal of sperm receptors (Vacquier et al., 1973; Carroll and Epel, 1975; Figs 6.4, 6.16).

In addition, other changes of the egg cortex are believed to be a result of proteolysis at fertilization. For example, limited proteolytic cleavage of egg surface components other than the vitelline layer occurs at fertilization and artificial activation (Shapiro, 1975). What role, if any, this limited proteolysis may serve in development has not been determined. Separation and lifting of the vitelline layer from the egg surface is accomplished by the hydration of sulfated mucopolysaccharides released from the cortical granules (Schuel et al., 1974).

Following its elevation, the fertilization membrane undergoes a

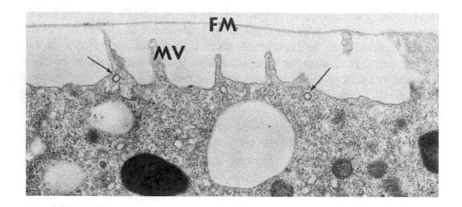

Fig. 6.16 Fertilized sea urchin (*Arbacia*) egg having completed the cortical granule reaction. The fertilization membrane (FM), elongate microvilli (MV) and endocytotic vesicles (arrows) are depicted.

sequence of changes resulting in a hardened glycoprotein coat that is resistant to most denaturing agents (Vernon *et al.*, 1977; Shapiro, 1981). The sequence of assembly in the sea urchin *Strongylocentrotus* proceeds from a pliable fertilization membrane that contains casts of the microvilli, which are shaped like inverted Us, to a final hardened structure in which the microvillar casts are converted to pyramidal tent-like forms (Fig. 6.17).

During the 2 min it takes to transform from the inverted U to the pyramidal form, the microvillar casts become coated with an orderly arrangement of repeating macromolecular units which spread to regions between these elevations (Chandler and Heuser, 1980). This coating is believed to play a role in the hardening of the fertilization envelope and acts to protect the embryo. A secretory product from cortical granules has been isolated that self-associates in a calcium-dependent manner to form sheets that have a pattern similar to that seen on the fertilization membrane (Bryan, 1970a, b). Deposition of this material may be involved in hardening the fertilization membrane.

Hardening of the fertilization membrane is affected by the formation of dityrosine crosslinks due to components present in the cortical granules, ovoperoxidase and proteoliaisin (Foerder and Shapiro, 1977; Hall, 1978; Somers *et al.*, 1989; Shapiro *et al.*, 1989; Fig. 6.17). With cortical granule exocytosis there is the release of a peroxidase (ovoperoxidase), a 70 kDa protein which becomes incorporated into the fertilization membrane via a cross-linking of tyrosine residues in proteoliaisin (235 kDa) using hydrogen peroxide from molecular oxygen. Peroxide synthesis occurs as a burst and accounts for two-

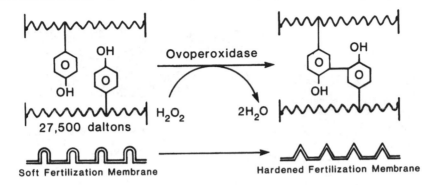

Fig. 6.17 Mechanism of fertilization membrane hardening. The soft fertilization membrane contains igloo-shaped projections which represent reflections of the vitelline layer over the microvilli of the egg. The hardened fertilization membrane contains tent-shaped projections. See Shapiro, 1981.

thirds of the oxygen taken up by the egg during the first 15 min after insemination. Peroxide is also toxic to sperm and in addition to hardening the fertilization membrane this system may provide an additional block to polyspermy. The interaction of ovoperoxidase with proteoliaisin is Ca^{2+}-dependent: together they form a 1:1 complex (Kay and Shapiro, 1985, 1987; Shapiro et al., 1989). Ovoperoxidase is tethered to proteoliaisin, which in turn is bound to the vitelline layer (Weidman and Shapiro, 1987). Crosslinking of proteoliaisin and other less characterized components lead to hardening of the fertilization envelope. Structural changes in the vitelline layer and jelly coat of sea urchin eggs have also been carried out using rapid-freezing and deep-etching to visualize macromolecular structure (Larabell and Chandler, 1991).

In *Xenopus* two major biochemical modifications in the extracellular layers surrounding the egg are brought about by the cortical granule reaction. First, the jelly layer (J1), which is apposed to the vitelline layer, lifts from the egg with cortical granule exocytosis (Shapiro et al., 1989). Some 70% of the protein from the cortical granule is composed of a Ca^{2+}-dependent galactose-specific agglutinin (43 kDa). J1 serves as a ligand for the cortical granule lectin and their interaction results in the formation of a Ca^{2+}-dependent precipitation and a fertilization layer (F layer; Wyrick et al., 1974; Fig. 6.18).

The fertilization layer is believed to act as a permeability barrier and may facilitate chemical conversion of the vitelline layer by trapping other modifying agents in the perivitelline space (Shapiro et al., 1989), as well as altering the receptivity of the vitelline layer to sperm (Wyrick et al., 1974). Second, the vitelline layer, which consists of at least seven proteins (Gerton and Hedrick, 1986), undergoes limited hydrolysis.

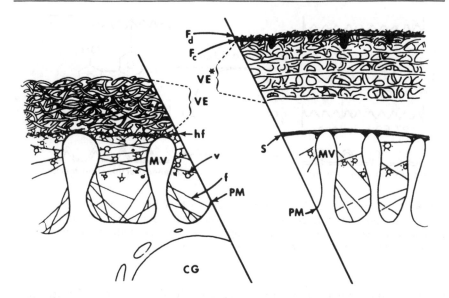

Fig. 6.18 Diagram of the extracellular matrix of the *Xenopus* egg before and after fertilization. The unfertilized egg (**left**) with a cortical granule (CG) beneath the plasma membrane PM and microvilli (MV) extending into the perivitelline space. Filaments (f) connect MV to each other, to the plasma membrane and to vesicles (v) seen in this region. The vitelline layer lies on the tips of the MV and is attached to the MV by a layer of horizontal filaments (hf) seen at the innermost portion of the vitelline envelope (VE). After fertilization (**right**), the S layer (S) encircles the egg, lying on the tips of the MV. The VE has elevated and has been chemically and structurally altered (VE*); the F layer (F_d and F_c) has been deposited on the outer surface of the altered VE. The jelly layer lies just superiorly to the VE. Reproduced with permission from Larabell and Chandler, 1988a.

These biochemical changes are accompanied by significant alterations in the vitelline layer as determined by quick-freezing, deep-etching and rotary shadowing (Larabell and Chandler, 1988a, b). New structural components of the extracellular layers have been observed (e.g. the S-layer – Fig. 6.18; Larabell and Chandler, 1988a, b) and their associations provide an indication of the number and complexity of molecular interactions taking place in the formation of the fertilization layer in *Xenopus*.

6.4.2 ZONA REACTION

Austin and Bishop (1957) recovered eggs from oviducts of mated mammals and quantified the incidence of multiple sperm penetration of the egg and zona pellucida. They found that blocks to polyspermy in mammals occurred at the zona pellucida and/or egg plasma membrane. In monospermic rabbit eggs, sperm were found in the perivitelline space, indicating that polyspermy is blocked at the plasma membrane. Experiments have demonstrated that parthenogenetically activated mouse eggs are capable of a transitory plasma membrane block that requires both cortical granule exocytosis and meiosis resumption (Tatone et al., 1994). In hamsters, dogs and sheep, sperm were not seen in the perivitelline space, suggesting that a polyspermy block can also occur at the level of the zona pellucida. The zona reaction represents changes in the zona pellucida that (1) reduce sperm binding and penetration and (2) alter its physiochemical properties, e.g. increasing its resistance to dissolution (Wolf and Hamada, 1977). Sperm receptor activity has been demonstrated in solubilized preparations of zonae pellucidae from mouse eggs by competition assays (Bleil and Wassarman, 1983). The fact that zonae from two-cell mouse embryos have no sperm-receptor activity and a decreased solubility is referred to as the zona reaction.

The mechanism responsible for the zona reaction is believed to be similar to that described for sea urchins, i.e. it involves the enzymatic modification of zona pellucida proteins, ZP2 and ZP3 (Bleil et al., 1981; Moller and Wassarman, 1989; Wassarman, 1993). ZP2 undergoes a loss in solubility due to limited proteolysis by a 21–34 kDa protease from the cortical granules (Moller and Wassarman, 1989). This brings about an alteration in the zona pellucida, such that it is more resistant to solubilization (Gulyas and Schmell, 1980; Schmell and Gulyas, 1980). Zona modification may also occur as a result of the release of ovoperoxidase, in a manner similar to that demonstrated for sea urchin eggs (see Hoodbhoy and Talbot, 1994).

Interestingly, the number of cortical granules in immature mouse oocytes is greater than in mature eggs (Ducibella et al., 1988a, b). This difference is due to the formation of a cortical-granule-free domain in the area of the metaphase II spindle (Fig. 6.19).

In vitro experiments in serum-free medium to study the depletion of cortical granules during meiotic maturation have revealed that the formation of the cortical-granule-free domain is correlated with the conversion of ZP2 to ZP2$_f$. This conversion is inhibited when oocytes are matured in serum-containing medium (Ducibella et al., 1990). The inhibitory effect of serum on ZP2 conversion may represent a mechanism to prevent precocious modification of the zona that would result in a premature block to polyspermy and, hence, inhibit fertilization.

ZP3 also undergoes limited digestion of its carbohydrate component and this in turn results in the loss of its sperm receptor activity (Wassarman, 1993). Studies by Miller et al. (1993) have shown that the

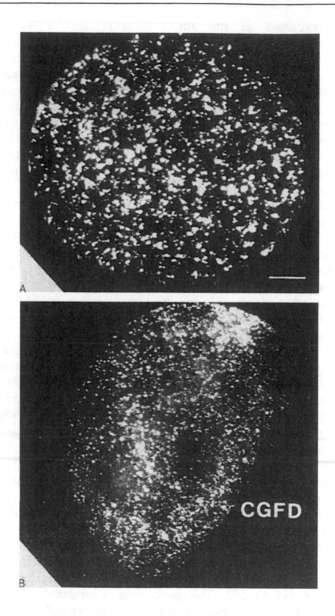

Fig. 6.19   Fluorescence photomicrographs of mouse oocyte cortices stained with *Lens culinaris* agglutinin to demonstrate the presence of cortical granules. **(A)** Cortical granules are distributed throughout the oocyte cortex. **(B)** With oocyte maturation cortical granules are found only associated with a limited portion of the egg cortex. The cortical-granule-free domain (CGFD) overlies the cytoplasmic region containing the second meiotic spindle. Reproduced with permission from Ducibella *et al.*, 1990.

loss in ZP3's sperm-binding activity is due to the removal of the GalTase binding site by N-acetylglucosaminidase, a cortical-granule-derived enzyme. PUGNAC, an inhibitor of N-acetyl-glucosaminidase, and antibodies to N-acetyl-glucosaminidase inhibit the zona block to sperm binding (Miller *et al.*, 1993; see Miller and Shur, 1994).

Evidence for cortical granule exudates effecting a block to poly-spermy at the plasma membrane level in mammals is equivocal. Exposure of zona-free eggs to crude preparations of cortical granules indicated involvement in a plasma membrane polyspermy block (Gwatkin, 1977). However, observations indicating that cortical granule secretions have no role in establishing a plasma membrane block have also been presented. For example, the fertility of zona-free mouse eggs having undergone a loss of cortical granules induced by the calcium ionophore A-23187 was identical to that of controls (Wolf *et al.*, 1979). In connection with these results in mammals, sea urchin embryos removed from their fertilization membranes and hyaline layers can be refertilized, indicating the absence of a permanent polyspermy block at the plasma membrane level in this organism as well (Longo, 1984).

In animals with large eggs, such as selachians, urodeles, reptiles and birds, where physiological polyspermy is common, only one male pronucleus becomes associated with the female pronucleus. In the urodele *Triton* Fankhauser (1948) showed that supernumerary male pronuclei regress in the presence of the cleavage (mitotic) spindle with proximity to the spindle determining the order of regression. More recent experiments by Iwao *et al.* (1993) in the newt indicate that a factor (maturation promoting factor?) associated with the female pronu-cleus rescues male pronuclei adjacent to the female pronucleus, while male pronuclei further from the female pronucleus are targeted for destruction. In mammals, pronuclear suppression reminiscent of that in *Triton* has been described in polyspermic eggs (Hunter, 1967a, b). A process of sperm abstraction in *in vitro* zona-free polyspermic mouse eggs has been described in which incorporated sperm are removed from the egg cytoplasm by a blebbing process that may restore the monospermic condition (Yu and Wolf, 1981).

Alterations in the egg cortex and cytoskeleton 7

7.1 PLASMA MEMBRANE

The cortical granule reaction results in a dramatic structural reorganization of the egg plasma membrane. The resultant membrane of the fertilized sea urchin egg has been referred to as a mosaic, indicating that it is derived from several sources, i.e. the egg plasma membrane, the cortical granule membrane and the sperm plasmalemma. The following discussion considers changes in the egg plasma membrane as the result of fertilization primarily involving the egg plasma membrane itself and contributions from the cortical granule membrane. Aspects concerning the integration of the sperm plasma membrane have been considered previously (Chapter 4) and are mentioned here only as they relate to broader questions of egg plasma membrane changes (see Longo, 1989a).

Formation of the mosaic plasma membrane and concomitant physiological changes in activity of the egg prompt the following questions:

1. Is the formation of the mosaic membrane related to physiological and biochemical changes characteristic of the activated ovum?
2. Do identifiable domains exist in the plasma membrane of the fertilized egg that are derived from the cortical granules?

It has been speculated that the contents of cortical granules serve other roles regulating the metabolism of the egg, e.g. they elicit perturbations of the plasma membrane that may be critical for activation (Schuel, 1978). In connection with this possibility it is pertinent to note that echinoid eggs can be parthenogenetically activated by trypsin and other proteases. Release of surface protein from the plasma membrane of sea urchin eggs at fertilization has been reported and it has been suggested that proteolytic processing of surface proteins may be an important aspect of activation (Shapiro, 1975).

Although activation of transport systems for specific metabolites occurs at fertilization there is no evidence linking this change with the insertion of cortical granule membrane into the plasma membrane. That transport systems develop in ammonia-activated sea urchin eggs, when the cortical granule reaction does not occur, suggests that alterations in the properties induced by the cortical granule reaction may not be essential for permeability changes characteristic of fertilization (Epel, 1978). Hence, the role of the cortical granule reaction in later developmental events is uncertain.

Because so many aspects of fertilization are membrane-mediated events it has been suspected that a change in the state of the plasma membrane is an obligatory step in cellular activation. Using electron spin resonance spectroscopy Campisi and Scandella (1978, 1980a) demonstrated an increase in bulk membrane fluidity of sea urchin eggs after fertilization. However, because the spin label (fatty acid) was equilibrated among all of the subcellular membrane fractions, it could not be determined whether: (1) ovum activation is accompanied by a change in total cellular membranes to a more fluid state or (2) more specialized membranes (such as the plasmalemma) entered a more fluid state and the probe was showing the average change experienced by altered and unaltered membranes. The structural changes of membrane lipids accompanying activation is probably not a result of the cortical granule reaction, as eggs partially activated by ammonia showed a similar effect. In experiments with cortical fractions it has been shown that the fluidity of the fertilized egg cortex is less than that of the unfertilized cortex (Campisi and Scandella, 1980b). Adding calcium to cortical fractions from unfertilized eggs resulted in a fluidity decrease *in vitro*. It has been suggested that this change may represent an alteration in membrane structure rather than a direct interaction of calcium with phospholipid groups. Experiments discussed earlier (Chapter 4) indicate that labeled sperm and egg plasma membrane components are able to move within the plasma membrane of the fertilized egg (Longo, 1989a, b).

Analysis of membrane lipid changes in sea urchin and mouse eggs using fluorescence photobleaching recovery suggest that fertilization is not accompanied by a change in bulk membrane viscosity; rather it is associated with alterations in the ensemble of lipid domains (Wolf *et al.*, 1981a, b). The different lipid analogs employed by Wolf *et al.* (1981a, b) indicated the existence of lipid domains differing in composition or physical states from the average for the plasma membrane. These results suggest that gel and fluid lipid domains exist within the egg plasma membrane, the proportion and composition of which change upon fertilization. At fertilization there may be a reordering of lipid domains, which release inactive proteins from gel regions of the plasma membrane into fluid regions, where they would become active.

Changes in lipid composition and extent of gel-fluid regions at fertilization could then act as a switch which would rapidly activate protein
functions that do not require the synthesis or insertion of new material
into the membrane.

Studies with sea urchin (*Arbacia*) eggs treated with filipin (Carron and
Longo, 1983) demonstrate alterations in membrane sterols at activation
(see also Decker and Kinsey, 1983). The plasma membranes of treated
unfertilized eggs possess numerous filipin–sterol complexes while
fewer complexes are associated with membranes delimiting cortical
granules, demonstrating that the plasma membrane is relatively rich in
β-hydroxysterols. Following fusion with the plasmalemma, membranes
formerly delimiting cortical granules undergo a dramatic alteration in
sterol composition – a rapid increase in the number of filipin–sterol
complexes. In contrast, portions of the fertilized egg plasma membrane,
derived from the original plasma membrane of the unfertilized egg,
display little change in filipin–sterol composition. Other than regions
involved in endocytosis, the plasma membrane of the zygote possesses
a homogeneous distribution of filipin–sterol complexes and appears
structurally similar to that of the unfertilized ovum.

The absence of patches within the plasma membrane of the
fertilized egg, which are relatively devoid of filipin–sterol complexes
and correspond morphologically to membrane formerly delimiting
intact cortical granules, indicate that the cortical granule membrane is
significantly altered when it fuses with the plasmalemma. How the
cortical granule membrane acquires an increase in filipin–sterol
complexes has not been determined. Lateral displacement of sterols
from membranous regions derived from the original egg plasma
membrane may be involved. Sterols have been shown to diffuse rapidly
in bilayers of somatic cells, which is consistent with an extremely rapid
lateral displacement of sterols into membranous patches derived from
cortical granules.

Fluorescence photobleaching recovery experiments have been
performed with mouse eggs using protein probes, suggesting that interactions with cytoskeletal components may regulate membrane protein
diffusion (Wolf and Ziomek, 1983). As with membrane lipids, the
proteins probed demonstrated a heterogeneous distribution. Moreover,
although membranes, i.e. cortical granule and sperm plasma
membranes, are added to the egg plasmalemma at fertilization, there is
no generalized effect on the diffusion of membrane protein in the
mouse egg.

Binding studies using plant lectins have been used in an effort to
demonstrate possible membrane changes between fertilized and unfertilized eggs. Investigations with mouse and hamster eggs have shown
that concanavalin A binding sites change quantitatively following ferti-

lization (Yanagimachi and Nicolson, 1976). In ascidian eggs both the agglutinibility and number of concanavalin A receptors increased following activation (O'Dell *et al.*, 1973). Qualitative changes in lectin binding following fertilization may reflect modifications in the nature and/or structure of the binding sites themselves and may also be influenced by membrane fluidity and functional states of the cytoskeleton. In the sea urchin *Strongylocentrotus* two classes of concanavalin A binding site have been identified: a high affinity site associated with the vitelline layer and a low affinity site associated with the plasma membrane. The number of low affinity sites doubles at fertilization, apparently as a result of the insertion of cortical granule membrane (Vernon and Shapiro, 1977). Although the increase in low affinity binding sites may be due to the appearance of cryptic sites, there is no doubling when eggs are activated with ammonia. This suggests that the increase in number of sites is caused by the addition of cortical granule membrane to the egg plasmalemma.

Examination of freeze-fracture replicas of unfertilized sea urchin eggs demonstrates a significant difference in the number of intramembranous particles within the plasmalemma and the cortical granule membrane: in *Arbacia* the number of intramembranous particles within the P-face of the cortical granule membrane is about 30% of that in the P-face of the cortical granule membrane (Longo, 1981a). Studies have been carried out to determine what happens to this dichotomy following cortical granule exocytosis, i.e. whether there is the appearance of localized areas, corresponding to patches of cortical granule membrane, within the plasma membrane of the fertilized egg, or whether particles within the plasma membrane of the activated egg are homogeneously distributed. The mosaic pattern of the fertilized egg plasmalemma, in terms of intramembranous particles, is temporary; recognizable differences between the original egg plasma membrane and cortical granule membrane are lost soon after cortical granule exocytosis. Patches are not found which contain a reduced number of intramembranous particles and correspond to the cortical granule membrane. This indicates a rapid alteration in the composition of cortical granule membrane following its fusion with the plasma membrane, i.e. an intermixing of components within the mosaic membrane. By 4 min postinsemination the density of intramembranous particles in the P-face of the plasma membrane of the fertilized egg is slightly reduced from that of the membrane of the unfertilized egg, suggesting a possible flow of intramembranous particles from the oolemma into membrane derived from the cortical granules. This suggestion is in keeping with fluid character of membranes and is consistent with schemes reported for other cell types (Frye and Edidin, 1970; Singer and Nicolson, 1972).

It has been shown that the total surface area of cortical granule membrane in sea urchin (*Strongylocentrotus*) eggs is greater than that of the egg plasmalemma (Schroeder, 1979; Vacquier, 1981). Hence, if all the cortical granule membrane is incorporated into the egg plasmalemma there would be at least a twofold increase in membrane delimiting the activated egg. However, by 16 min postinsemination the surface area of the activated egg is only slightly larger than that of the unactivated ovum, indicating a rapid accommodation in surface membrane. The microvillar elongation that occurs following insemination may be one means of accommodating a surface increase in the activated sea urchin eggs (Figs 6.4, 7.1).

However, elongated microvilli cannot compensate for all the cortical granule membrane that might be incorporated and membrane internalization has been proposed as a mechanism to quantitatively modify the surface area of activated eggs (Schroeder, 1979). Rapid elongation of microvilli is believed to occur primarily in areas occupied by the original plasma membrane (Chandler and Heuser, 1981). Microvillar elongation may take place only at sites on the egg surface where cortical granules have exocytosed and involves a reorganization of the cortical cytoskeletal system (Vacquier, 1981).

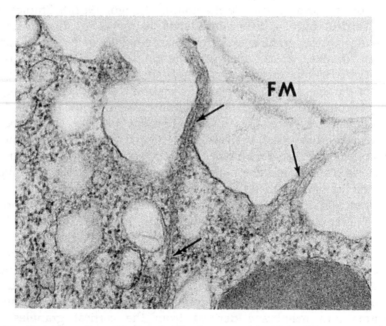

Fig. 7.1   Elongate microvilli of a fertilized sea urchin (*Arbacia*) egg containing bundles of microfilaments (arrows). FM = fertilization membrane. See Carron and Longo, 1982.

7.1.1 ENDOCYTOSIS

Following the cortical granule reaction and concomitant with the elongation of microvilli is the development of endocytotic pits and vesicles (Donovan and Hart, 1982; Fisher and Rebhun, 1983; Carron and Longo, 1984; Sardet, 1984). Endocytosis in sea urchin eggs commences as a burst 3–5 min postinsemination in which portions of the zygote plasma membrane are taken into the cytoplasm (Fig. 6.4). Whether or not portions of the original plasmalemma or the cortical granule membrane are preferentially endocytosed has not been determined. In the light of observations demonstrating significant changes in the composition of the egg plasma membrane at fertilization it seems unlikely that discrete patches of membrane persist intact to be selectively endocytosed.

That the mosaic membrane does, in fact, undergo endocytosis is demonstrated in studies employing fluid phase and absorptive tracers (e.g. horseradish peroxidase and cationized ferritin), which become internalized within vesicles of activated eggs. The fact that endocytosis follows the cortical granule reaction suggests a mechanism for both surface area reduction and cell surface remodeling which may be relevant to physiological changes characteristic of fertilized eggs. Its presence is consistent with observations in secretory cells where, after exocytosis, excess membrane may be removed from the cell surface. The extent of membrane internalized by endocytosis beginning at fertilization appears to be extensive and persists up to the time of cleavage in sea urchins (Fisher and Rebhun, 1983; see also Bernardini *et al.*, 1986). Whether endocytosis remains constant over this period has not been established, but it has been estimated that, in *Strongylocentrotus*, about 26 300 μm^2 of surface membrane per egg is readsorbed by endocytosis during the first 4 min of fertilization. This represents approximately 46% of the membrane presumably added to the egg surface by cortical granule exocytosis. The relationship between cortical granule exocytosis and endocytosis, in terms of the quantity of membrane in flux, is unclear since:

1. the rate of membrane interiorization is unknown;
2. the amount of cortical granule membrane added to the zygote surface has not been established and may be less than 100%;
3. mechanisms other than endocytosis that may contribute to the reduction of surface area have not been elucidated.

Following the appearance of tracer in pinocytotic vesicles of fertilized *Arbacia* eggs, label has been observed in lyosomes (Carron and Longo, 1984). This transition indicates that the tracer travels from one cellular compartment to another. That tracer was localized to lysosomes of zygotes examined up to 60 min postinsemination also suggests that surface membrane may be degraded or modified. Membrane components may then re-enter cytoplasmic precursor pools by traversing the lysosomal membrane to be utilized at later stages of embryogenesis.

The interiorization of zygote plasma membrane indicates the existence of a pathway from the cell surface to the cytoplasm and has significant biological implications. In a variety of cell types nutritional and regulatory molecules are selectively incorporated from the extracellular milieu. Receptor-mediated uptake and degradation systems may serve to modify hormones and surface receptors and may represent a mechanism for alteration of the physiologically response of cells to their environment (Goldstein *et al.*, 1979).

7.2 ACTIN AND THE CYTOSKELETON

Actin is present in high concentration in the cortices of unfertilized sea urchin and starfish ova (Otto and Schroeder, 1984; Bonder *et al.*, 1989). Much of the actin in the unfertilized sea urchin egg is in a monomeric form and surrounds the cortical granules; filamentous actin is concentrated predominately in the numerous microvilli that line the egg's surface (Spudich and Spudich, 1979; Bonder *et al.*, 1989). In addition to actin and myosin, a variety of actin-binding proteins have been found in sea urchin eggs (Vacquier, 1981; Mabuchi *et al.*, 1985; Bement *et al.*, 1992). A profilin-like protein may prevent actin from polymerizing in the unfertilized egg (Mabuchi, 1981; Hosoya *et al.*, 1982). The pool of unpolymerized actin present in the unfertilized sea urchin egg is believed to function in accommodation of the surface area change of the egg brought about by the cortical granule reaction. Monomeric actin is recruited into the formation of elongate microvilli, thereby mitigating potentially unfavorable consequences of the massive insertion of cortical granule membrane into the egg plasma membrane (Bement *et al.*, 1992).

Investigations, with both intact eggs and isolated cortices exposed to different ionic conditions, demonstrate that microvillar elongation is stimulated by the Ca^{2+} flux characteristic of egg activation (Carron and Longo, 1982; Begg *et al.*, 1982). Microvillar elongation does not occur when eggs are incubated in media, such as ammonia, that induce an increase in intracellular pH. However, actin filament bundle formation is triggered by an increase in intracellular pH. Formation of actin filament bundles is not necessary for microvillar elongation but is required to provide a rigid support for the microvilli. Hence, the events of activation prior to the intracellular pH increase induce the formation of cortical microfilamentous networks and microvillar elongation. The microfilaments may provide the structural and/or contractile framework for support of the egg surface, which is undergoing extensive rearrangement. Microfilament organization within the micro-villi, i.e. bundle formation, may then be a consequence of cytoplasmic alkalinization. Hence, actin filament bundle formation in the cortex of the sea urchin fertilized egg appears to be a two-step process:

1. the polymerization of actin filaments;
2. the association of filaments to form bundles.

The mechanisms of cortical reorganization are not known but are likely to involve actin-binding proteins as described in other systems (Pollard and Craig, 1982). Aggregation of actin filaments and their asso-ciation with bundling protein, e.g. fascin, may give rise to microfila-ment bundles in egg microvilli (Otto et al., 1980). Although fascin is found in the unfertilized sea urchin egg and is localized in microvilli of fertilized ova, its interaction with actin has not been shown to be calcium- or pH-sensitive (Bryan and Kane, 1982). Hence, other actin-binding proteins may be instrumental in microvillar elongation; cyto-plasmic alkalinization may give rise to microfilament bundle formation by promoting actin–actin-binding-protein interactions.

In addition to changes in microvillar conformation, the eggs of a number of different animals undergo changes in cortical rigidity and contraction that involve the cytoskeleton (Vacquier, 1981; Bement et al., 1992). Within minutes of insemination in sea urchin eggs there is an increase in cortical rigidity due presumably to changes in its actin–myosin system. In starfish eggs changes in cortical rigidity are coordi-nated with meiotic maturation of the maternal chromosomes. When oocytes are treated with 1-methyladenine to induce meiotic maturation there is a decrease in cortical stiffness. After germinal vesicle breakdown cortical rigidity remains low but increases during polar body formation. *Chaetopterus* oocytes possess a cortical cytoskeleton that localizes mRNA and undergoes morphological rearrangement during early development, taking attached mRNA with it (Sawalla et al., 1985; see Bement et al., 1992). Such changes in the cytoskeleton may be important in the locali-zation and segregation of specific mRNAs to different parts of the egg cytoplasm and for differentiation. An actin cytoskeletal network is a prominent feature of *Tubifex* (oligochaete) eggs (Shimizu, 1988), which in coordination with microtubules may be responsible for the furrow-like depressions that appear along the animal vegetal axis before polar body extrusion and for organelle translocation (Shimizu, 1979). Cyclical changes in surface tension and contraction have been correlated with cytoskeletal alterations and also occur in anucleate egg fragments with the same cycle as in normal embryos (Yoneda et al., 1978; Yamamoto and Yoneda, 1983). These observations indicate that egg activation initiates processes that are autonomous of the nucleus, and regulate, in a cyclical manner, cytoskeletal components and cytoplasmic contraction.

Ascidian eggs undergo extensive surface and internal rearrangements when activated. In *Styela* the myoplasm (a portion of the egg cytoplasm that ultimately gives rise to mesenchyme and muscles) and its associated mRNA become concentrated in the cortex of the vegetal hemisphere following fertilization and then move to the subequator to form the yellow crescent (Fig. 7.2; Bement *et al.*, 1992).

These changes in the localization and segregation of cytoplasmic components are brought about by alterations in the cytoskeleton, involving both actin and microtubules formed in conjunction with the development of the sperm aster (Bement *et al.*, 1992; Sawada and Schatten, 1988). In *Phallusia* extensive rearrangements of the cytoplasm and cortical contractions have also been described (Sardet *et al.*, 1989) that are brought about by a cortical actin network. Elevation of internal Ca^{2+} by activation triggers a contraction wave, the direction of which is only partially determined by the animal–vegetal axis (Speksnijder *et al.*, 1990). The direction of ooplasmic segregation may be determined by the point of sperm entry.

Actin filaments, intermediate filaments and microtubules all undergo extensive rearrangements at fertilization in *Xenopus* eggs (Bement *et al.*, 1992; Houliston and Elinson, 1991). The reorganizations are involved with fertilization events *per se*; axis determination is believed to reflect specializations of the cytoplasm that allow the egg to cope with its large size (Bement *et al.*, 1992). Cortical contraction is prominent in fertilized frog eggs (Elinson, 1980; Kirschner *et al.*, 1980). At sperm entry

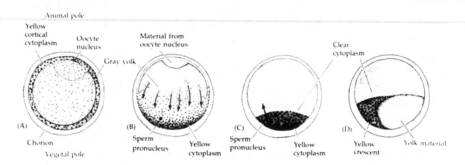

Fig. 7.2 Cytoplasmic rearrangement in the egg of the tunicate *Styela*. **(A)** Before fertilization, yellow cortical cytoplasm surrounds gray yolky cytoplasm. **(B)** After sperm entry, the yellow cortical cytoplasm and the clear cytoplasm derived from the breakdown of the oocyte nucleus stream vegetally toward the sperm. **(C)** As the sperm pronucleus migrates towards the egg pronucleus, the yellow and clear cytoplasms move with it. **(D)** The final positions of the clear and yellow cytoplasms. These mark the positions where the cells give rise to the mesenchyme and muscles, respectively. Reproduced with permission from Gilbert, 1991.

there is a contraction in the animal hemisphere, called the activation wave. After the first wave ceases a second wave, the post-fertilization wave, starts at the site of sperm incorporation. Other calcium-sensitive waves follow. This contractile activity is believed to reduce the size of the animal hemisphere, thereby assisting the movements of the male pronucleus and pulling the egg surface from the forming fertilization envelope. The egg then is free to rotate within the fertilization envelope. Changes in components comprising the cytoskeleton of *Xenopus* eggs and surface changes involving processes such as microvillar elongation, localization of mRNA and rotation of the egg cortex relative to the underlying cytoplasm have been reviewed (Bement *et al.*, 1992; see also Houliston and Elinson, 1991).

In mammals changes in the egg cytoplasm and cytoskeleton in association with cortical granule exocytosis have not been extensively investigated. Major alterations in the localization of actin and microtubules occur in conjunction with fertilization cone formation, meiotic maturation, and pronuclear migration and association in mammalian zygotes and are discussed in Chapters 4, 8 and 13.

Resumption of meiotic maturation

8

Oocyte maturation is accomplished by two successive meiotic divisions during which the chromosome number is halved. This reduction is achieved by the formation of two polar bodies – two small cells that are formed during the meiotic divisions and serve as 'receptacles' for extra chromosomes which the egg/zygote eliminates. The haploid chromosomes that remain within the egg/zygote become a part of the female pronucleus and eventually 'one half' of the embryonic genome. Early studies established the nomenclature and sequence of meiotic maturation; contemporary studies have been concerned with chromosome structure and mechanisms involving the regulation of this process (Masui and Clarke, 1979; Murray and Hunt, 1993).

As previously indicated, eggs of various mollusks and annelids are normally inseminated during meiotic prophase (the germinal vesicle stage) while eggs of some organisms, such as starfish, are fertilized during germinal vesicle breakdown, just prior to the formation of the first meiotic apparatus. The sequence of events in eggs fertilized at the germinal vesicle stage differs from that of ova fertilized at later stages of meiosis. In the former, this variation primarily involves germinal vesicle breakdown, formation of a meiotic spindle and its movement to the periphery of the zygote. Morphogenesis of the maternally derived chromatin of eggs belonging to classes I–III (see Fig. 5.1), involves the later stages of meiotic maturation, e.g. polar body formation, and are comparable in all other aspects. Figure 8.1 illustrates some of the major chromosomal events occurring during meiosis.

8.1 GERMINAL VESICLE BREAKDOWN AND FORMATION OF THE MEIOTIC SPINDLE

One of the most dramatic indications of activation in eggs fertilized at meiotic prophase is the breakdown and disappearance of the germinal

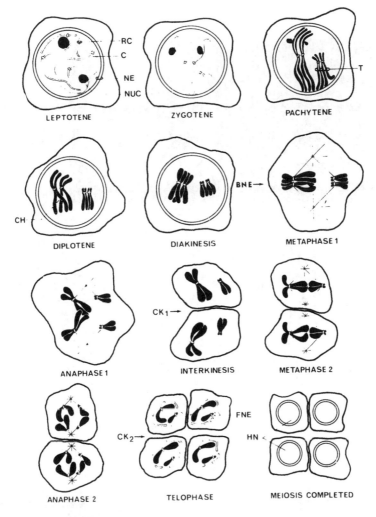

Fig. 8.1 Structures and stages of meiosis. RC = replicated leptotene chromosomes; C = centromere; NE = nuclear envelope; NUC = nucleolus; T = tetrad; CH = chromosomes; BNE = breakdown of nuclear envelope; CK$_1$ and CK$_2$ = cytokinesis 1 and 2; FNE = formation of nuclear envelope; HN = haploid nuclei. Reproduced with permission from Longo and Anderson, 1974.

vesicle. The germinal vesicle is a large, euchromatic spheroid nucleus that often contains a large nucleolus with several distinct regions (Fig. 8.2(A), (B).

The nuclear envelope is distinguished by its smooth contour and numerous pores. Granular aggregations have been observed within the germinal vesicle in a number of different species. The composition of

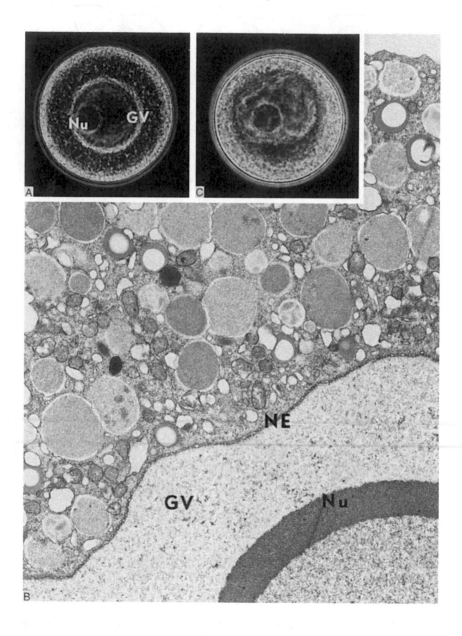

Fig. 8.2 **(A)** Germinal vesicle (GV) of the surf clam (*Spisula*) oocyte. Nu = nucleolus. **(B)** Germinal vesicle (GV) of the starfish (*Asterias*) oocyte. NE = nuclear envelope; Nu = nucleolus. **(C)** Germinal vesicle breakdown in a surf clam (*Spisula*) zygote. See Longo and Anderson, 1970a.

these structures has not been determined but they are thought to contain RNA, possibly in transit to the cytoplasm.

Breakdown of the germinal vesicle is a fairly rapid event; in the surf clam, *Spisula*, it occurs 10–15 min after insemination at 20°C (Chen and Longo, 1983). This process has been examined in a number of different organisms using electron microscopy and appears to be morphologically similar in each organism examined (Calarco *et al.*, 1972). Initiation of germinal vesicle breakdown is recognized when its surface becomes highly irregular and plicated (Fig. 8.2(C)). Internally, the chromosomes condense and the nucleolus disappears. Subsequently, openings are seen along the nuclear envelope, owing to multiple fusions between the inner and outer membranes of the nuclear envelope. These breaks lead to the formation of cisternae which may outline the chromosomes for a brief period but then are scattered into the cytoplasm. In some instances, e.g. human oocytes matured *in vitro*, cisternae derived from the nuclear envelope outline the condensing chromosomes as they are arranged along the metaphase plate; later the cisternae disappear.

During nuclear envelope breakdown in *Spisula* eggs, a 67 kDa lamin protein is extensively phosphorylated concomitant with its solubilization (Dessev and Goldman, 1988). Nuclear envelope breakdown in this case is believed to be controlled by a mechanism involving lamin phosphorylation similar to that which occurs in mitotic cells (Newport and Spann, 1987). A protein kinase activity ($p34^{cdc2}$-H1 kinase, MPF) phosphorylates the lamins and this, in turn, is coupled to breakdown of the nuclear envelope (Dessev *et al.*, 1991). Nuclear envelope breakdown in *Spisula* oocytes *in vitro* also requires ATP and Mg^{2+}, but not Ca^{2+} and is unaffected by protease inhibitors (Dessev *et al.*, 1989). Somewhat different results have been obtained with the oocytes of other species. Studies with starfish oocytes (Stricker and Schatten, 1989) indicate that phosphorylation of nuclear proteins is accompanied by structural changes of the nuclear envelope and nuclear lamins during germinal vesicle breakdown and that nuclear envelope disassembly occurs before depolymerization of nuclear lamins is completed. Corresponding changes in chromatin and lamin B localization have been described for maturing mouse oocytes (Albertini *et al.*, 1993).

Organelles identified as lysosomes aggregate along the germinal vesicle of rat oocytes undergoing meiotic maturation (Ezzell and Szego, 1979). It has been suggested that this mobilization of lysosomes is a specific response to factors promoting the resumption of meiotic maturation that leads to germinal vesicle breakdown. Meiotic maturation in mouse oocytes can be reversibly blocked at discrete stages prior to metaphase II when oocytes are cultured in the presence of various drugs (Wassarman *et al.*, 1976). Germinal vesicle breakdown does not take place in the presence of dibutyryl cAMP; chromosome condensation is initiated but then aborts. Similarly, chloroquine, an inhibitor of lysosomal function, blocks germinal vesicle breakdown. These findings suggest that germinal vesicle breakdown may occur by a cAMP-controlled protease-activated mechanism analogous to that proposed for a variety of polypeptide hormones mediating biological phenomena.

8.1.1 ACTIN AND TUBULIN CHANGES

Concomitant with the breakdown of the germinal vesicle, asters appear in the egg/zygote cytoplasm. Early cytologists recognized that fertilized eggs may be initially associated with the formation of one or two asters. More recent studies with fluorescently labeled tubulin antibody have demonstrated the presence of an aster or centrosphere in association with the germinal vesicle of starfish eggs (Schroeder and Otto, 1984; Otto and Schroeder, 1984). In surf clam (*Spisula*; Kuriyama *et al.*, 1986) and oyster (Longo *et al.*, 1993) zygotes and starfish (*Asterias*) oocytes, two asters become situated on either side of the disrupting germinal vesicle. In mammalian eggs/zygotes, centrioles have not been observed in the asters of meiotic and cleavage spindles (Zamboni, 1971; Szöllösi *et al.*, 1972; Schatten *et al.*, 1988b). Dense accumulations of fine textured material and vesicles are observed at the polar regions of the meiotic spindle that are reminiscent of material observed in microtubule organizing centers of somatic cells (McIntosh, 1983). The amorphous material is believed to be involved in the nucleation, polymerization and organization of the microtubules that are a part of the asters and constitute a microtubule organizing center (MTOC).

At the time of germinal vesicle breakdown in many species two centrosomes, composed of electron-dense fibrillar material from which microtubules radiate, appear near the nucleus. These aggregates form small asters and become situated at the poles of the spindle during the meiotic divisions. Fascicles of microtubules project from the central portion of the asters and are separated by areas filled with yolk bodies, mitochondria and cisternae of endoplasmic reticulum. Microtubules appear within breaks of the nuclear envelope, become associated with condensing chromosomes and orient to the poles of the developing spindle. Concomitantly, the two asters move opposite one another and their centers define the poles of the meiotic spindle (Longo and Anderson, 1970a). In starfish (Kato *et al.*, 1990) and *Mytilus* (Longo and Anderson, 1969a) each pole of the first meiotic spindle has two centrioles, whereas each pole of the second meiotic spindle contains only one centriole. These observations indicate that the centrioles do not duplicate between meiosis I and II and that the egg possesses only one centriole immediately following its maturation (Sluder *et al.*, 1989).

In maturing oocytes the germinal vesicle and/or the meiotic spindle as it develops moves to the cortex. Eventually, the meiotic spindle becomes anchored to the animal pole. In contrast to observations of some invertebrate eggs (Longo et al., 1993), cytochalasin B prevents the localization of the meiotic spindle to the cortex of maturing mouse oocytes, suggesting that microfilaments are involved in this movement (Longo and Chen, 1985; Maro et al., 1986; Alexandre et al., 1989). Actin has been demonstrated in nuclei and the meiotic spindle and has been implicated in force production of chromosome movements during mitosis (Zimmerman and Forer, 1981). In light of these investigations and studies demonstrating the disruptive effects of cytochalasin B on actin, an actin-based system may be responsible for the cortical localization of the meiotic spindle in mouse oocytes. Experiments with *Xenopus* oocytes have demonstrated that translocation of metaphase I spindles and assembly of metaphase II spindles are not blocked by cytochalasin B (Gard et al., 1995; see also Ryabova et al., 1986). Filamentous actin, however, is required for anchoring and rotation of the spindle.

The role of microtubules in movements of the meiotic spindle is less clear, although microtubular changes appear to be a prerequisite for germinal vesicle migration and are involved in chromosome translocation (see below). In *Xenopus* eggs, the meiotic spindle is barrel-shaped and within its centrosomes are found centrioles (Gard, 1992). At both meiosis I and II the spindle is initially oriented perpendicular to the animal/vegetal axis (or perpendicular to the surface of the animal pole). Similar movements of the meiotic spindle have also been observed in starfish and oyster oocytes (Shirai et al., 1990; Longo et al., 1993). The basis for this rotation is unknown but in *C. elegans* embryos and leech oocytes microtubules have been implicated in similar movements of the spindle. In most invertebrates the spindle of the unfertilized egg is usually positioned with its long axis perpendicular to the surface of the ovum/zygote (Fig. 8.3); in the unfertilized eggs of some mammals, however, the spindle is oriented tangential to the egg surface (Figs. 8.4, 8.5).

Placement of the meiotic apparatus within the periphery of invertebrate eggs frequently results in an asymmetry in its structure. The aster located in the cortex of the egg (the peripheral aster) is reduced in size in comparison to the centrally located aster.

In many mammalian eggs (e.g. mouse, hamster and rat) the region that overlies the meiotic spindle is distinguished by the absence of microvilli and cortical granules and the presence of a dense layer of actin filaments (Nicosia et al., 1977; Longo and Chen, 1985; Maro et al., 1986). This specialized cortical region in mouse ova is referred to as the microvillus-free area (Figs 8.4, 8.5). Although a meiotic spindle is not formed in mouse oocytes treated with colchicine, the chromosomes move to the egg cortex and a microvillus-free area forms in the region of the cortex associated with the chromosomes. Moreover, when the

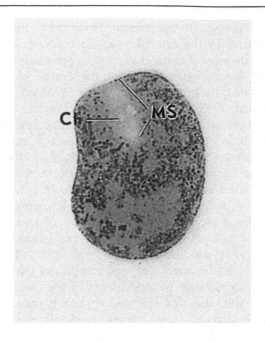

Fig. 8.3 Unfertilized mussel (*Mytilus*) egg at the first metaphase of meiosis. The long axis of the meiotic spindle (MS) is orientated perpendicular to the egg surface. Ch = chromosomes. See Longo and Anderson, 1969a.

meiotic spindle or the chromosomes are prevented from moving to the egg cortex a microvillus-free area does not develop. These observations indicate that interaction of the meiotic chromosomes with the egg cortex brings about the formation of the microvillus-free area. A similar type of interaction may be manifest when the sperm nucleus enters the egg cortex and with the formation of polar bodies (see below).

The tubulin content of surf clam (*Spisula*) eggs has been shown to be slightly greater than 3% of the total protein of the ovum (Burnside *et al.*, 1973). Since unfertilized *Spisula* eggs lack morphologically distinguishable microtubules, tubulin is apparently 'stored' until activation, when it is presumably used in the formation of the meiotic spindle. Isolated tubulin from *Spisula* eggs is capable of assembling into microtubules *in vitro* (Weisenberg and Rosenfeld, 1975). In contrast to unfertilized ova, activated *Spisula* eggs contain centrioles and granules that are capable of organizing microtubules into asters *in vitro*. This change in the ability to polymerize and assemble microtubule-containing structures appears to be a general characteristic of activated eggs.

In the unfertilized mouse egg microtubules are located predominantly

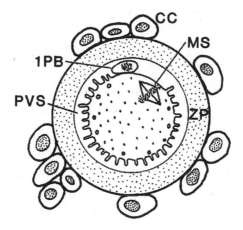

Fig. 8.4 Diagrammatic representation of an unfertilized mouse egg at the second metaphase of meiosis. The long axis of the meiotic spindle (MS) is oriented tangential to the egg surface. CC = cumulus cells; 1PB = first polar body; PVS = perivitelline space; ZP = zona pellucida.

in the meiotic spindle. At fertilization approximately 15 centrosomes appear (Schatten *et al.*, 1985b, 1986), each of which organizes an array of microtubules around itself to form an aster. The asters increase in size during the course of meiotic maturation and become associated with the male and female pronuclei during their movements to the center of the zygote (Fig. 8.6).

Modifications of tubulin, such as acetylation, as well as the presence of different tubulin isoforms, have been examined in maturing and fertilized mouse oocytes (Schatten *et al.*, 1988a; Palacios *et al.*, 1993). In somatic cells there appears to be a correlation between acetylation and microtubule stability (Webster and Borisy, 1988): the 'oldest', more stable microtubules appear to be acetylated while the dynamic, newly assembled microtubules are not. Examination of mouse eggs with monoclonal antibodies that bind acetylated α-tubulin suggests a cell-cycle-specific pattern of acetylation (Schatten *et al.*, 1988a) where acetylated tubulin is located at the poles of the meiotic spindle of unfertilized eggs. Acetylated microtubules are not present in the cytasters. At anaphase the spindle is labeled and by the formation of the second polar body only the mid-body microtubules are acetylated.

Gamma-tubulin is found in the pericentriolar material of somatic cells and is thought to play a role in the nucleation, growth and polarity of

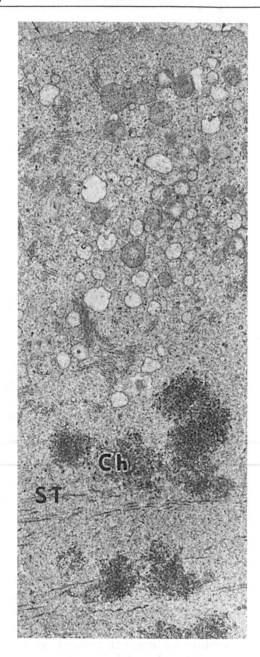

Fig. 8.5 Cortical region of a mouse egg containing a portion of the meiotic spindle. The surface of the egg associated with the meiotic spindle lacks microvilli immediately subjacent to the plasma membrane (arrows) is a layer of actin filaments. Ch = chromosomes; ST = spindle microtubules. Reproduced with permission from Longo and Chen, 1985.

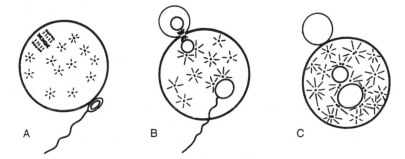

Fig. 8.6 Microtubular patterns during fertilization of the mouse egg. **(A)** Mouse sperm lack centrosomes and the unfertilized oocyte has 16 cytoplasmic aggregates of centrosomal antigen as well as centrosomal bands at the meiotic spindle poles. **(B)** Each centrosome organizes an aster and, after sperm incorporation, some foci along with their asters begin to associate with the developing male and female pronuclei. **(C)** When the pronuclei are closely apposed at the egg center, several foci are found in contact with the pronuclei and typically a pair reside between the adjacent pronuclei. Dots = centrosomes; lines = microtubules. Reproduced with permission from Schatten *et al.*, 1986.

microtubules. The localization of γ-tubulin in fertilized mouse eggs is consistent with the idea that it functions as a component of the MTOC and in spindle organization (Palacios *et al.*, 1993). It is concentrated at the spindle poles of the second meiotic spindle and at the centers of cytasters in fertilized mouse eggs. At later stages of fertilization γ-tubulin foci coalesce in the perinuclear microtubule organizing region surrounding the apposed male and female pronuclei. The presence of γ-tubulin with the basal body of the sperm is rather puzzling for, as indicated above (see also Schatten *et al.*, 1985b, 1986), this organelle is not believed to be a MTOC.

The role of kinetochores in fertilized mouse eggs has also been investigated by immunofluorescent methods (Schatten *et al.*, 1988b). Injection of antikinetochore antibodies into eggs interferes with chromosome congression. Interestingly, anaphase I and II are not blocked nor are microtubule capture and spindle formation inhibited (Simerly *et al.*, 1990).

Experiments examining phosphorylation of protein in the centrosomes of somatic cells have stimulated similar studies in mouse oocytes (Messenger and Albertini, 1991; Wickramasinghe and Albertini, 1992). Two populations of centrosomes have been noted, one of which appears to mediate nuclear and the other cytoplasmic events. In other words, some are consistently associated with chromatin throughout meiotic maturation and appear to function in the formation of spindle poles, while the remainder exhibit a maximum number and a heightened capacity for nucleation at prometaphase and anaphase of meiosis I (Messenger and Albertini, 1991). The fact that in both cases the centrosomes remain phosphorylated throughout metaphase I and II suggests that phosphorylation may not play a direct role in the regulation of the microtubule dynamics each group exhibits. Further efforts examining the potential role of centrosome phosphorylation suggest that the G2/M transition that occurs when mouse oocytes undergo meiotic maturation involves a phosphorylation of the centromeres (Wickramasinghe and Albertini, 1992). Maintenance of phosphorylation is required for *in vitro* expression of meiotic competency. Components of maturation promoting factor ($p34^{cdc2}$ + cyclin) are localized to centrosomes of somatic cells (Pines and Hunter, 1991) and it has been suggested that centrosomes play a key role in the entry of cells into metaphase. In light of these observations Wickramasinghe and Albertini (1992) suggest that centrosome phosphorylation in mouse oocytes is related to the functional expression of meiotic competency.

8.2 REGULATION OF MEIOTIC COMPETENCY

Questions concerning mechanisms by which eggs become arrested and the means by which this inhibition is alleviated such that oocytes become competent to resume meiosis have been of major concern to cell and developmental biologists. However, only within the past 10–15 years have the methodologies and the techniques been developed to allow investigators to answer such questions. The flurry of research activity and the insights obtained in the past 10 years with respect to determining the regulation of the cell cycle in a wide variety of cell types and the convergence of biochemical analyses and embryology has led to a resurgence in the use of oocytes and cleaving eggs as sources for the study of the cell cycle. The following section deals with the regulation of meiotic competency of the oocyte and factors which play a role in meiotic maturation (Murray and Hunt, 1993).

In most animals oocyte maturation is dependent upon ovarian functions. Generally, only fully grown oocytes mature in response to ovarian stimuli and cease development until inseminated. The systemic factors controlling ovarian egg maturation are gonadotropins which in turn stimulate the synthesis and release of hormones that are directly active on the egg (Masui and Clarke, 1979). In only a few groups of animals is there direct evidence of a chemically defined substance acting as a maturation-inducing substance. In amphibians, progesterone appears to be the natural maturation-inducing substance, although other steroids are known to be effective. In starfish, 1-methyladenine has been shown to be the natural maturation-inducing substance

(Kanatani *et al.*, 1969). Considerable controversy exists as to the precise roles of gonadotropins, steroids, prostaglandins, purines and cyclic nucleotides and inhibin during oocyte maturation in mammals (Channing *et al.*, 1982; Eppig and Downs, 1984; Downs, 1995).

Morphological differences between meiotically competent and incompetent mouse oocytes have been demonstrated (Wickramasinghe *et al.*, 1991; Zuccotti *et al.*, 1995). Meiotically competent oocytes are larger, possess a nucleolus which is surrounded by chromatin and undergo a progressive diminution of cytoplasmic microtubules with the appearance of MTOC. Use of the antibody, MPM-2, specific to M-phase phosphoprotein, indicates that the appearance of cytoplasmic positive foci is directly correlated with the appearance of nuclear changes characteristic of meiotically competent oocytes. Wickramasinghe *et al.* (1991) suggest that failure of mouse oocytes to progress beyond meiosis I may be due to a quantitative or qualitative deficiency in protease activity that degrades cyclin or other metaphase arresting factors that appear to trigger anaphase onset (Sagata *et al.*, 1989).

8.2.1 CALCIUM

In some animals, e.g. the surf clam, *Spisula*, the induction of oocyte maturation is known to be dependent on the presence of external calcium (Masui and Clarke, 1979). Oocytes of the starfish *Asterias*, however, can be induced to mature in sea water lacking calcium. Chelation of internal free calcium from oocytes by injection of EGTA inhibits maturation, whether or not calcium is present in the external medium. These observations indicate that transients of internal calcium may play a role in the initiation of oocyte maturation and are supported by experiments in amphibians (*Xenopus*) and echiuroid worms (*Urechis*), which demonstrate the release of calcium into the external medium when oocytes are stimulated to mature (O'Connor *et al.*, 1977; Johnston and Paul, 1977).

Consistent with an active role for calcium in the initiation of oocyte maturation are the results of experiments employing agents known to interfere with its physiological action. Agents such as D-600, which blocks calcium channels, and procaine inhibit 1-methyladenine-induced maturation in starfish oocytes as well as suppressing the luminescent discharge from 1-methyladenine-stimulated, aequorin-injected oocytes (Moreau et al., 1978). In the mollusk *Barnea* there is a calcium-dependent, D-600-sensitive period that lasts 3–4 min postactivation, which corresponds to the period in which there is an influx of calcium. A-23187, an ionophore that facilitates the transport of calcium across membranes, induces oocyte maturation in a wide variety of animals if external calcium levels are appropriate (Steinhardt *et al.*, 1974). Maintenance of internal Ca^{2+} and pH above threshold levels is required for germinal vesicle breakdown in the mollusks *Mactra* and *Limaria* (Deguchi and Osani, 1994b).

The role of calcium in the resumption of meiosis in mouse oocytes has been explored (Tombes *et al.*, 1992). Germinal vesicle breakdown is independent whereas meiosis I is dependent on external Ca^{2+}. In the absence of external Ca^{2+} formation of the first meiotic spindle is delayed and the first polar body fails to form. In the ascidian (*Phallusia*) egg postactivation Ca^{2+} waves are required for the completion of meiosis I and II (McDougall and Sardet, 1995). Interestingly, the Ca^{2+} waves originate from a region within the egg which is filled with endoplasmic reticulum and is believed to function as a pacemaker. Although *Spisula*, starfish, *Xenopus* and mouse oocytes are arrested at different stages in the cell cycle (G2, G2/MI, and MII, respectively) all employ a Ca^{2+} flux to resume the cell cycle, i.e. meiosis. In starfish the total amount of releasable Ca^{2+} does not appear to vary during maturation, but the pattern of the Ca^{2+} transients can differ depending on the stage of maturation and/or the type of Ca^{2+} releasing agent (Stricker *et al.*, 1994). There is considerable evidence that Ca^{2+} transients trigger entry into meiosis of mammalian oocytes (see Homa, 1995). Additionally, post-transitional modifications of maturation promotion factor, which regulate its activity, are believed to be controlled by Ca^{2+} signals. It has been proposed that different eggs have utilized the reactions induced by Ca^{2+} to overcome different stages of arrest (Murray and Hunt, 1993).

One possible mechanism for calcium action at maturation is that it binds to calmodulin. Injection of calmodulin triggers meiosis reinitiation in amphibians (Maller and Krebs, 1980). Although amphibian calmodulin stimulates bovine brain phosphodiesterase, it has no apparent effect on amphibian oocyte phosphodiesterase, an enzyme involved in the regulation of oocyte maturation (Masui and Clarke, 1979). Additional problems with this model include observations that calmodulin antagonists, chlorpromazine or fluphenazine, trigger maturation.

In the mouse oocyte Ca^{2+} is thought to trigger the resumption of meiosis by the destruction of cytostatic factor (CSF; Masui and Shibuya, 1987), a component that has not been characterized but arrests cells in metaphase (Masui and Markert, 1971). The *ser/thr* protein kinase, c-mos, is believed to play a key role in the generation and maintenance of CSF activity and is discussed further below.

In addition to changes in internal Ca^{2+}, sodium and potassium ion permeability and electrophysiological properties of oocytes have been shown to change during maturation. These alterations are necessary steps in the initiation of this process by maturation-inducing substances. Specific investigations of these properties are reviewed by Masui and Clarke (1979).

Reinitiation of meiotic maturation may also be accompanied by an increase in intracellular pH (Lee and Steinhardt, 1981). The role of internal pH changes in the maturation process has not been fully elucidated. Since maturation still occurs when the internal pH of progesterone-treated amphibian oocytes is constant, it has been suggested that a threshold value of internal pH does not exist and that internal pH change during maturation is of little importance as a regulating parameter (Cicirelli *et al.*, 1983). Similarly, starfish oocytes do not appear to require a significant increase in internal pH in order to mature in response to 1-methyladenine (Johnson and Epel, 1982).

8.2.2 PROTEIN SYNTHESIS

Germinal vesicle breakdown is not inhibited by anaerobic conditions or uncouplers of phosphorylation in mollusks. Such conditions and agents, however, prevent the formation of the meiotic apparatus. Spontaneous maturation of rat oocytes can be prevented by hypoxia. Germinal vesicle breakdown and chromosome condensation take place in mouse oocytes in the presence of protein synthesis inhibitors such as puromycin; however, nuclear progression is blocked at the circular bivalent stage when oocytes are cultured continuously in the presence of the drug (Clarke and Masui, 1983).

In general, changes in protein synthesis activity occur during meiotic maturation in all species studied thus far (Masui and Clarke, 1979). An increase in protein synthesis occurs during amphibian oocyte maturation (Shih *et al.*, 1978). Removal of the germinal vesicle from *Rana pipiens* oocytes affects neither the rate nor the pattern of protein synthesis. Protein synthesis inhibitors block germinal vesicle breakdown and oocyte maturation in amphibians and fish; however, a substantial reduction in the level of protein synthesis occurs under anaerobic conditions that does not prevent oocyte maturation, provided the eggs have undergone germinal vesicle breakdown (Smith and Ecker, 1970). With inhibition of protein synthesis in starfish (*Asterias* and *Piasterias*) oocytes, germinal vesicle breakdown occurs and maturation is not arrested until metaphase I. Fusidic acid blocks protein synthesis and stops maturation before germinal vesicle breakdown in *Chaetopterus* (annelid) oocytes, whereas puromycin inhibits protein synthesis by 50% but does not affect maturation, implying that some of the protein synthesized during maturation is actually required for this process.

Further experiments have examined what portion of the protein synthesized by the maturing oocyte is responsible for the progression of maturation and when the proteins necessary for each step of maturation are synthesized (Masui and Clarke, 1979; Murray and Hunt, 1993). In general, meiosis I in frogs requires protein synthesis, whereas nascent protein is not required for starfish and clam oocytes to progress through meiosis I. Protein synthesis, however, is required for the oocytes of all three groups to progress through meiosis II. Clarke and Masui (1983) have demonstrated the existence of critical transition points requiring protein synthesis between the germinal vesicle stage and metaphase I in mouse oocytes. Further experiments on this aspect of the regulation of meiosis (Hampl and Eppig, 1995) have shown that there is a translation-dependent mechanism regulating the activation of $p34^{cdc2}$ kinase from the germinal vesicle stage to metaphase I. The requirement for nascent protein appears at germinal vesicle breakdown and is correlated with a surge in cyclin B synthesis. Modifications (poly-adenylation) and translation of cyclin mRNA in immature and mature starfish oocytes have been examined (Standart *et al.*, 1987; Tachibana *et al.*, 1990).

Much of our knowledge concerning the regulation of meiosis, including its arrest and resumption, stems from investigations employing oocytes of frogs, starfish and mammals. A number of studies have also been carried out with the oocytes of mollusks. The following describes the regulation by maturation promoting factor (MPF) of meiotic maturation in each of these groups. The cell cycle of blastomeres is essentially biphasic and includes S1 and M phases. Unlike replicating somatic cells, eggs do not possess or have very abbreviated G1 and G2 phases of the cell cycle. One of the major objectives of fertilization is the initiation of the cell cycle from cells (sperm and egg) that have ceased to replicate. How the maturing oocyte and fertilized egg accomplish this task has been of considerable interest to cell and developmental biologists for the past 10 years (for reviews see Murray and Hunt, 1993; Eckberg, 1988; Matten and VandeWoude, 1994; Maller, 1994; Meijer and Mordet, 1994) and may be summarized as follows for the eggs of some organisms. In frogs the secretion of progesterone from surrounding follicle cells activates protein synthesis and initiates the meiotic cycle; the oocyte progresses from G2 to metaphase II. A similar series of events take place in mammalian eggs. In starfish, 1-methyladenine triggers the oocyte to undergo meiosis I and II and, unless fertilized, the mature egg ceases to develop. It has been shown for both amphibian and starfish oocytes the maturation-inducing substance (progesterone and 1-methyladenine, respectively) are effective only when applied to the oocyte surface (Gurdon, 1967; Dorée and Guerrier, 1975; see Masui and Clarke, 1979). In clams, G2-arrested oocytes are spawned and fertilization activates resumption of meiosis, events of fertilization per se and the mitotic cycle. In sea urchins, oocytes are induced to complete meiotic maturation prior to spawning. The eggs are spawned and fertilized at G0, an early state of the G1 phase of the cell cycle. Fertilization initiates the mitotic cycle.

Unlike dividing somatic cells, factors regulating the cell cycle in oocytes are, for the most part, present in the cytoplasm. For example, transplantation of nuclei derived from cells that normally do not divide (e.g. neurons) into frog oocytes stimulated with progesterone triggers the transplanted nucleus to undergo division. Conversely, transplantation of nuclei of dividing cells into immature oocytes results in the cessation of division of the transplanted nucleus. Serial transfer of cytoplasm, as shown in Fig. 8.7, demonstrates that whatever promotes maturation in frogs and starfish oocytes is a component of the oocyte rather than progesterone or 1-methyladenine (Masui and Markert, 1971). These experiments also indicate that the generation of the active factor is brought about by a catalytic process.

8.2.3 MATURATION PROMOTING FACTOR: FROGS

The substance responsible for oocyte maturation is maturation promoting factor (MPF; Masui and Markert, 1971), also referred to as meiosis or mitosis promoting factor (Murray and Hunt, 1993). In frogs progesterone induces MPF activation only when the oocyte is allowed to synthesize protein. When cycloheximide (a protein synthesis inhibitor) treated oocytes are microinjected with MPF containing cytoplasm, they undergo maturation suggesting that MPF exists in a

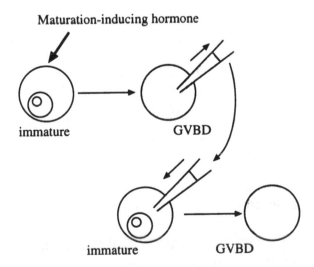

Fig. 8.7 Demonstration of the presence of MPF in the cytoplasm of maturing oocytes. Injection of cytoplasm taken from a maturing oocyte after treatment with maturation-inducing hormone induces maturation in an intact immature oocyte. Reproduced with permission from Kishimoto, 1988.

preactivated form that can be converted to activated MPF by post-translational modification. The presence of the oocyte nucleus is not necessary for MPF activation since enucleated frog oocytes can produce active MPF when treated with progesterone (Kishimoto et al., 1981). Unlike the situation in somatic cells, cytoplasmic reactions activate an inactivate MPF in a cyclical fashion and this oscillation of MPF is independent of DNA synthesis and meiotic spindle assembly (Fig. 8.8).

MPF has been purified from frog and starfish oocytes (Lohka et al., 1988; Labbe et al., 1989) and has been shown to be a heterodimer consisting of a 34 kDa protein that is homologous to the yeast gene product $p34^{cdc2}$ and cyclin, a 45 kDa protein. Cyclins are a family of proteins similar to the yeast protein $p56^{cdc13}$. They are synthesized throughout the cell cycle, accumulate during S and are degraded at the end of each mitosis (Swenson et al., 1986). Contrary to mitotic cell regulation, meiotic division in Xenopus does not appear to be controlled by cyclin synthesis (Minshull et al., 1991). In its active form, MPF ($p34^{cdc2}$ + cyclin), a kinase, is responsible for germinal vesicle breakdown and chromosome condensation (see Smith, 1989; Sagata et al., 1989a, b; Nebreda and Hunt, 1993). In keeping with these activities, among MPF's substrates are histone H1 and nuclear lamins. It has been suggested that phosphorylation of histone may play a role in chromosome condensation and it has been shown that phosphorylation of nuclear lamins leads to their disassembly.

Changes in the meiotic cell are coupled to fluctuations in MPF (Fig. 8.9).

In frog oocytes progesterone induces the activation of MPF, leading to meiosis I. After a brief decline, MPF activity rises and induces meiosis II. The egg remains at metaphase II until fertilization, at which time MPF activity falls and remains low until first mitosis. This sequence of events is triggered by progesterone. Progesterone binding to the oocyte surface induces a decrease in cAMP and a reduction in activity of cAMP-dependent protein kinase activity (PKA: Maller and Krebs, 1977; Maller, 1985). PKA is a negative regulator (Fig. 8.10).

1. It prevents protein synthesis, specifically in this case it blocks c-mos synthesis.
2. It blocks the activation of cdc25 phosphatase, an enzyme that catalyzes one of the last steps in the activation of MPF (Matten and VandeWoude, 1994; Maller, 1994).

The proto-oncogene product p39 c-mos kinase has been proposed to transform pre-MPF to active MPF (Sagata et al., 1989b). It is translated de novo upon progesterone stimulation and injection of c-mos mRNA or c-mos protein into oocytes induces germinal vesicle breakdown. Results of Daar et al. (1993) and Yew et al. (1992) indicate that the synthesis of

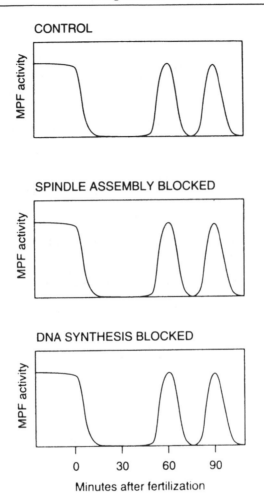

Fig. 8.8 MPF oscillates independently of DNA synthesis and spindle assembly. The fluctuation of MPF activity in normally fertilized embryos (control) is compared with that of embryos fertilized in the presence of an inhibitor of DNA polymerization (aphidicolin) or spindle assembly (nocodazole). The oscillations in the three sets of embryos are identical, showing that the failure to complete DNA replication does not influence the reactions that periodically activate and inactivate MPF. Reproduced with permission from Murray and Hunt, 1993.

c-mos is not only necessary, but is sufficient for germinal vesicle breakdown (Fig. 8.11).

Additional nascent proteins are also required for the progression from meiosis I to meiosis II. As suggested by Matten and VandeWoude (1994), progesterone exposure induces a dephosphorylation at thr 14 and

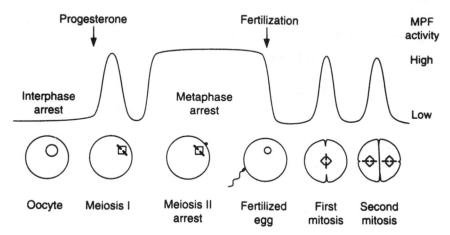

Fig. 8.9 MPF fluctuations in meiotic and mitotic cell cycles. Oocytes have low levels of MPF activity. Progesterone induces the activation of MPF, leading to meiosis I. After a brief decline, a second rise in MPF activity induces meiosis II and the oocytes remain arrested in metaphase of meiosis II with high levels of MPF. This arrest is overcome by fertilization, which leads to a precipitous decline in MPF activity. Interphase of the first mitotic cell cycle lasts about 60 min, while that of cycles 2 through 12 last about 15 min each. Each mitosis is initiated by the activation of MPF and is terminated by its inactivation. Reproduced with permission from Murray and Hunt, 1993.

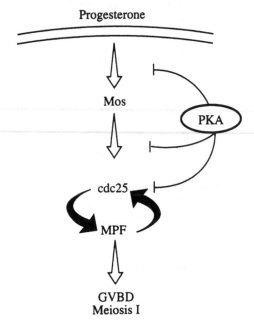

Fig. 8.10 Points of action of the cAMP-dependent protein kinase (protein kinase A, PKA). PKA blocks premature initiation of maturation at two critical points: synthesis of Mos protein and activation of cdc25. The unfilled arrows represent incomplete identification of the pathway participants. Reproduced with permission from Matten and VandeWoude, 1994.

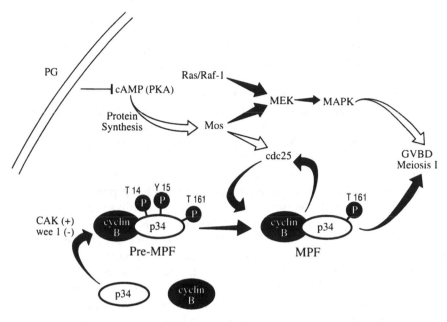

Fig. 8.11 A partial model of the signal transducing pathway in *Xenopus* oocyte maturation. Unfilled arrows reflect uncertainty regarding the pathway participants. Filled arrows indicate direct regulatory interactions. PG = progesterone; PKA = cAMP-dependent protein kinase; MEK = MAP/Erk kinase; CAK = the p34^{cdc2} activating kinase which phosphorylates p34^{cdc2} at threonine 161 (T161); (+) indicates this is an activating phosphorylation and (−) indicates that Wee1 performs the inhibitory phosphorylation at tyrosine 15 (Y15). Reproduced with permission from Matten and VandeWoude, 1994.

tyr 15 of p34^{cdc2} and activation of MPF. Cdc25 and Wee1 are believed to be involved in this activation, although this has not been conclusively established (Murray and Hunt, 1993). Phosphorylation of the thr 161 on MPF, which is catalyzed by CAK (*cdc2*-activation kinase), is essential for kinase activity. Mitogen activated protein kinase (MAPK) is activated at the same time as MPF; this is accomplished by a kinase referred to by a variety of terms, including MAPK kinase or MEK (Map/ERK kinase; Matten and VandeWoude, 1994). c-Mos is believed to phosphorylate and, hence, activate MEK *in vivo*. The exact function of MAPK in meiosis I and II has not been fully established; it is suggested to be involved in spindle formation and may prevent chromosome dispersion and nuclear assembly during the period between meiosis I and II (Murray and Hunt, 1993; Matten and VandeWoude, 1994). The latter suggestion has been proposed to account for the decline in MPF activity at the end of meiosis I and the absence of nuclear assembly.

8.2.4 CYTOSTATIC FACTOR

When cytoplasm from unfertilized frog eggs is injected into one cell of a two-cell embryo (Masui and Markert, 1971), the injected cell arrests in metaphase (Fig. 8.12).

The component responsible for this arrest is referred to as cytostatic factor (CSF; Masui and Clarke, 1979). CSF has not been isolated and characterized but three proteins have been shown to be involved in its activity: c-mos, MAPK and cyclin dependent kinase 2 (Haccard *et al.*, 1993; Matten and VandeWoude, 1994). Cytostatic factor-like activity has also been observed in mammalian oocytes (Balakier and Czolowska, 1977). It is believed to stabilize MPF by preventing cyclin degradation. With the wave of Ca^{2+} release at fertilization, metaphase II arrest is abolished by the inactivation of MPF and CSF. The c-*mos* proto-oncogene product, p39^{mos}, undergoes selective proteolysis by the Ca^{2+}-dependent cysteine protease calpain (Watanabe *et al.*, 1989). Calmodulin-dependent protein kinase II has also been shown to mediate the inactivation of MPF and CSF at fertilization of *Xenopus* eggs (Lorca *et al.*, 1993). Cyclin proteolysis, which is apparently independent of c-mos degradation, is also induced by elevated free Ca^{2+} (Lorca *et al.*, 1992; Watanabe *et al.*, 1991). That c-mos degradation takes longer than cyclin degradation is difficult to reconcile with the

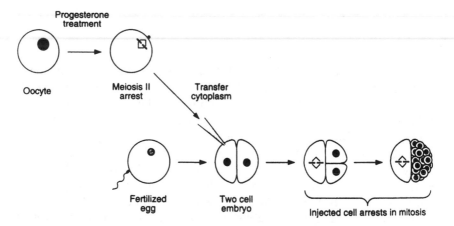

Fig. 8.12 Demonstration of cytostatic factor (CSF). Cytoplasm removed from unfertilized eggs and injected into one cell of a two-cell embryo induced the injected cell to enter a prolonged metaphase arrest at the next mitosis. This experiment shows that unfertilized eggs contain a factor capable of inducing cell cycle arrest. Reproduced with permission from Murray and Hunt, 1993.

idea that CSF (c-mos) keeps the *Xenopus* egg arrested in metaphase by actively preventing cyclin degradation. Clearly other factors are involved here.

8.2.5 MATURATION PROMOTING FACTOR: STARFISH

In starfish oocytes mechanisms regulating meiotic maturation are similar to those found for frog oocytes (Meijer and Mordet, 1994). There is good evidence that 1-methyladenine (1-MA), the natural inducer of meiotic maturation of starfish oocytes and produced by surrounding follicle cells, interacts with a receptor in the oocyte's plasma membrane. The receptor, which has not been identified, is believed to be coupled to G protein and when activated, the G protein dissociates into α and $\beta\gamma$ subunits. Although the roles of the subunits and their effectors have not been determined, recent immunolocalization experiments demonstrate that the $\beta\gamma$ subunit coexists with cytokeratin filaments in starfish oocytes. Stimulation of oocyte maturation by 1-MA induces the disappearance of the cytokeratin filaments and, presumably, the disassembly of $\beta\gamma$ subunits (Chiba *et al.*, 1995). As in frog oocytes, a drop in cAMP (induced by 1-MA) is necessary, but not sufficient, for induction of maturation.

A number of mechanisms have been shown to be involved in steps leading to MPF activation in starfish oocytes (Meijer and Mordet, 1994). For example, cAMP and its dependent kinase are believed to inhibit a step leading to the activation of MPF (Meijer and Mordet, 1994). A chymotrypsin-like protease activity associated with a large 650 kDa structure, referred to as a proteosome, has been suggested to be involved in mediating 1-MA activity and MPF activation (Sawada *et al.*, 1992). The mechanism of action and the site where the proteosome functions have not been identified, although there is evidence suggesting that the proteosome is activated by intracellular Ca^{2+} and is responsible for the ubiquitin-dependent proteolysis of cyclin and/or other proteins during the metaphase–anaphase transition (Kawahara and Yokosawa, 1994). There is also evidence for tyrosine phosphorylation and protein kinase C activity involvement in 1-MA induced maturation in starfish oocytes (Meijer and Mordet, 1994).

Activation of p34^{cdc2}/cyclin B in starfish oocytes is believed to occur in a manner similar to that described for frog oocytes (Meijer and Mordet, 1994). In addition, the translocation of MPF at the time of meiotic maturation has been followed by immunolocalization (Ookata *et al.*, 1992). In immature oocytes inactive MPF is located within the cytoplasm; with exposure to 1-MA, a fraction of MPF moves to the nucleus and binds the nuclear envelope and chromosomes. Another fraction becomes associated with the developing meiotic spindle. The changes in MPF localization are considered essential in order to bring active MPF in contact with its substrates.

8.2.6 REGULATION IN SURF CLAM OOCYTES

Mechanisms involved in the regulation of meiosis in surf clam oocytes have been investigated (Eckberg, 1988; Swenson et al., 1989). Events occurring in the first 3–4 min after insemination in Spisula commit oocytes to germinal vesicle breakdown, which occurs about 10 min postinsemination (Allen, 1953). Investigations carried out to determine what occurs during the 3–4 min period of commitment supports the idea that protein kinase C and Ca^{2+}-dependent processes are involved in biochemical pathways leading to germinal vesicle breakdown (Eckberg et al., 1987; Bloom et al., 1988; Dubé et al., 1991). Protein kinase C is activated at about 3 min postinsemination, i.e. near the end of the commitment period, by diacylglycerol (Bloom et al., 1988). Unlike the situation in frogs and starfish, IP_3 induces germinal vesicle breakdown in Spisula oocytes. The hydrolysis of phosphotidylinositol trisphosphate occurs before germinal vesicle breakdown, suggesting that both products of inositol lipid hydrolysis act synergetically to elicit germinal vesicle breakdown (Eckberg, 1988). The increase in protein phosphorylation that occurs in activated Spisula oocytes is required for germinal vesicle breakdown and is affected by alterations in Ca^{2+} influx. Investigations by Longo et al. (1991b) have shown that A-23187 is able to override the effects of protein synthesis inhibition on pronuclear development in Spisula, suggesting that both nascent protein and Ca^{2+} signals are involved in regulating events controlling the status of the maternal and paternal chromatin. Studies with frog oocyte extracts (Lohka and Masui, 1984a) and in sea urchin eggs (Twigg et al., 1988) indicate that Ca^{2+} transients regulate chromatin condensation and progression of the cell cycle. It has been postulated that Ca^{2+} transients function as pacemakers, governing the timing of cell cycle transitions (Whitaker and Patel, 1990).

Increases and decreases in the amount of cyclins A and B during meiosis I and II have been observed in Spisula oocytes (Swenson et al., 1986; Fig. 8.13).

The proteins have been shown to be synthesized continuously throughout the cell cycle and degraded at specific times. High levels of the cyclins do not correlate directly with nuclear envelope breakdown and degradation of cyclins does not result in chromatin dispersion and nuclear assembly as both cyclins A and B are degraded between meiosis I and II yet nucleus formation fails to occur (Longo and Anderson, 1970a, b; Longo, 1983). Low doses of the protein synthesis inhibitor emetine, which reduces but does not abolish protein synthesis,

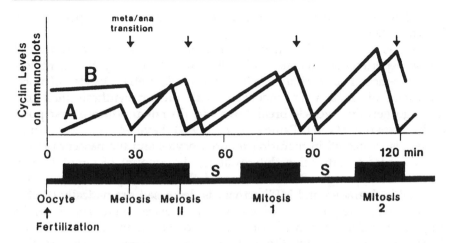

Fig. 8.13 Diagrammatic representation of the levels of clam cyclin A and B during meiosis and mitosis. This diagram summarizes the behavior of the cyclins as seen in several types of experiments: continuous label, pulse-chase, immunoblot and immunofluorescence. The periodic rise and fall in cyclin levels is due to continuous synthesis across the cell cycle, followed by periodic destruction near the end of each M phase. Reproduced with permission from Swenson *et al.*, 1989.

cause a lengthening of the cell cycle, suggesting that cells must accumulate a threshold amount of one or more inducers (cyclin plus other proteins?) allowing them to enter mitosis (Swenson *et al.*, 1989).

8.2.7 REGULATION IN MAMMALIAN OOCYTES

Extracts of maturing mouse oocytes induce germinal vesicle breakdown in immature starfish and frog oocytes (Kishimoto *et al.*, 1984), indicating that mammalian oocytes possess MPF activity. MPF activity, as indicated by histone H1 kinase activity, increases upon entry into M-phase (Rime and Ozon, 1990) and fluctuates with the cell cycle, i.e. it declines at anaphase I and increases at metaphase II (Hashimoto and Kishimoto, 1988; Fulka *et al.*, 1992). Inhibitors of phosphorylation (6-dimethylaminopurine) block maturation whereas the phosphatase inhibitor okadaic acid induces germinal vesicle breakdown (Rime and Ozon, 1990). It has been proposed that p34^{cdc2} in its phosphorylated state is inactive and, as described for frog and starfish oocytes, when dephosphorylated, MPF is activated in mammalian oocytes (Rime and Ozon, 1990; see also Gavin *et al.*, 1994). Phosphorylation and dephosphorylation of other proteins are, no doubt, important in meiotic maturation in mammals but have not been ascertained. Protein kinase A maintains denuded oocytes in meiotic arrest, presumably through phosphorylations that may or may not directly involve p34^{cdc2} (see Downs, 1995).

CSF has also been demonstrated in mouse oocytes indirectly by cell fusion experiments (Kubiak *et al.*, 1993) and directly by the identification of c-mos (Paules *et al.*, 1989; Weber *et al.*, 1991). Oocytes of mice, like those of frogs, when incubated with protein synthesis inhibitors undergo germinal vesicle breakdown but do not produce a first polar body. Experiments by Paules *et al.* (1989) have shown that c-mos protein is expressed in maturing mouse oocytes and the nascent protein is required for maturation. Interestingly, the amount of c-mos is very high in mouse eggs after formation of the second polar body, while cyclin B is degraded and MPF activity is decreased dramatically (Weber *et al.*, 1991). Degradation of c-mos takes place later. These results suggest that degradation of c-mos may not be involved in the release from metaphase II arrest. Evidence has been presented to suggest that protein kinase C is involved in the release of the metaphase II block in mouse oocytes and, together with the increase in internal Ca^{2+}, may lead to chromosome decondensation and pronuclear formation (Colonna and Tatone, 1994). The possible interplay between the cell cycle and microtubule networks in maturing mouse oocytes has been discussed by Maro *et al.* (1994), who propose that MAP kinase is related to CSF and that this enzyme regulates changes in microtubule organization and chromatin condensation. In support of this proposal MAP kinase has been localized to the meiotic spindle and the MTOCs present at the spindle poles of maturing mouse oocytes (Verlhac *et al.*, 1993).

8.3 POLAR BODY FORMATION

Formation of the first polar body has been described in eggs and zygotes of a number of different organisms and involves the elongation of the spindle and movement of dyad chromosomes to their respective poles (Zamboni, 1971; Longo, 1973). Topoisomerase II is required during chromosome segregation at meiosis I for resolution of recombined chromosomes (Rose *et al.*, 1990). Concomitant with these events is the production of a cytoplasmic protrusion at the animal pole which becomes the first polar body (Fig. 8.14). At anaphase I, the more peripherally positioned chromosomes become localized within this protrusion (Fig. 8.15).

At the base of the protrusion a cleavage furrow is formed. The cleavage furrow associated with the formation of the first polar body in mouse eggs develops paratangentially following the rotation of the equatorial plate. Eventually the projection containing the peripheral aster and chromosomes becomes separated from the zygote. Separation

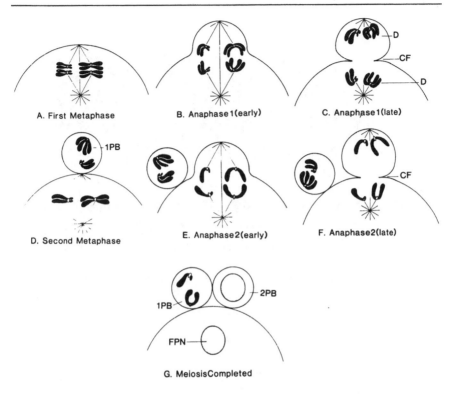

Fig. 8.14 Events involved in polar body formation. 1PB and 2PB = first and second polar bodies; FPN = female pronucleus; CF = cleavage furrow; D = dyad chromosomes. See Longo and Anderson, 1974.

is mediated by the formation of a ring of microfilaments that is located at the base of the projection. In the case of mammalian eggs (human and mouse) a dense layer of microfilaments is also located at the distal end of the projection that becomes the polar body (Lopata *et al.*, 1980). By immunofluorescent procedures this layer, as well as the cleavage furrow, has been shown to stain intensely for actin (Maro *et al.*, 1984; 1986).

The morphology of the cleavage furrow during polar body formation is similar to that which develops during cytokinesis of mitotic cells (Schroeder, 1975). The electron-opaque material which lines the furrow during polar body formation is actin and may function in the manner of a 'purse string' for partitioning the cytoplasm. Progression of the cleavage furrow eventually yields a short-lived cytoplasmic bridge which joins the egg/zygote and the first polar body. A causal role for actin in polar body formation has been demonstrated in surf clam (*Spisula*) and mouse zygotes, where furrow constriction and polar body

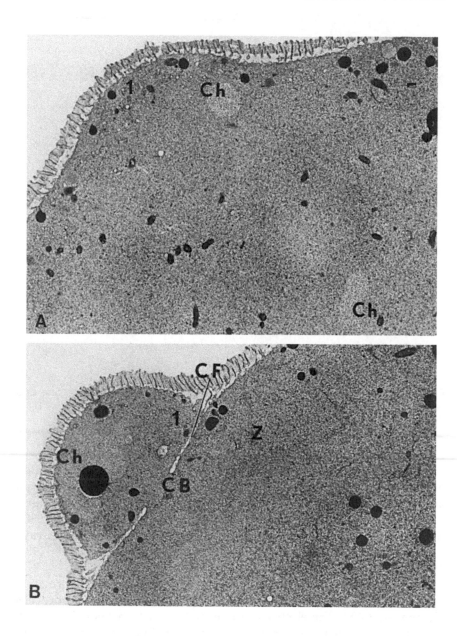

Fig. 8.15 Early **(A)** and late **(B)** stages in the development of the first polar body in the mussel *Mytilus*. 1 = developing first polar body; Ch = chromosomes; CB = cytoplasmic bridge joining the first polar body and the zygote (Z); CF = cleavage furrow. See Longo and Anderson, 1969a.

extrusion are prevented by cytochalasins (Longo, 1972; Maro *et al.*, 1984, 1986). In *Spisula* and oyster (Longo *et al.*, 1993) zygotes, inhibition of polar body formation with cytochalasin B occurs at anaphase I or II with similar results. The chromosomes that would normally be emitted are retained, thereby altering zygote ploidy. When zygotes are incubated in cytochalasin B only at meiosis I or meiosis II triploid zygotes result (Longo *et al.*, 1993; Guo *et al.*, 1992a, b); when incubated in cytochalasin B throughout the course of maturation, i.e. both meiosis I and meiosis II, all the chromosomes normally emitted with the polar bodies remain within the zygote and become organized into a variable number of pronuclei. In such cases the zygote is pentaploid.

The first polar body appears as a miniature cell and has been observed in a variety of invertebrates and mammals (Fig. 8.16).

A distinguishing feature of the first polar body is the presence of compacted chromatin which is usually not associated with a nuclear envelope. In many of the organisms studied, particularly mammals, the first polar body may contain cortical granules and variable amounts of

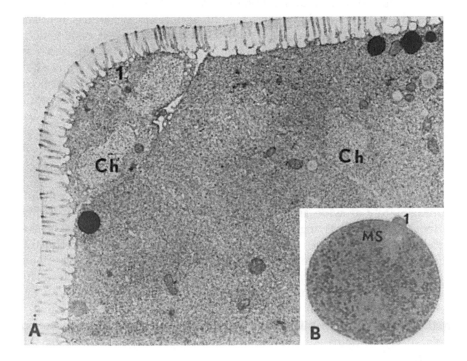

Fig. 8.16 (A) and **(B)** Second meiotic spindle (MS) at metaphase II and first polar body of the surf clam *Spisula*. Ch = chromosomes; 1 = first polar body. See Longo and Anderson, 1970a.

endoplasmic reticulum and mitochondria (Zamboni, 1971). Large quantities of cytoplasmic inclusions are usually found in the first polar body of the mussel, *Mytilus*, while lesser amounts are observed in the surf clam, *Spisula*. Centrioles and Golgi complexes have been observed in the first polar body of *Mytilus*. As to the possible cleavage of the first polar body, a human pronucleate zygote has been observed to be associated with three polar bodies, two of which were interpreted as being division products of the first (Zamboni *et al.*, 1966). In *Spisula* and *Mytilus*, the first polar body becomes highly electron-opaque during later stages of development. This increase in electron-opacity may be an indication of necrosis. The first polar body of some mammals, e.g. the rat, may disintegrate soon after its formation.

Following the formation of the first polar body, the chromosomes remaining in the egg/zygote become aligned on the equatorial plate of the second meiotic apparatus (Longo *et al.*, 1993; Fig. 8.16). As indicated earlier, this usually takes place without intervening telophase and prophase stages. The meiotic apparatus formed is structurally similar to the first and occupies an area relatively devoid of large cytoplasmic constituents (Longo, 1973; see Longo *et al.*, 1993). Centrioles have been found in the asters of the second meiotic apparatus in mollusk (*Spisula* and *Mytilus*) and starfish zygotes but they have not been observed in mammals. In the oligochaete *Tubifex* the second meiotic apparatus is formed and positioned perpendicular to the egg surface 40 min after the formation of the first. The meiotic apparatus appears to be tethered to the egg surface by structural connections between filamentous elements in the cortex and microtubules of the peripheral aster (Shimizu, 1981a).

Movement of the chromosomes and elongation of the spindle at anaphase II are similar to events occurring during anaphase I (Fig. 8.17).

The chromosomes located most peripherally move into a protrusion, approximately as wide as the peripheral aster, which is formed during spindle elongation (Longo, 1973; Shimizu, 1981b). At the base of the protrusion, a cleavage furrow develops which is morphologically similar to that observed during the formation of the first polar body; it too is associated with a band of filamentous actin along its leading edge. Progression of the cleavage furrow at the base of the developing second polar body yields a cytoplasmic bridge containing a midbody. The cytoplasmic bridge connecting the zygote and the second polar body remains at least until the first cleavage division of the zygote.

The second polar body contains a variable number of cytoplasmic constituents. Very few structures, such as mitochondria, endoplasmic reticulum, lipid droplets and yolk bodies, are observed in the surf clam,

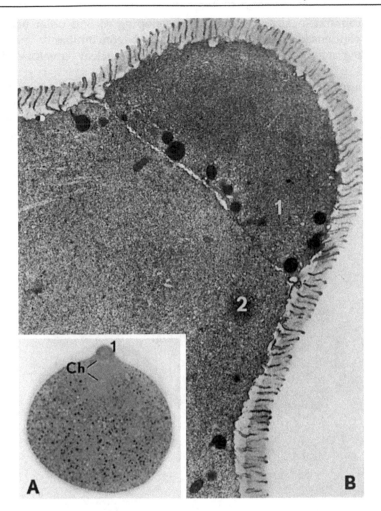

Fig. 8.17 Forming second polar body (2) in the mussel *Mytilus*. 1 = first polar body; Ch = chromosomes at anaphase II of meiosis. See Longo and Anderson, 1969a.

Spisula. In mammals the second polar body contains relatively more endoplasmic reticulum, mitochondria and yolk bodies than observed in *Spisula*. Few or no cortical granules are observed in the second polar body of the rat, hamster and human (Zamboni *et al.*, 1966; Lopata *et al.*, 1980). This is consistent with the formation of the second polar body following the release of the cortical granules in these species. The same kinds of cytoplasmic organelle and inclusion are observed in the second body of the mussel, *Mytilus*, as in the first.

The chromatin localized within the developing second polar body is initially condensed. Later, it disperses and concomitantly vesicles aggregate along its margin, fuse and form a nuclear envelope (Fig. 8.18).

Chromatin delimited by a continuous nuclear envelope has been observed in the second polar body of invertebrates and mammals (Longo, 1973). Apart from its reduced size and the absence of a male pronucleus, the second polar body is structurally similar to the zygote.

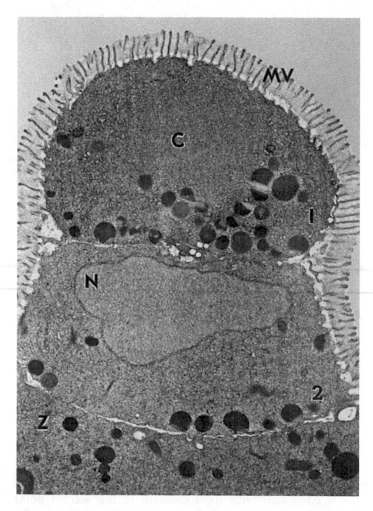

Fig. 8.18   First (1) and second (2) polar bodies of the mussel *Mytilus*. C = chromatin of first polar body; N = nucleus of second polar body; Z = zygote cytoplasm; MV = microvilli. See Longo, 1983.

Structural differences in the chromatin of the first and second polar bodies have been attributed to the lack of organelles that normally participate in nuclear envelope formation and the size difference of the zygote and polar bodies. However, the polar bodies in many species appear to contain the same kinds and number of membranous structures, and their sizes in many instances are essentially the same. Chromatin differences between the first and second polar bodies may depend upon the sequential appearance of substances regulating chromatin morphogenesis during the course of polar body formation. For example, if the egg/zygote develops the property for nuclear envelope formation and chromosome condensation after the first meiotic division, only the second polar body would be capable of forming a nucleus. Other than acting as receptacles for redundant chromatin, additional roles, if there are any, for the polar bodies during fertilization and embryogenesis have not been established. In humans there have been cases described of chromosomal mosaic individuals who are believed to be derived, in part, from the retention, replication and differentiation of polar bodies.

Polar body formation may be viewed as an extreme case of unequal cleavage and represents one of the few instances where the plane of cleavage does not bisect the metaphase spindle midway between its poles (Rappaport, 1971). The cleavage furrow, which separates the polar body from the zygote, forms at the base of the prospective polar body and not at a plane in register with the metaphase plate. Asymmetrical cleavage has been shown to be associated with an inequality of aster size, for the pole containing the smaller aster is the one at which the smaller blastomere is formed (Dan and Nakajima, 1965). A similar relation may also be involved during the formation of the polar bodies. Unequal aster size is characteristic of the meiotic apparatus in the mollusks *Spisula* and *Mytilus* (Longo, 1973). The peripheral aster appears restricted in dimension due to its proximity to the plasma membrane. A similar situation has also been observed in dividing zygotes of *Spisula* where one aster, located in that region of the cell which becomes the smaller blastomere, is flattened on its polar surface as if it were being pushed or pulled toward the cell's surface.

Centrifugation studies of molluskan eggs have indicated that the inequality of the meiotic division is not due exclusively to an inherent property of the maturation spindle but to the orientation of the meiotic apparatus within the egg (Raven, 1966). Centrifugation of gastropod (*Lymnaea*) eggs prior to the first and second meiotic divisions results in the formation of large polar bodies due to the movement of the meiotic spindle from the egg cortex. Normal embryonic development of the zygote occurs only when the giant polar body contains less than 25% of the egg volume.

Characteristic features of polar body formation – production of a cytoplasmic projection and the failure of the cleavage furrow to bifurcate the metaphase plate – may be due to the aster's ability to affect the cell cortex adjacent to it, causing it to become distensible. Internal pressure might then induce the cytoplasm along this region to evaginate, thereby forming a cytoplasmic mass which develops into a polar body. Before polar body formation in starfish eggs the cell surface expands at the animal pole and contracts at the vegetal pole. There is an accompanying movement of endoplasm from the vegetal pole to the animal pole (Hamaguchi and Hiramoto, 1978). With the formation of the polar body, the cell surface around the extruded region contracts and the surface at the vegetal pole expands with a movement of cytoplasm from the animal pole to the vegetal pole. Following polar body extrusion, the cell surface at the animal pole expands with a movement of cytoplasm towards the animal pole.

8.4 DEVELOPMENT OF THE FEMALE PRONUCLEUS

Following anaphase II, the chromosomes remaining in the zygote disperse. Concomitantly, vesicles aggregate along the edge of dispersing chromosomes and progressively fuse to form a bilaminar envelope (Fig. 8.19).

This results in the formation of chromosome-containing vesicles, i.e. the chromosomes are delimited by two parallel membranes structurally similar to the nuclear envelope (Fig. 8.19). Subsequently, the chromosome-containing vesicles coalesce. This coalescence involves the fusion of the inner and outer laminae of the chromosome-containing vesicles, thereby forming an irregularly shaped female pronucleus. These events, which are similar to those occurring during the formation of the nucleus during telophase of mitotic cells, have been observed in the eggs of a wide variety of species, including mollusks, starfish, amphioxus and mammals (Longo, 1973, 1983; Holland and Holland, 1992).

Subsequent to its formation, the female pronucleus becomes spheroid and may acquire nucleoli, intranuclear annulate lamellae or crystalline structures (Longo, 1973). In many forms, it is difficult to distinguish the female pronucleus from the male and identification is based primarily on the proximity of one pronucleus to the site of polar body formation. In some mammals the pronuclei may be identified by size difference or by association with the sperm tail (Austin, 1961). In the sea urchin, *Arbacia*, the male pronucleus is distinguished by the presence of a centriolar fossa, its relatively more electron-opaque chromatin and smaller diameter.

During the formation of the female pronucleus, the aster and portion of the meiotic spindle remaining in the zygote regress and may

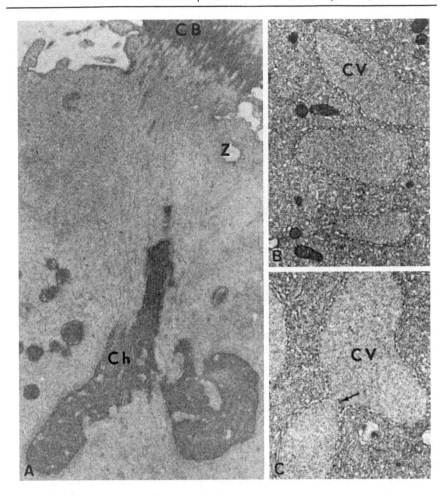

Fig. 8.19 Formation of the female pronucleus. **(A)** Chromosomes (Ch) remaining in a rabbit zygote (Z) following completion of the second meiotic division. CB = cytoplasmic bridge connecting the zygote to the second polar body. **(B)** Formation of chromosome-containing vesicles (CV) in a surf clam (*Spisula*) zygote. **(C)** Fusion (arrow) of chromosome-containing vesicles (CV) to form a female pronucleus in a *Spisula* zygote. See Longo, 1973.

disappear. In the mollusks *Spisula* and *Mytilus* the female pronucleus is usually associated with an area relatively devoid of large cytoplasmic components that contains endoplasmic reticulum and some microtubules. Occasionally, a centriole is observed in this region which is believed to be a remnant of the inner aster of the second meiotic apparatus (Fig. 8.20).

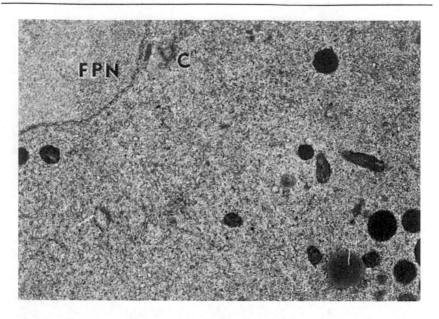

Fig. 8.20 Portion of a female pronucleus (FPN) adjacent to a centriole (C) in a mussel (*Mytilus*) zygote. See Longo and Anderson, 1969a.

8.5 CYTOPLASMIC REARRANGEMENTS DURING FERTILIZATION/MEIOTIC MATURATION

During fertilization/meiotic maturation, and usually in concert with polar body formation, zygote modifications occur involving the development of crenations along the vegetal pole, polar lobes or the redistribution of cytoplasmic components (Raven, 1966; Conrad and Williams, 1974a, b; Dohmen and Van Der Mey, 1977). These morphogenetic events are often part of the general phenomenon of cytoplasmic localization/ooplasmic segregation (Davidson, 1976). The eggs of many animals do not demonstrate these modifications; the most dramatic examples are seen in eggs that develop polar lobes during meiotic maturation (e.g. the mud snail, *Ilyanassa*).

Many studies consider polar lobe formation only at the time of cleavage. Nevertheless, polar lobes (i.e. protrusion of the vegetal region of the oocyte) may also occur during meiotic maturation. In *Ilyanassa*, the first polar lobe forms when the first meiotic apparatus moves to the animal pole; it recedes upon formation of the first polar body. Just prior to the formation of the second meiotic apparatus, the second polar lobe forms and recedes following the formation of the second polar body.

Cytoplasmic streaming and the redistribution of cellular inclusions in zygotes undergoing meiotic maturation have been described for a number of animals. In general, the movement of cytoplasmic inclusions is believed to be influenced by components in the egg cortex (Raven, 1966; Luchtel, 1976). Aspects involving how these changes are brought about, their correlation with maturation events of the egg, the localization of ooplasmic components and possible cytoskeletal alterations have been reviewed (Jeffrey, 1984).

Metabolic alterations at egg activation

9

Numerous physiological changes occur at fertilization that profoundly affect the activity of the egg, e.g. changes in permeability to small molecules, oxygen uptake, carbohydrate metabolism and synthesis of DNA, RNA and protein. In review of these areas consideration is often given to changes that occur not only at fertilization but throughout embryogenesis (Giudice, 1973; VanBlerkom, 1977; Raff, 1980); the descriptions presented here relate primarily to changes in the fertilized egg.

9.1 PERMEABILITY CHANGES

Investigators have demonstrated that the sea urchin egg undergoes permeability changes to different molecules following activation, e.g. the uptake of amino acids and nucleosides increases after fertilization (Giudice, 1973). Active mechanisms for the transport of amino acids and nucleosides are expressed at fertilization and, with respect to glycine and thymidine, are sodium-dependent. The mechanism of transport activation may require both early (possibly cortical granule exocytosis) and late events (possibly gradients of Na^+ and K^+) of egg activation, involving increased energy metabolism (Schneider, 1985). Similar permeability changes do not necessarily occur in organisms other than sea urchins and may differ significantly in their mode of activation. In the oyster, *Crassostrea*, fertilized eggs take up uridine at the same rate as unfertilized ova. In the surf clam, *Spisula*, there is an increase in amino acid uptake of ten- to 12-fold at the completion of meiotic maturation, approximately 50 min postinsemination (Bell and Reeder, 1967).

In mice the rate of amino acid uptake is low and relatively constant from the one-cell stage to the blastocyst stage *in vivo*; there are no apparent qualitative or quantitative differences between unfertilized and fertilized eggs (Holmberg and Johnson, 1979). Amino acid uptake in mouse eggs is a carrier-mediated process and that which is accumulated is available for exchange via a carrier-mediated system. Apparently, there is no insertion or activation of amino acid transporting enzymes of similar or novel kinetic activity at fertilization as appears to be the case with sea urchins. In mouse embryos the uptake of uridine and adenine increases with development but the rates are strikingly different for each nucleoside (Daentl and Epstein, 1971; Epstein and Daentl, 1971; Epstein et al., 1971). In the fertilized egg adenine is taken up about 350 times more efficiently than uridine and its uptake increases about 20-fold by the blastocyst stage. Uridine uptake during the same period, however, increases 300-fold. This difference in uptake rates of the two nucleosides suggests that they are transported by different systems which are regulated independently.

9.2 OXYGEN UPTAKE AND CARBOHYDRATE METABOLISM

At fertilization the rate of oxygen consumption increases rapidly in sea urchins (Giudice, 1973). In other organisms there is little change at fertilization, e.g. the annelids Sabellaria and Nereis, or respiration is reduced, as in Cumingia (bivalve mollusk) and Chaetopterus (annelid). In Paracentrotus lividus (sea urchin) oocytes the respiration rate is slightly higher than in newly fertilized eggs and much higher than in mature unfertilized eggs (Fig. 9.1). In fish and amphibian eggs there is reportedly no change in respiration at fertilization.

In sea urchin eggs the respiratory burst that occurs during fertilization membrane elevation is associated with the production of peroxide, which is the substrate for ovoperoxidase (Foerder et al., 1978). Ovoperoxidase joins tyrosyl residues in dityrosyl linkages which serve to crosslink polypeptide chains of the nascent, soft fertilization membrane resulting in its hardening.

What conditions limit respiration in unfertilized sea urchin eggs and how they are reversed at fertilization has not been fully elucidated. The respiratory change is preceded by a several-fold increase in coenzyme NADPH which is apparently generated by a phosphorylation of NAD (Giudice, 1973). This suggests either an activation of the NAD kinase or that the enzyme and its substrate are compartmentalized and prevented from interacting prior to fertilization. In view of the correlation between the NADPH level and synthetic cellular activities, phosphorylation may be important in initiating and controlling biosynthetic processes of the egg. As pointed out by Swezey and Epel (1995), a major use of NADPH at fertilization is to produce H_2O_2, which is used as a substrate for hardening of the fertilization membrane. Some 30–70% of the NADPH produced by the pentase shunt is required for generation of H_2O_2.

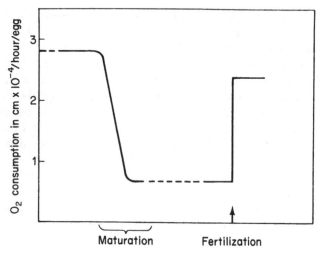

Fig. 9.1 Changes in the rate of oxygen consumption during maturation and at the time of fertilization in the sea urchin *Paracentrotus*. Reproduced with permission from Monroy, 1965.

The level of glycolytic intermediates and hexose phosphate is low in the eggs of some sea urchins, which may limit the oxidative breakdown of carbohydrates and respiration. This, together with the presence of large amounts of glycogen-like material in the unfertilized egg, indicates that a block exists in the pathway leading from glycogen to glucose-6-phosphate. This block is apparently reversed at fertilization, as a several-fold increase of glucose-6-phosphate is observed soon after insemination and shortly before the increased oxygen consumption. These observations suggest that at fertilization there is a mobilization of substrates, possibly due to an activation of the enzyme glycogen phosphorylase. However, the activity of glycogen phosphorylase has been found to be at the same level in homogenates of unfertilized and fertilized eggs. The enzyme is released at fertilization from a particulate fraction and may be held in a cellular compartment before fertilization and then released when required.

Intermediates of the tricarboxylic acid cycle are rapidly oxidized by homogenates of sea urchin (*Arbacia*) eggs and isolated mitochondria are capable of oxidative phosphorylation (Giudice, 1973). Pentose cycle activity is enhanced at fertilization and tends to predominate through fertilization and early cleavage (Swezey and Epel, 1995).

Metabolic changes in mollusk eggs at fertilization appear to be quite variable depending upon the species examined (Raven, 1972). Respiratory activity increases in the oyster *Crassostrea virginica*, decreases in the bivalve *Cumingia* and does not change in the oysters *Crassostrea commer-*

cialis and *Crassostrea gigas*. Glycolysis is regulated at steps of phosphorylase, phosphofructokinase and pyruvate kinase in fertilized *Crassostrea* eggs. Phosphorylase is activated first, followed by pyruvate kinase and phosphofructokinase. Because this pattern is similar to that found in sea urchin eggs it is possible that these rate-limiting steps are regulated in the same manner, and carbohydrate utilization is enhanced at fertilization by an activation of glycolysis (Yasumasu *et al.*, 1975). Activation of NAD kinase, as inferred from the elevation of total NADP levels in surf clam (*Spisula*) zygotes, occurs immediately following insemination and is regulated by calcium and calmodulin *in vitro* (Epel *et al.*, 1981).

Although respiration and energy utilization increase at fertilization in the sea urchin *Strongylocentrotus*, there is reportedly no change in the levels of ATP, ADP and AMP during the period of maximum respiratory activity (Epel, 1969). These results do not support the hypothesis that respiration in eggs is limited by ADP. ATPase activity reportedly increases after fertilization in sea urchins and is calcium-activated. A magnesium-activated ATPase whose activity increases slightly following fertilization has been described (Monroy, 1957).

9.3 DNA SYNTHESIS

One of the most dramatic consequences of fertilization is stimulation of the egg to undergo DNA replication and cell division (Fig. 9.2).

Nuclear transplantation experiments showed that brain cell nuclei injected into mature amphibian eggs undergo DNA synthesis which is replicative rather than repair (Gurdon *et al.*, 1969; Laskey and Gurdon, 1973). In similar experiments with immature oocytes, injected nuclei do not undergo DNA synthesis. These observations correlate with the overall increase in DNA polymerase activity during *Xenopus* maturation.

Critical events in the initiation of DNA synthesis at fertilization are unknown. The enzymatic machinery required for synthesis is present and the deoxynucleotide pool, although small, does not appear to be limiting (DePetrocellis and Rossi, 1976). All four deoxyribonucleotide kinases have been directly assayed in the sea urchin, *Strongylocentrotus* (Fansler and Loeb, 1969). Their activity was not found to vary significantly in unfertilized and fertilized eggs. The enzymatic activity of DNA polymerase from homogenates of *Strongylocentrotus* does not undergo variations with phases of the cell cycle. The enzymes (kinase and polymerase) appear to be present in both the nucleus and cytoplasm of the unfertilized egg. In *Strongylocentrotus* the percentage of DNA polymerase activity associated with the nuclear fraction increases progressively while activity in the cytoplasm declines. By the late blastula stage most of the DNA polymerase activity is associated

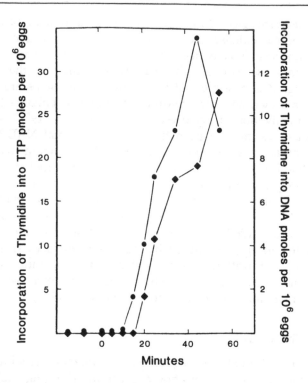

Fig. 9.2 Incorporation of [³H]-thymidine into thymidine triphosphate (black circles) and DNA (diamonds) of sea urchin (*Arbacia*) zygotes. The sperm suspension was added at 0 min. Reproduced with permission from Longo and Plunkett, 1973.

with the nucleus. This change in DNA polymerase activity has been interpreted as representing a transfer of the enzyme from the cytoplasm to the nucleus as development proceeds (Fansler and Loeb, 1972; Loeb *et al.*, 1969; Loeb and Fansler, 1970).

DNA polymerase activity is not detected in sea urchin spermatozoa, suggesting that the ability of male pronuclei to undergo DNA synthesis is due to the association of DNA polymerase of maternal origin with the paternally derived DNA during pronuclear development. In mammals, RNA and DNA polymerase activities have been shown to be associated with the mitochondria of bull sperm (Hecht, 1974; Hecht and Williams, 1979).

DNA synthesis in sea urchin eggs has been experimentally inhibited by a number of different agents with delays or complete cessation of cleavage (Giudice, 1973). DNA synthesis is inhibited by substances that interfere primarily with protein synthesis, such as puromycin and cycloheximide, suggesting that some proteins associated with DNA replication are synthesized *de novo*. Even though these drugs block cell division they appear to have little effect on pronuclear development and association.

Although the unfertilized sea urchin egg can take up thymidine, it is unable to phosphorylate the nucleoside to thymidine triphosphate (TTP; Fig. 9.2). Phosphorylation of thymidine to TTP begins about 10 min postinsemination in *Arbacia* zygotes (Longo and Plunkett, 1973). Since homogenates of unfertilized eggs are unable to carry out thymidine phosphorylation, there may be a compartmentalization of phosphorylation enzymes, thereby preventing their interaction with appropriate substrates. In *Arbacia*, DNA synthesis normally follows pronuclear fusion and occurs in the zygote nucleus about 16 min postinsemination at 20°C (Fig. 9.3).

In sand dollar, mouse and rabbit zygotes DNA synthesis occurs in both pronuclei during their migration (Simmel and Karnofsky, 1961; Oprescu and Thibault, 1965; Luthardt and Donahue, 1973; Howlett and Bolton, 1985). There is a distinctive localization of silver grains over the female pronucleus in autoradiographs of rabbit zygotes (Fig. 9.3). The grains are distributed in a polarized fashion and are located along that region of the female pronucleus proximal to the male pronucleus.

There is little doubt that DNA synthesis is controlled by cytoplasmic factors in the egg (Naish *et al.*, 1987a). When DNA synthesis begins before pronuclear fusion, it occurs simultaneously in both pronuclei. If

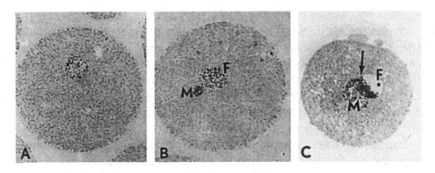

Fig. 9.3 Autoradiographs of sea urchin and rabbit zygotes incubated in [³H]-thymidine. **(A)** Autoradiographic grains over the zygote nucleus of a sea urchin (*Arbacia*) zygote. In this species DNA synthesis normally follows pronuclear fusion. **(B)** *Arbacia* zygote treated with colchicine to inhibit pronuclear fusion. Both the male (M) and the female (F) pronuclei have synthesized DNA. **(C)** Closely apposed male (M) and female (F) pronuclei of a rabbit zygote that have synthesized DNA. Autoradiographic grains in the female pronucleus are located at the pole (arrow) closest to the male pronucleus. See Longo and Plunkett, 1973 and Longo, 1976c.

eggs are made polyspermic or injected with accessory sperm, DNA synthesis begins at the same time in all pronuclei (Graham, 1966; Longo and Plunkett, 1973). When ascidian (*Ascidia*) eggs are cut into animal and vegetal halves and then fertilized, DNA synthesis begins at the same time in both halves (Ortolani et al., 1975). In crossfertilization studies using hamster eggs and human sperm, incorporated human sperm undergo DNA synthesis at the same time as would incorporated hamster sperm, i.e. 3–8 hours postfertilization (Naish et al., 1987b). Human sperm incorporated into human eggs normally undergo DNA synthesis about 12 hours postfertilization (Tesarik and Kopecny, 1989). These results suggest that fertilization triggers a chain of reactions which rapidly propagate throughout the cytoplasm and allow male pronuclei to undergo DNA synthesis.

The observation that DNA synthesis can be initiated and maintained by exposure of unfertilized sea urchin eggs to ammonia gave rise to the idea that DNA synthesis is activated by an increase in intracellular pH (Mazia and Ruby, 1974). Although eggs activated in ammonia and calcium-free sea water undergo an increase in internal pH, they do not initiate DNA synthesis. This suggests that ammonia activation is more than just an elevation of internal pH. It appears that both calcium and pH elevation are required for the initiation of DNA synthesis in sea urchin eggs (Whitaker and Steinhardt, 1981).

9.4 RNA AND PROTEIN SYNTHESIS

The existence of informational macromolecules in the egg cytoplasm is suggested by studies of marine embryos that demonstrate the presence of morphogenetic factors and that various regions of the uncleaved ovum are not equivalent in their developmental potential (Davidson, 1976). In many species the role of nuclear expression at fertilization and early development is minimal. Physical or chemical (actinomycin; Gross and Cousineau, 1964) enucleation studies have shown that a nucleus is not necessary for cleavage. Harvey (1956) produced egg fragments by centrifugation and showed that parthenogenetic merogones (enucleate egg fragments) were capable of developmental changes. In crossfertilization studies, hybrid embryos follow the maternal pattern of development. Later studies using biochemical rather than morphological criteria demonstrated that the eggs of sea urchins possess a store of messenger RNA that is translated subsequent to fertilization. In other words, enucleated eggs could synthesize protein at a rate comparable to its nuclear counterpart (Denny and Tyler, 1964; Craig and Piatigorsky, 1971). Similar results were also obtained using frog eggs (Smith and Ecker, 1965; Ecker and Smith, 1971).

Incorporation of amino acid into protein by non-nucleate fragments

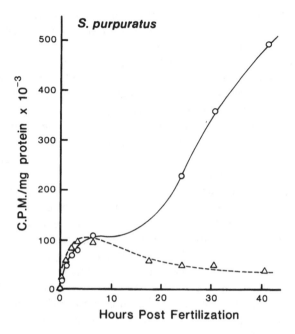

Fig. 9.4 Effect of actinomycin D on protein synthesis of sea urchin (*Strongylocentrotus*) embryos determined by [^{14}C]-valine incorporation. Open circles = controls; triangles = embryos exposed continuously to 20 µg/ml actinomycin D from before fertilization. The second rise in controls begins after the hatching blastula stage and does not occur in specimens treated with actinomycin D. Rates of actinomycin-D-treated specimens are slightly higher for the first 2–3 hours than the controls. Reproduced with permission from Gross, 1964.

of sea urchin eggs can be stimulated by artificial activation to the same extent as by fertilization, which suggests the presence of mRNA in the egg cytoplasm (Brachet *et al.*, 1963; Denny and Tyler, 1964). Inhibition of transcription in sea urchin embryos treated with actinomycin D does not prevent DNA or protein synthesis and cleavage (Fig. 9.4) and treated embryos are capable of developing to the hatching blastula stage (Gross and Cousineau, 1964).

An implicit assumption of these studies is that actinomycin D affects only transcription (Sargent and Raff, 1976). Actinomycin D experiments have also been carried out with embryos of snails and tunicates, with results comparable to those in sea urchins. Mammalian embryos continue protein synthesis in the presence of actinomycin D with an inhibition of RNA synthesis (Manes, 1973; Golbus *et al.*, 1973). Similar results have been obtained in mammalian embryos with α-amanitin, which inhibits RNA synthesis but neither blocks cleavage nor inhibits protein synthesis.

The sea urchin embryo can reach the blastula stage in the absence of RNA synthesis, but this does not imply that RNA is not normally synthesized during this early period. In addition to an incorporation of precursors into the pCpCpA sequence of tRNA, the synthesis of heterogeneous RNA does, in fact, take place in the period from fertilization to hatching in sea urchins (Gross et al., 1965), as well as histone mRNA (Poccia et al., 1985; Santiago and Marzluff, 1989). Moreover, there is evidence that the paternally-derived genome may be active in RNA synthesis (Longo and Kunkle, 1977; Poccia et al., 1985).

The synthesis of RNA upon fertilization has been investigated in mollusks (Collier, 1979; Kidder, 1976; McLean, 1976). Fertilized oyster (Crassostrea) eggs take up uridine at the same rate as unfertilized eggs and the rate at which uridine is incorporated into high-molecular-weight RNA is not altered at insemination (McLean and Whiteley, 1974). Hence, development of Crassostrea during cleavage is apparently controlled by information stored in the unfertilized egg. Although RNA synthesis is not markedly increased at fertilization in the snail Lymnaea, all major forms of RNA are synthesized from oogenesis throughout fertilization. In contrast to these results, eggs and early embryos of the gastropod Acmaea reportedly do not incorporate uridine until late cleavage (Karp, 1973).

In fertilized frog eggs there is no or very little new transcription until the midblastula transition (Santiago and Marzluff, 1989). Incorporation of the nucleosides uridine, cytosine and adenine into pronuclei of mouse zygotes was demonstrated by Mintz (1964), although later studies using [^{3}H]-uridine failed to detect RNA polymerase activity in vitro. However, transcriptional activity at the pronuclear stage in mouse zygotes has been demonstrated using [^{3}H]-adenine, possibly due to the more efficient uptake of adenine versus uridine (Clegg and Pikó, 1982). The absolute rate of RNA synthesis increases about twofold as mouse embryos progress from the one- to the eight-cell stage of development; much of this newly synthesized RNA is attributable to enhanced synthesis of ribosomal RNA and the production of ribosomes. Correlated with the onset of detectable rRNA synthesis is the modification in nucleolar structure. With the appearance of granule elements and the progressive reticulation of the nucleolar matrix, ribosomes and polysomes become more abundant within the cytoplasm (Van Blerkom and Motta, 1979). Messenger RNA synthesis has been demonstrated in mouse zygotes and two-cell embryos (Chen et al., 1986; Schultz, 1986).

Unfractionated RNA from unfertilized sea urchin eggs has been translated in cell-free systems indicating that 3–4% of the egg RNA is mRNA (Raff, 1980). RNA extracted from 20–40 S particles of sea urchin eggs directed the synthesis of histones in vitro (Gross et al., 1973). Similar results have also been obtained from surf clam (Spisula) embryos. Sea urchin eggs contain poly(A)$^+$ mRNA, poly(A)$^-$ histone mRNA and poly(A)$^-$ nonhistone mRNA. Amphibian oocytes contain poly(A)$^+$ mRNA, poly(A)$^-$ and poly(A)$^+$ histone mRNA. Observations of Ruderman and Pardue (1977) suggest that sea urchin embryos, in contrast to eggs, contain relatively few prevalent nonhistone poly(A)$^-$ mRNAs and that newly synthesized nonhistone poly(A)$^-$ mRNAs contribute less (at least quantitatively) to the embryonic program than do the poly(A)$^+$ mRNA and histone mRNA components. Sequence complex analysis has been carried out for eggs of sea urchins, amphibians (*Xenopus*) and the echuroid worm *Urechis* (Davidson, 1976). RNAs corresponding to the single copy portion of the genome are present in oocytes of these organisms, equivalent to about 25 000 different mRNAs of 1500 nucleotides in length and reflect an enormous developmental potential.

Not all stored mRNAs are uniformly distributed within the egg. Rebagliati *et al.* (1985) have shown that some mRNAs are localized to the animal or vegetal poles of *Xenopus* oocytes. The mRNAs and proteins that are unevenly distributed are believed to be regulators of different developmental fates that correlate with the animal/vegetal axis (Weeks *et al.*, 1995). Histone mRNA is localized within the female pronucleus of sea urchin eggs (Showman *et al.*, 1982; DeLeon *et al.*, 1983). The basis for this compartmentalization of histone mRNA has not been established.

Changes in the rate of protein synthesis for eggs inseminated at different stages of meiosis are summarized in Table 9.1.

In sea urchin eggs a large spectrum of proteins is synthesized at a comparatively low rate (Brandhorst, 1976). However, fertilization in sea urchins and surf clams is followed by a large increase in the rate of protein synthesis, which indicates the release of the block restricting mRNA translation in the unfertilized egg (Fig. 9.5; Santiago and Marzluff, 1989).

In sea urchins the protein synthesis increase is accompanied by an increase in polysomes at the expense of single 80 S ribosomes, as stored maternal mRNA is recruited in a linear manner following fertilization. The recruitment of mRNA into polysome is estimated to be responsible for a 50-fold increase in protein synthesis. There is also a twofold increase in the rate of translation, such that the overall increase in the rate of protein synthesis is 100-fold (Santiago and Marzluff, 1989).

The low rate of protein synthesis in sea urchin eggs does not appear to be due to a lack of ribosomes, mRNA or energy sources. Studies by various investigators using sea urchin eggs provide evidence that the dramatic increase in protein synthesis at fertilization is a result of the following (see Raff 1980; Santiago and Marzluff, 1989; Standart, 1992 for reviews):

Table 9.1 Comparison of changes in the rates of protein synthesis and respiration at fertilization for selected species of vertebrates and invertebrates (see Houk and Epel, 1974)

Species	Maturation state when fertilized	Increase in rates	
		Protein synthesis	*Respiration*
Urechis caupo (echiuroid)	Intact germinal vesicle (GV)	At fertilization, 2 ×	At fertilization, 1.2 ×
Spisula solidissima (mollusk)	Intact GV	At fertilization, 3–4 ×	Not at fertilization
Sea urchins, many species	Pronucleate egg	At fertilization, 6–30 ×	At fertilization, 6–10 ×
Rana pipiens (amphibian)	Second metaphase	At GV breakdown, 10 ×	Not at fertilization
Asterias forbesii (asteroid)	After GV breakdown	After GV breakdown	At completion of meiosis
Patiria miniata (asteroid)	After GV breakdown	Before GV breakdown, 5 ×	At completion of meiosis

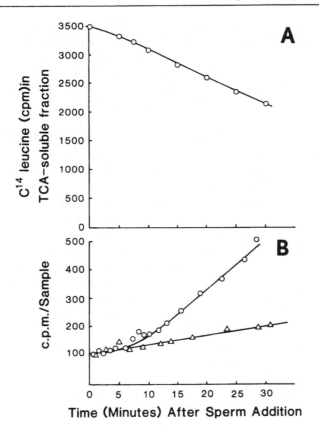

Fig. 9.5 **(A)** Incorporation of [^{14}C]-leucine in fertilized sea urchin (*Lytechinus*) eggs. Sperm was added at time zero. Unfertilized eggs were preloaded with [^{14}C]-leucine; one part of the sample was fertilized (open circles), the other was left unfertilized (triangles). Samples collected at time intervals were analyzed for incorporation of the label in hot trichloroacetic acid-insoluble material. **(B)** Loss of [^{14}C]-leucine from the free amino acid pool in fertilized eggs. Reproduced with permission from Epel, 1967.

1. unmasking of mRNA (Infante and Nemer, 1968; Gross *et al.*, 1973; Raff *et al.*, 1971); in other words, mRNAs are modified as they are recruited into polysomes;
2. processing of mRNA for more efficient translation, e.g. modification of the 3′ or 5′ ends;
3. Changes in translational efficiency, i.e. modifications of the translational apparatus of the egg such that it becomes more efficient in protein synthesis (Hille *et al.*, 1985; Danilchik *et al.*, 1986); for example, changes in pH activation are believed to unmask mRNA and/or initiation factors (Winkler *et al.*, 1985);

4. Sequestration of messages and/or compartmentalization of the protein synthesis apparatus of the egg, thereby preventing the association of mRNA with ribosomes (Moon et al., 1982); examples of such a situation include the localization of histone mRNA within the female pronucleus (DeLeon et al., 1983) and the association of mRNAs with the cytoskeleton (Moon et al., 1983).

As suggested above, the principal mechanism affecting the increase in protein synthesis at fertilization in sea urchin eggs is due to recruitment of stored mRNAs into polysomes and enhancement in the activity of the translation machinery (Santiago and Marzluff, 1989; Standart, 1992). Fertilization/artificial activation triggers unmasking of mRNAs. Consistent with this hypothesis are studies demonstrating that the mRNAs found in presumptive cytoplasmic messenger ribonuclear particles (mRNPs) exhibit S values greater than those of the same mRNAs after purification. For example, histone mRNPs typically have S values of 9–12, but sediment at 20–60 S in egg homogenates. In addition, purified poly(A)$^+$ mRNAs, which have model S values of 20–30, are found to sediment at 30–70 S when detected in egg homogenates.

The existence of mRNPs in the cytoplasm of unfertilized and fertilized eggs, although consistent with the hypothesis of masked mRNA, is not sufficient to substantiate the existence and developmental role of masked mRNA. Messenger RNPs isolated from polysomes or cytoplasm of various systems are as effective as the isolated mRNAs in stimulating heterologous in vitro protein synthesis (Raff, 1980). Moreover, demonstration of the validity of the masking hypothesis requires that egg mRNPs, isolated in their native state, be nontranslatable by an in vitro protein synthesizing system until they have been modified. This modification allows the contained mRNA to be translated.

The translatability of poly(A)$^+$ mRNPs from sea urchin eggs has been examined (Jenkins et al., 1978; Young and Raff, 1979). Cytoplasmic poly(A)$^+$ mRNPs fail to stimulate translation in a wheatgerm system, whereas mRNA extracted from nontranslatable particles was shown to be as template-active as mRNA extracted from whole eggs. The lack of translation of egg mRNPs may not be due to the presence of an inhibitor, since addition of mRNPs to deproteinized mRNA has no effect on the translation efficiency of the latter.

As a working hypothesis, Humphreys (1969, 1971) suggested that the rate of protein synthesis may be increased at fertilization by a heightened efficiency of translation and/or the translation of additional mRNA molecules. The efficiency of translation, defined as the number of protein molecules produced per mRNA molecule per unit time, is similar in sea urchin eggs and embryos. Protein synthesis is believed to be accelerated at fertilization by the translation of additional mRNA molecules. Measurements of mRNA entering polysomes in fertilized sea urchin eggs revealed that ribosomes in polysomes increase about 30-fold from 0.75% to 20% following fertilization, thereby substantiating Humphreys's speculation. These results suggest that the translational control mechanism in the egg acts directly at the level of the mRNA molecule.

The modification or processing of mRNA at fertilization has not been extensively studied. In sea urchins a significant portion of the egg mRNAs possesses an inverted 'cap' of 7-methylguanosine at the 5′ end (Raff, 1980). Capping of mRNAs could provide a store of nontranslatable message since mRNA becomes translatable after methylation. A second type of modification found on the 3′ polyadenylation of mRNA with codycepin has no apparent effect on the rise of protein synthesis or polysome formation. Histone mRNA, which lacks poly(A) segments in sea urchin eggs, participates in the postfertilization rise in protein synthesis. Thus, polyadenylation may not represent either the factor that controls the interaction of ribosomes with mRNA or the condition that makes the maternal mRNA translatable. Experiments comparing the rate of translation of poly(A)$^+$ and poly(A)$^-$ globin mRNA injected into amphibian (*Xenopus*) oocytes demonstrate that the initial rate of globin synthesis is similar in both cases (Huez *et al.*, 1974). With longer incubation, the rate of globin synthesis directed by poly(A)$^+$ mRNA is considerably lower than that directed by poly(A)$^+$ mRNA. These observations suggest that poly(A)$^+$ sequences may increase the stability of mRNA.

Relatively little is known about the involvement of various factors controlling initiation, elongation and termination of polypeptide chains during protein synthesis in fertilized eggs. A translation inhibitor has been isolated from egg ribosomes with salt solutions and suppresses the translation of poly(U) templates *in vitro*, suggesting that this factor may act to inhibit translation in unfertilized eggs. However, Hille (1974) showed that the inhibitor could be isolated from active embryo ribosomes and egg ribosomes that presumably possess the inhibitor can translate globin mRNA as well as embryo ribosomes (Clegg and Denny, 1974).

By studying the translational activity of sea urchin ribosomes in a fractionated reticulocyte cell-free system Danilchik and Hille (1981) have determined that unfertilized egg ribosomes differ from activated egg ribosomes. In comparison to ribosomes from embryos, those from eggs are less active in polymerizing amino acids and slowly increase this activity over the course of incubation. Furthermore, differences between egg and blastula ribosome activity may be involved in postactivation of protein synthesis. Consistent with Danilchik and Hille's observations are those showing a quantitative loss of high-molecular-weight protein from *Strongylocentrotus* egg ribosomes following fertilization (Unsworth and Kalenas, 1975). Investigations demonstrating that the ribosome transit time (the time necessary to traverse a mRNA during translation) decreases by more than one-half in fertilized sea urchin eggs offers another potential challenge to the view that the protein synthesis increase at fertilization results solely from recruitment of stored messenger RNA (Brandis and Raff, 1978; Hille and Albers, 1979). However, the magnitude of such changes can only account for a small proportion of the increased synthetic activity. Hence, it appears that the rate of protein synthesis in eggs and zygotes is primarily controlled by availability of mRNA. A pool of masked mRNA may be maintained in the egg and upon fertilization unmasking begins and translational efficiency rises. The result is the recruitment of mRNA into polysomes and a concomitant increase in protein synthesis. In sea urchins this process may involve modification or substitution of proteins in the mRNPs.

In the mud snail *Ilyanassa* protein synthesis occurs in the unfertilized egg; approximately 15 min postfertilization, amino acid uptake and incorporation increase. When the percentage of incorporation is calculated, protein synthetic activity of the zygote is about 2.5 times greater than the unfertilized ovum (Mirkes, 1970). Rapid and dramatic changes in the pattern of protein synthesis occur in fertilized surf clam (*Spisula*) eggs (Rosenthal *et al.*, 1980; Standart *et al.*, 1985). There is a synthesis reduction of prominent oocyte-specific proteins and a synthesis increase for at least three proteins (cyclins A and B and the small subunit of ribonucleotide reductase). Labeling of the latter three proteins dominates the pattern of protein synthesis at fertilization in *Spisula* and is due to stage-specific utilization of different subsets of mRNA from a common maternal pool. Discrimination of mRNAs may be achieved by a selective repression of availability by a phenol-soluble component of the egg. Although changes in internal pH and/or activation of protein kinase are believed to be involved in the unmasking of mRNAs in *Spisula* zygotes (Standart, 1992), in general, mRNAs active in translation undergo an extension of their (poly)A tail (Rosenthal *et al.*, 1983). This suggests that mRNA modification may play a major role in the selection of specific mRNAs for translation. Recent studies with surf clam oocytes demonstrate the phosphorylation of an 82 kDa protein at fertilization that selectively binds the masking portions of ribonucleotide reductase and cyclin A mRNAs (Walker *et al.*, 1996). This alteration may be an important step in the regulation of translational unmasking.

A small number of polysomes have been shown to be active in protein synthesis in *Spisula* eggs (Firtel and Monroy, 1970). After fertilization, there is a progressive increase in the number of ribosomes that become associated with polysomes, and by 30 min postinsemination the specific activity of polysomes is 2.5 times greater than in unfertilized ova. By the time the pronuclei are formed there are four to five times more polysomes than were present in the egg prior to maturation. These results suggest that the increase in the number of polysomes at fertilization is due, at least in part, to the activation of stored maternally derived mRNA. Eggs of the clam *Mulinia* also undergo an increase in polysomes by 45 min postinsemination (Kidder, 1976). Furthermore, there does not seem to be a shift in polysome size and distribution at fertilization. Protein synthesis in the mollusks *Spisula* and *Mulinia*, therefore, is similar to that observed in sea urchins, suggesting that gene transcription patterns between early mosaic and regulative embryos are not significantly different.

Studies with amphibian (*Rana pipiens*) oocytes have demonstrated that an increased rate of protein synthesis (70%) follows the onset of maturation; fertilization results in a further increase (50%) and by the blastula stage the rate has increased an additional twofold (Shih *et al.*, 1978). In *Xenopus* eggs the polysome content increases an additional twofold shortly after fertilization. In frog and starfish eggs, specific changes in protein synthesis occur at fertilization (see Standart, 1992); mRNAs for cyclins and DNA synthesis/cell proliferation are activated while those involved in housekeeping functions, such as actin, tubulin and elongation factors, are masked.

Accompanying the final meiotic division of the mammalian oocyte are qualitative and quantitative changes in the pattern of protein synthesis (Van Blerkom, 1977; Sherman, 1979; Wassarman *et al.*, 1981). The majority of polypeptides synthesized by mature, unfertilized mouse eggs are also made in the zygote and some show a change in the relative rate of synthesis following insemination. In addition, during the immediate postfertilization period of both rabbit and mouse embryos, major changes in the pattern of protein synthesis take place (Chen *et al.*, 1980). Fertilization of mouse eggs is accompanied by a 40% increase in the absolute rate of protein synthesis.

Certain proteins, e.g. tubulin and ribosomal, are synthesized in large amounts by growing mouse oocytes and continue to be synthesized at similar or even greater rates during early embryogenesis (Wassarman *et al.*, 1981). In rabbit eggs there is a translation of a population of stage-specific polypeptides that is autonomous of fertilization and appears to follow a timed translational schedule initiated with germinal vesicle breakdown (Van Blerkom, 1979). While tubulin is synthesized from mRNA stored in unfertilized sea urchin eggs, it is unclear to what extent maternally derived tubulin mRNA directs tubulin synthesis during mouse embryogenesis (Wassarman *et al.*, 1981). Like tubulin, ribosomal protein synthesis represents a major portion of the total protein synthesized in mouse eggs. In addition, at least five new proteins are synthesized in mouse eggs following insemination (Howlett and Boulton, 1985). Although these proteins have not been thoroughly characterized, they may be involved in the generation of the cell cycle (Howlett, 1985). A significant increase in protein synthesis occurs just prior to and at the two-cell stage (Lathan et al., 1991), i.e. at about the time of the transition from maternal to zygotic control of transcription (Howlett and Boulton, 1985).

Initially, direct comparisons between intracellular pH and protein synthesis in eggs and early sea urchin embryos indicated that changes in internal pH might play a direct role in the increase in protein synthesis at egg activation (Epel et al., 1974; Grainger et al., 1979; Dubé et al., 1985; see Epel, 1989, for a critical review of the importance of pH changes at egg activation). More recent studies, however, demonstrate that internal pH is only one of several signals involved in turning on protein synthesis at fertilization (Rees et al., 1995). Calcium release in the absence of an intercellular pH increase does not stimulate protein synthesis, while a pH increase in the absence of calcium release yields a partial stimulation. Only in the presence of intracellular pH increase and calcium release is the rate of protein synthesis in experimentally treated ova comparable to that of fertilized eggs (Winkler et al., 1980). The mechanism by which these ionic signals activate protein synthesis is unknown. A cell-free system derived from the sea urchin Lytechinus pictus exhibits the ionic controls found in vivo and is capable of initiation, elongation and termination of the normal spectrum of proteins synthesized in vivo at physiological ion concentrations. However, this system has a relatively low and variable activity which suggests that it is incomplete.

Calcium may also play a significant and wide-spread role in biosynthetic changes at fertilization (see Epel, 1989). Ca^{2+} is an activator of calmodulin-regulated NAD-kinase (Epel et al., 1981) and because this enzyme regulates the redox potential of cells it indirectly affects a myriad of biosynthetic processes. Ca^{2+} induces other enzymes such as lipoxygenase and protein kinase C (Epel, 1989). Activation of the latter enzyme may yield an array of protein phosphorylations, which in turn affect changes in regulatory pathways involved with biosynthetic processes.

Development of the male pronucleus

10

Although specific details involving the development of the male pronucleus vary from one organism to another, there are three basic features of this process that are common to the species that have been studied thus far:

1. breakdown of the sperm nuclear envelope;
2. dispersion of the condensed sperm chromatin;
3. development of a nuclear envelope, the pronuclear envelope.

These events are accompanied by a dramatic transformation in the shape, volume, chromatin conformation, nucleoprotein content and activity of the incorporated sperm nucleus (Longo, 1973, 1981b; Longo and Kunkle, 1978).

10.1 BREAKDOWN OF THE SPERM NUCLEAR ENVELOPE

The sperm chromatin of some animals is not delimited by a nuclear envelope (Baccetti and Afzelius, 1976). Following gamete fusion in these instances the paternally derived chromatin is placed in direct association with the egg cytoplasm, without an intervening membranous boundary, and is free to undergo changes leading to its transformation into a male pronucleus. However, the sperm of most organisms contain a nuclear envelope. Immediately following sperm incorporation in these cases, the inner and outer laminae of the sperm nuclear envelope fuse at multiple sites, thereby forming vesicles that initially outline the condensed sperm chromatin but are then scattered throughout the surrounding cytoplasm (Fig. 10.1).

The vesicles lack distinguishing features and are soon lost among other membranous elements. As a result of the breakdown of the sperm nuclear envelope, the condensed sperm chromatin is directly exposed to the zygote cytoplasm (Longo, 1973).

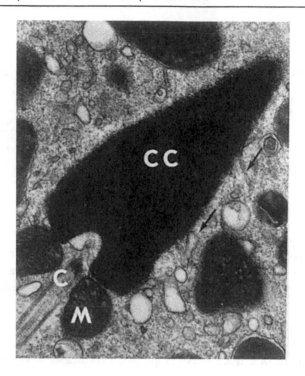

Fig. 10.1 Incorporated sea urchin (*Arbacia*) sperm nucleus that has undergone sperm nuclear envelope breakdown. Vesicles (arrows) adjacent to the condensed sperm chromatin (CC) may be derived from the vesiculation of the sperm under nuclear envelope. M = sperm mitochondrion; C = centriole.

A number of observations indicate that breakdown of the sperm nuclear envelope at fertilization is a highly regulated event. For example, in the sea urchin *Arbacia* vesiculation of the sperm nuclear envelope is not complete (Longo, 1973). Segments of the sperm nuclear envelope formerly associated with the acrosome and centrioles are left intact; subsequently they become incorporated into the nuclear envelope of the male pronucleus. Retention of specific regions of the sperm nuclear envelope has also been observed in mammalian zygotes (Yanagimachi and Noda, 1970a). In *Arbacia*, breakdown of the sperm nuclear envelope sometimes occurs in the immediate vicinity of the female pronucleus. In this instance the nuclear envelope of the female pronucleus fails to undergo similar changes, indicating the specificity of this process. Incorporated sperm of immature mouse and rabbit oocytes (Berrios and Bedford, 1976; Szöllösi *et al.*, 1990) retain an intact nuclear envelope, suggesting that a factor formed or activated during the course of oocyte maturation/activation is responsible for disruption of the sperm nuclear envelope. Such observations are not universal among mammals, for sperm incorporated into hamster and cattle immature oocytes undergo sperm nuclear envelope breakdown (Yanagimachi, 1994).

Although factors regulating the disappearance of the sperm nuclear envelope have not been determined, it is possible that processes similar to nuclear envelope disruption in mitotic cells are involved. Mature amphibian ova contain a cytoplasmic factor that can induce germinal vesicle breakdown when injected into immature oocytes (Masui and Clarke, 1979). The activity of this factor is quickly lost after fertilization but later reappears and cycles with cleavage (Wasserman and Smith, 1978). Its appearance coincides with the onset of mitosis in cycling cells, suggesting that the factor may not be restricted to maturing oocytes but may play a more general role in regulating nuclear envelope breakdown.

In cells with open mitosis breakdown of the nuclear envelope at the end of prophase is accompanied by phosphorylation of the nuclear lamins, proteins of the intermediate filament family (Gerace and Burke, 1988). Although phosphorylation is necessary to dissolve the nuclear lamina, nuclear envelope fragmentation is not dependent on complete dissolution of the lamina (Heald and McKeon, 1990). The status of nuclear lamins in sperm has not been determined for many animals (Stricker et al., 1989) and in the case of mammals is controversial (McPherson and Longo, 1993). Sperm of Xenopus possess a specific lamin, L_{IV} (Benavente and Krohne, 1985). In sea urchin sperm, human autoimmune serum that recognizes A/C and B nuclear lamins of mammalian cells reacts with a substance(s) located at the acrosomal (apex) and centriolar fossa (base) of the sperm nucleus (Schatten et al., 1985a). In mammalian sperm Longo et al. (1987) found no evidence for lamin proteins, but a nuclear matrix (Bellvé et al., 1992) and nuclear lamins (Maul et al., 1986; Moss et al., 1987) have been demonstrated under specific preparative conditions and fixation (see McPherson and Longo, 1993). The fate of sperm nuclear lamins at fertilization in those species that possess such a component and the trigger for nuclear envelope dissolution have not been determined (see Fisher, 1987).

Intact sea urchin sperm mechanically injected into eggs in which the nuclei apparently remain surrounded by membrane do not undergo chromatin dispersion and do not transform into male pronuclei (Hiramoto, 1962). The failure of sperm nuclei to metamorphose in this instance may be due to the presence of the nuclear envelope and plasma membrane and an inability of cytoplasmic factors to enter the sperm nucleus. A similar situation is observed in inseminated immature sea urchin eggs, where incorporated sperm nuclei often remain surrounded by an intact nuclear envelope and fail to undergo chromatin dispersion

(Longo, 1978b). Evidence that sperm nuclear envelope breakdown is required for development of a male pronucleus also comes from observations of gynogenetic fish (Yamashita *et al.*, 1990). When *Carassius auratus langsdorfii* eggs are fertilized by sperm from *C. a. cuvieri*, the sperm nuclear envelope does not break down and male pronuclei do not form. However, if isolated *C. a. cuvieri* sperm nuclei lacking a nuclear envelope and plasma membrane are injected into *C. a. langsdorfii* eggs, the sperm nuclei decondense and form male pronuclei. It is speculated that eggs of *C. a. langsdorfii* possess a factor for breakdown of the sperm nuclear envelope but that it is not activated at fertilization with *C. a. cuvieri* sperm (Yamashita *et al.*, 1990). In this case sperm of *C. a. cuvieri* merely serve the purpose of activating *C. a. langsdorfii* eggs and make no contribution to the genome of resultant embryos. These observations differ from those in mammals, which indicate that decondensation does not require breakdown of the nuclear envelope (Kopecny and Pavlok, 1984; Szöllösi *et al.*, 1994). At present, the basis for these apparent differences has not been determined.

Intracytoplasmic sperm injection (ICSI) is a strategy of *in vitro* fertilization in humans (Van Steirteghem *et al.*, 1993). In such cases the injected whole sperm undergoes pronuclear formation with a success rate of up to 51% and the delivery of healthy infants (Lanzendorf *et al.*, 1988; Palermo *et al.*, 1992, 1993). In addition to their importance as a means of fertility control, these results relate to a number of basic issues.

1. Apparently events involved with sperm entry into the egg are unnecessary for pronuclear development. This has also been demonstrated by experiments showing that isolated sperm nuclei are capable of pronuclear development when injected into hamster eggs (Uehara and Yanagimachi, 1976).
2. In the light of what was discussed above regarding the importance of the 'removal' of the plasma membrane and nuclear envelope for pronuclear development, how microinjected human sperm with presumably intact membranes are able to develop into pronuclei has not been determined.

In addition, the success of ICSI may have a bearing on previously published experiments attempting to use sperm as foreign DNA vectors for transgenics (Lavitrano *et al.*, 1989; see also Arezzo, 1989; Brackett *et al.*, 1971). Mature mouse sperm were incubated with cloned DNA (pSV2CAT plasmid) and the sperm were then used to fertilize eggs. Remarkably, some of the offspring (< 30%) possessed the plasmid. Although these experiments have not been repeatable, how whole, live sperm might be capable of bringing plasmids into eggs is a mystery (Patil and Khoo, 1996). That whole human sperm injected into eggs are capable of developing into male pronuclei suggests that the plasma membrane of human sperm, as well as other mammals, is unique. This uniqueness may be responsible for the success of ICSI in humans and the results observed by Lavitrano *et al.* (1989).

10.2 CHROMATIN DISPERSION

With few exceptions (e.g. decapod crustaceans) the presence of condensed chromatin in sperm is a distinctive feature of this cell type. Although reasons for the presence of condensed chromatin in sperm nuclei includes the following, all have been shown to be deficient in some way (Risley, 1990).

1. The hydrodynamically efficient shape of some sperm nuclei is believed to be due to the presence of condensed chromatin, but the shape of some mammalian sperm appears to be poorly designed for swimming.
2. Condensed chromatin has been suggested to protect the DNA from mutagens, although the protection afforded by condensation is relatively weak.
3. Condensation has been suggested to be a manifestation of a genetically blank slate; however, there is evidence for imprinting (Groudine and Conkin, 1985).

Transformation of the condensed sperm chromatin into the dispersed form of the male pronucleus is a dramatic alteration that has been examined in a variety of organisms (Longo, 1973). Morphological changes usually occur first along the periphery of the condensed, incorporated sperm nucleus; dense-staining chromatin grades into a more dispersed and lightly staining mass (Fig. 10.2).

As this process continues, the peripheral dispersion zone increases in volume, while the central dense portion gradually decreases until it disappears. One interpretation of such a pattern of morphogenesis is that the agent(s) responsible for dispersion initiates the process at the periphery; once the outer chromatin is dispersed the more central chromatin is free to disperse also. Eventually, all the sperm chromatin becomes a morphologically homogeneous mass of dispersed chromatin (Fig. 10.3).

In some cases, chromatin dispersion appears to occur uniformly and simultaneously throughout the sperm nucleus, without the formation of particular regions with different densities and conformations. As a result of dispersion, the paternal chromatin undergoes a significant increase in volume.

Changes in the sperm chromatin consistent with development into a male pronucleus may occur prior to gamete membrane fusion. In the sea urchin *Arbacia* exposure of sperm to egg jelly induces the phosphor-

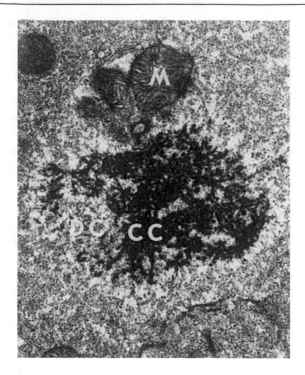

Fig. 10.2 Incorporated surf clam (*Spisula*) sperm nucleus undergoing chromatin dispersion. DC and CC = dispersed and condensed chromatin; M = sperm mitochondria. See Longo and Anderson, 1970b.

ylation of histone H_3, which contrasts with phosphorylation of histone H_1 in *Strongylocentrotus* sperm (Vacquier *et al.*, 1989). Egg jelly induces the degradation of histones in starfish sperm (Amano *et al.*, 1992). A product derived from the jelly layer, ARIS, is involved in the induction of the acrosome reaction (Ikadai and Hoshi, 1981) and in structural changes of the sperm nucleus (Longo *et al.*, 1995). The latter alterations may be a morphological manifestation of histone degradation demonstrated by Amano *et al.* (1992).

The pattern of chromatin dispersion appears to partially govern the initial structure of the male pronucleus. For example, in the sea urchin *Arbacia* the retention of the nuclear envelope along the apical and basal regions of the sperm nucleus and the lateral dispersion of sperm chromatin yields a heart-shaped mass which, following the development of the pronuclear envelope, becomes a spheroid male pronucleus. In mammals the dispersed chromatin profile is ellipsoidal and is reminiscent of the original shape of the sperm nucleus.

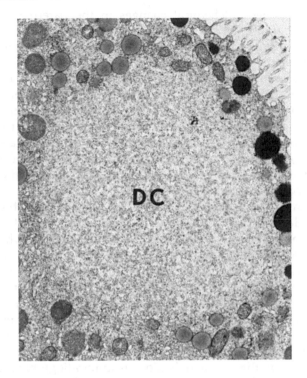

Fig. 10.3 Dispersed sperm chromatin (DC) of a surf clam (*Spisula*) zygote prior to the formation of the male pronuclear envelope.

It has been speculated that sperm chromatin dispersion in mammals is the opposite of nuclear condensation during spermatogenesis (Szöllösi and Ris, 1961; Bedford, 1970). If this were the case, it would be suggestive of a reversible process regulating the paternal chromatin at spermatogenesis and at fertilization. However, morphological and biochemical analyses of paternal chromatin during pronuclear development and spermiogenesis indicate that this is not the case.

Dispersion of sperm nuclei in egg extracts (see below) suggests that the process consists of two stages: (1) a rapid but limited decondensation of the sperm nucleus followed by (2) a slower, membrane-dependent dispersion (Lohka and Maller, 1987; Philpott *et al.*, 1991). Sperm nuclear decondensation is independent of the female pronucleus, nuclear or mitochondrial DNA synthesis and protein synthesis (Poccia, 1989). It is not unreasonable to suggest that dispersion of the condensed sperm chromatin is a morphological manifestation of changes in nucleoprotein content; in fact, chemical alterations are coincident with sperm chromatin dispersion. Formidable problems are inherent in studies of chemical changes in the paternal chromatin at fertilization because of technical difficulties; for example, isolation of pronuclei is arduous because of an extremely high cytoplasm/nucleus ratio characteristic of the zygote. Despite these difficulties, a number of studies have been performed that have yielded interesting results.

In many organisms, during the differentiation of the spermatogonium into a spermatozoon, histones characteristic of somatic cells are replaced by a distinct group of basic nuclear proteins, often unique to the mature spermatozoon and more basic than those found in somatic cells (see Risley, 1990). It has been suggested that the complexing of sperm DNA to these basic proteins permits condensation of the chromatin and repression of the DNA (Bloch, 1969).

Cytochemical examinations of fertilized eggs demonstrate that the paternally-derived chromatin stains differently subsequent to the metamorphosis of the sperm nucleus into a male pronucleus, which suggests that the DNA brought into the egg with the spermatozoon acquires different basic nucleoproteins during the dispersion of the sperm chromatin (Bloch and Hew, 1960; Das et al., 1975). Support for the idea that the sperm basic nuclear proteins are removed from paternal DNA comes from experiments in situ employing autoradiographic analyses of incorporated sperm nuclei labeled with amino acids (Ecklund and Levine, 1975). During the differentiation of the sperm nucleus into a male pronucleus there is a reduction in autoradiographic grains associated with the dispersing chromatin, indicating that basic proteins unique to the sperm nucleus are not simply 'masked' but are dissociated from the DNA. Similarly, antibodies to mouse sperm basic proteins fail to detect antigenic sites in fertilized eggs (Rodman et al., 1981).

Biochemical studies of transforming, paternal chromatin at fertilization indicate that the basic proteins of the sperm nucleus are lost and the paternal DNA associates with basic proteins similar to those found within the female pronucleus (Carroll and Ozaki, 1979; Poccia et al., 1981; Poccia, 1989; Poccia and Green, 1992). That this transformation occurs in the absence of protein synthesis indicates the presence of components within the egg for the modification, removal and replacement of sperm nucleoproteins (Poccia, 1989; Poccia and Green, 1992). Immediately upon the incorporation of the sperm nucleus sperm histones H_1 and H_{2B} are phosphorylated (Fig. 10.4).

In concert with decondensation, nearly all of the phosphorylated sperm histone H_1 is lost with the assimilation of cleavage stage histone H_1 (Green and Poccia, 1985). Phosphorylated sperm histone H_{2B} persists for a period but is eventually lost (Fig. 10.4).

It has been postulated that the extended arms of sperm histone H_1 and H_{2B} act to stabilize or condense sperm nuclei by crosslinking

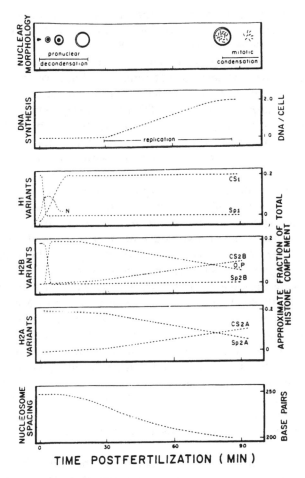

Fig. 10.4 Transitions in the sea urchin male pronucleus during the first cell cycle following fertilization. Transitions in H_1, H_{2A}, and H_{2B} histone variants are correlated with cell cycle parameters (replication and mitosis) and chromatin structural changes (nucleosomal spacing and pronuclear decondensation). Reproduced with permission from Poccia, 1986.

chromatin fibers (Poccia, 1989). At fertilization phosphorylation would neutralize the extended arms with a decrease in affinity for DNA. However, phosphorylation may be permissive for decondensation but is not sufficient, since it occurs even when decondensation is blocked (Poccia *et al.*, 1990). Additional evidence against phosphorylation as the sole mechanism controlling the state of condensation of sperm chromatin comes from studies of spermatogenesis (Poccia *et al.*, 1987). Phosphorylated forms of sperm-specific variants appear at the last stages of sperm development when the chromatin is condensed. As in the case of basic nucleoproteins, the nonbasic nucleoproteins of the sperm are apparently replaced by ones similar to those found within the female pronucleus (Kunkle et al., 1978a).

Investigations by Gurdon (1976) demonstrate that oocyte nuclei contain decondensation factors that remain in high concentration in the egg cytoplasm after germinal vesicle breakdown. Similar observations have also been carried out in mice (Thadani, 1979). Sperm nuclei fail to form pronuclei when exposed to cytoplasm from which the germinal vesicle has been removed prior to oocyte maturation (Katagiri and Moriya, 1976; Skoblina, 1976; Lohka and Masui, 1983a).

Injection of only soluble germinal vesicle contents is sufficient to restore the ability of enucleated eggs to induce pronuclear formation (Lohka and Maller, 1987). The use of an in vitro system employing permeabilized sperm incubated in egg extracts has been used to determine the active agent involved with sperm chromatin dispersion.

Sperm nuclear decondensation activity of eggs has been examined in a variety of organisms (sea urchins, amphibians, surf clams and fruit flies) by mixing permeabilized sperm with extracts of egg homogenates (Kunkle et al., 1978b; Eng and Metz, 1980; Lohka and Masui, 1983a, b; Ulitzer and Gruenbaum, 1989; Cameron and Poccia, 1994; Longo et al., 1994b). The cell-free system has a number of advantages, such as the availability of developing male pronuclei, and its usefulness for biochemical analyses has provided numerous insights into aspects of chromatin decondensation, nuclear envelope assembly and regulation of the cell cycle (Murray and Hunt, 1993). Moreover, it may be used as part of a heterologous system; for instance, Xenopus egg extract has been shown to induce decondensation of human sperm (Brown et al., 1987).

When permeabilized sperm nuclei of the amphibian Bufo are incubated in egg extract they undergo a loss of protamine accompanied by decondensation (Ohsumi and Katagiri, 1991a, b). Protamine removal and decondensation activities are also present in growing and mature oocytes and pregastrula embryos. There is no evidence for an enzymatic degradation of protamine, but the acid histone-binding protein (Laskey et al., 1978; Krohne and Franke, 1980) nucleoplasmin is implicated in protamine removal. It is important to note that nucleo-plasmin is localized in the germinal vesicle during oocyte maturation and becomes distributed throughout the animal hemisphere at germinal vesicle breakdown. Further observations by Philpott and co-workers (Philpott et al., 1991; Philpott and Leno, 1992) demonstrated that in Xenopus cell-free systems nucleoplasmin is both necessary and sufficient for the first stage of sperm nuclear decondensation, i.e. the initial, rapid decondensation. Sperm nuclear decondensation is accompanied by the loss of two specific sperm basic proteins with the acquisition of histones H_{2A} and H_{2B} and nucleosome formation. Immunodepletion of extracts with nucleoplasmin antibodies prevents these changes and addition of purified nucleoplasmin reverses this inhibition (Philpott and Leno, 1992). Hence, nucleoplasmin has been shown to be the active agent; such a protein may not only function in protamine removal and nucleosome assembly but in the cascade of events which lead to oocyte activation and germinal vesicle breakdown (Herlands and Maul, 1994). It is possible that a nucleoplasmin-like protein may exist in the eggs of mammals, as well as other organisms, to mediate sperm nuclear decondensation.

The possible roles of MPF, Ca^{2+} and protein phosphorylation in cell-free extracts to support nuclear envelope assembly and chromosome decondensation on the one hand, or nuclear envelope breakdown, chromosome condensation and spindle formation on the other, have been reviewed (Lohka and Maller, 1987). In summary, permeabilized *Xenopus* sperm nuclei incubated in egg extract undergo chromatin decondensation, nuclear envelope assembly and form pronuclei. Permeabilized sperm nuclei incubated in egg extract prepared with buffer containing EGTA form chromosomes on spindles. A 150 000 xg supernatant of the EGTA-extract does not support the dispersion of sperm nuclei into chromosomes but induces chromosome condensation and spindle formation as well as nuclear envelope breakdown when added to pronuclei that form in complete extract or isolated liver or brain nuclei. Similar changes are induced by MPF. The importance of Ca^{2+} in these processes is further illustrated by the addition of Ca^{2+} to extracts containing condensed chromosomes and spindles. Calcium induces the dissolution of spindles, chromosome decondensation and nuclear formation (Lohka and Maller, 1985). These observations are consistent with Masui's proposal (Masui *et al.*, 1977; Masui, 1985) that in metaphase-arrested eggs cytoplasmic factors promoting chromosome condensation are active only with low Ca^{2+}. The increase in Ca^{2+} at fertilization promotes chromosome decondensation and nuclear formation. Furthermore, these observations agree with results of *in vivo* experiments demonstrating that the increase in intracellular Ca^{2+} at fertilization in *Xenopus* eggs alters cytoplasmic factors that control the cell cycle and sperm chromatin structure (Kline, 1988; see also Wilding, 1996). Sperm chromatin decondensation and male pronuclear formation are prevented by microinjections of buffers that maintain a low internal Ca^{2+} level.

The initial decondensation of permeabilized sea urchin sperm nuclei incubated in egg extract requires ATP and is sensitive to kinase and metabolic inhibitors (Cameron and Poccia, 1994; Luttmer and Longo, 1987). Further nuclear swelling requires the presence of a nuclear envelope and is stimulated by GTP.

The extent of sperm nuclear decondensation in *Spisula* is related to the time egg extracts are prepared from cultured oocytes and is consistent with the sequence of fertilization events *in vivo* (Longo *et al.*, 1994b). About 30% and greater than 90% of sperm nuclei incubated in extract from eggs with or without germinal vesicles undergo chromatin dispersion, respectively. Nuclear envelope assembly occurs only in extracts from eggs taken about 65 min after activation, i.e. at a period when pronuclear envelopes normally assemble *in vivo*. These results suggest that the capacity of *Spisula* egg extracts to support pronuclear development is stage-dependent and correlated with the cell cycle (Longo, 1984, 1990).

Early investigations with mammalian sperm *in vitro* demonstrated that in most species exposure to strong acid, proteases, DNAse and detergents had little effect on the structural integrity of the nucleus. However, incubation of sperm nuclei in reducing agents such as β-mercaptoethanol or dithiothreitol induced their expansion, indicating that cleavage of disulfide bonds is required for decondensation, by facilitating the disruption of nucleoproteins bound to DNA (Calvin and Bedford, 1971). A combination of a reducing agent such as dithiothreitol and urea or trypsin has been employed to induce dispersion *in vitro*. The extreme conditions effected by such combinations may mimic, but undoubtedly do not reflect the biology of *in vivo* decondensation. Later studies verified the need for disulfide bond reduction *in vivo* and indicated the involvement of glutathione in this process (Zirkin *et al.*, 1989; Perreault, 1992). The importance of disulfide bond reduction *in vivo* has been demonstrated by injecting isolated sperm nuclei from different regions of the epididymis and testis into hamster eggs (Perreault *et al.*, 1987). Although newly formed testicular sperm possess nuclei with condensed chromatin, protamines are not linked to one another by disulfide bonds. Disulfide bond formation occurs progressively as sperm move through the epididymis. Hence, the time required for testicular sperm nuclei injected into hamster eggs to decondense was 5–10 min, whereas nuclei of sperm from the caudal portion of the epididymis required 45–60 min to decondense (Perreault *et al.*, 1987).

In general, mammalian germinal vesicle oocytes do not support sperm nuclear decondensation (Balakier and Tarkowski, 1980; Zirkin *et al.*, 1989; Yanagimachi, 1994). According to Yanagimachi (1994) this inability may be due to a lack or deficiency in a substance such as nucleoplasmin (Maleszewski, 1992) rather than low levels of glutathione (Perreault *et al.*, 1988; Perreault, 1990) because the glutathione concentration in immature eggs is only slightly lower than that present in mature ova (Wiesel and Schultz, 1981; Perreault *et al.*, 1988).

With the loss of protamines from the decondensing sperm nucleus there is the appearance of histones (Nonchev and Tsanev, 1990). This change occurs around anaphase II, whereas sperm chromatin dispersion is massive between anaphase II and telophase II (Garagna and Redi, 1988), the latter coinciding with the second stage of expansion which occurs during and following nuclear envelope assembly. The exact mechanisms involved in decondensation and how they relate to the removal of protamines and the addition of histones are not well understood (see Yanagimachi, 1994).

Including the possible involvement of a nucleoplasmin-like molecule (Philpott et al., 1991), protamine displacement from the mammalian sperm nucleus has been postulated to occur by a change in charge via, for example, phosphorylation and proteolysis (Zirkin et al., 1985). Phosphorylation is involved in the modification of basic nucleoproteins during spermatogenesis and, in combination with other enzymatic modifications of nucleoproteins, may be instrumental in dispersion of the condensed sperm chromatin. For example, high levels of phosphorylation in the mouse egg at fertilization and the phosphorylation of sperm protamines by egg extracts suggest that this process could lead to a charge modification and a destabilization of protamine–DNA interactions of the sperm chromatin during male pronuclear development (Young and Sweeney, 1978; Wiesel and Schultz, 1981). Such a scheme may be similar to that observed in fertilized sea urchin eggs, where sperm-specific histones H_1 and H_{2b} are phosphorylated (see above; Green and Poccia, 1985; Poccia, 1989).

Sulfhydryl-induced proteolytic activity has also been proposed to be involved in mammalian sperm nuclear decondensation in vivo (Zirkin and Chang, 1977; Zirkin et al., 1980). An acrosin-like protease associated with isolated rabbit sperm nuclei causes nucleoprotein degradation and decondensation in vitro. However, the normal role of this proteolytic activity in vivo has been questioned, since the decondensing activity intrinsic to isolated sperm nuclei may be of acrosomal origin and may become bound to the chromatin during sperm isolation (Young, 1979). In connection with the possible involvement of proteolytic agents in sperm chromatin dispersion, it is noteworthy that proteolytic activity is associated with the nuclei of somatic cells and processing of histone.

Topoisomerase II, an enzyme involved in alterations of the topological structure of DNA by transiently breaking and rejoining both strands of the DNA helix, may be important in chromatin changes at fertilization (Wright and Schatten, 1990; Perreault, 1992). Such an enzyme, or one similar to it, is likely to be required to assist in the unfolding of DNA from a condensed state to one organized around nucleosomes. Inhibition of topoisomerase II activity with teniposide in Spisula and hamster zygotes induces aberrant chromosome condensation, arrest of first polar body formation and the inhibition of DNA synthesis. Teniposide arrests nuclear assembly in Xenopus sperm nuclei incubated in egg extract (Newport, 1987), but this agent does not block sperm nuclear decondensation nor male pronuclear development in fertilized *Spisula* or hamster eggs (Wright and Schatten, 1990; Perreault, 1992). What role topoisomerase II may play in the development of the male pronucleus *in vivo* remains to be determined.

Models for the organization of chromatin and genes within sperm have been proposed (see Koehler *et al.*, 1983; Ward, 1993; Ward *et al.*, 1996) in order to better understand genome function and are particularly relevant to an understanding of sperm chromatin dispersion at fertilization. Sperm DNA is believed to be organized into loop domains which are anchored to a nuclear matrix (Fig. 10.5).

In amphibian sperm the loops are approximately 25 kb in length (Risley *et al.*, 1986) while in the hamster they are approximately 47 kb in length, i.e. about half the size of loops found in somatic cells (Ward *et al.*, 1989). A basic tenant of Ward's (1993) model is that the loops provide a means to transfer the paternal genome to the zygote in an organized but compact form. The proper three-dimensional DNA organization may function to ensure that the developing embryo can access the paternal genetic information rapidly and efficiently (Ward, 1993). This organization may also provide an orderly mechanism for decondensation and facilitate nucleoprotein exchange during the reorganization of the sperm nucleus into a male pronucleus.

Amphibian and sea urchin eggs have been shown to contain DNA binding proteins and pronuclei of these ova can concentrate cytoplasmic nonhistone proteins (Barry and Merriam, 1972; Kunkle *et al.*,

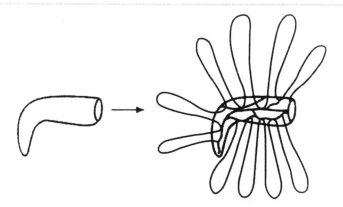

Fig. 10.5   DNA is organized into loop domains which are anchored to the sperm nuclear matrix (right). How the DNA loops are packaged in the fully condensed sperm nucleus has not been determined. The profile of the hamster sperm nucleus is depicted. Reproduced with permission from Ward, 1993.

1978b). A significant accumulation of label in male and female pronuclei of rabbit oocytes was found when eggs were incubated *in vitro* with [³H]-lysine and subsequently inseminated, providing additional evidence that proteins from the egg cytoplasm are incorporated into both pronuclei at fertilization (Motlik *et al.*, 1980).

Cytoplasmic proteins are recruited into nuclei before swelling and DNA synthesis occur, suggesting a causal relation (Merriam, 1969). The control of gene activity by elements within egg cytoplasm has been demonstrated, and it is not unreasonable to presume that changes exhibited by the sperm nucleus during pronuclear development also allow for its reprogramming (Gurdon and Woodland, 1968; Johnson and Rao, 1971). As indicated by Newrock *et al.* (1977), 'Egg and sperm are highly specialized cells which differ markedly in packing of their chromatin, and it would seem necessary to restore both genomes to the same state if remodeling processes at subsequent stages of development are to generate the same structural and functional qualities in the two chromosome sets.'

10.3 FORMATION OF THE MALE PRONUCLEAR ENVELOPE

Development of the male pronuclear envelope has been studied in a variety of organisms and is similar to the series of events described for nuclear envelope formation in mitotic and meiotic cells (Longo, 1973; Franke, 1974). It is not known to what extent remodeling of the sperm chromatin must proceed before nuclear assembly can occur. The timing of the formation of the nuclear envelope that surrounds the dispersed sperm-derived chromatin (male pronuclear envelope) varies in the different species that have been studied. For example, in the sea urchin *Arbacia* development of the male pronuclear envelope is initiated during chromatin dispersion. In the surf clam, *Spisula*, formation of the male pronuclear envelope occurs after chromatin dispersion, apparently in concert with the formation of the nuclear envelope of the female pronucleus. In either case morphological events involving the formation of the dispersed chromatin are similar. Vesicles coalesce along the periphery of the dispersed chromatin and fuse to form elongate cisternae that develop pores (Fig. 10.6).

The cisternae fuse to enclose the dispersed chromatin and form a nuclear envelope. In those cases where portions of the sperm nuclear envelope are incorporated into the structure of the male pronuclear envelope, the elongate cisternae fuse with the sperm derived membranes and the latter also become a part of the membranous boundary of the male pronucleus. Regions of the sperm nuclear envelope incorporated into the male pronuclear envelope frequently retain distinctive morphological features that allow for their identifica-

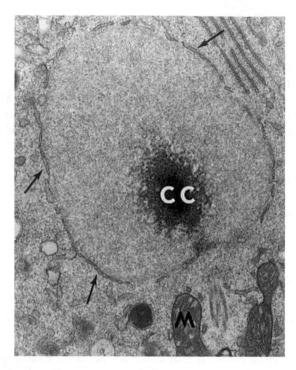

Fig. 10.6 Development of the male pronuclear envelope in a sea urchin (*Arbacia*) zygote. The vesicles (arrows) surrounding the dispersed sperm chromatin fuse to form a continuous lamina. CC = condensed chromatin; M = sperm mitochondrion.

tion at later stages of fertilization or embryonic development, e.g. following the fusion of the male and female pronuclei. What role portions of the sperm nuclear envelope that are incorporated into the male pronuclear envelope might serve has not been determined.

Investigations have been carried out to determine the source(s) of membrane that comprises the male pronuclear envelope (Longo, 1976b). Investigations with the sea urchin *Arbacia* demonstrate that portions of the sperm nuclear envelope, specifically the apical and basal regions, are incorporated into the male pronuclear envelope. This raises the question of whether the amount of nuclear envelope present within the incorporated sperm is adequate to completely enclose the dispersed paternally derived chromatin. In *Arbacia* the amount of membrane present in the sperm nuclear envelope is sufficient to delimit only about 20% of the surface of the male pronucleus. A likely source of membrane for the formation of the pronuclear envelope is the endoplasmic reticulum – because of its prevalence, continuity with the nuclear envelope and involvement in nuclear envelope formation and repair in other cells. Investigations of male pronuclear development in centrifuged eggs in which the endoplasmic reticulum is localized to a specific region of the ovum indicate that the time required to form a male pronuclear envelope is prolonged in areas lacking in and accelerated in areas rich in endoplasmic reticulum (Longo, 1976b). These results strongly suggest that the endoplasmic reticulum directly contributes to the formation of the pronuclear envelope.

Although the question of the involvement of membrane biosynthesis in the formation of the pronuclear envelope is a complex one, the following principles seem clear: newly synthesized proteins and lipids appear to be inserted into preexisting membranes and membrane components are frequently synthesized at or inserted into sites distinct from their ultimate destinations. Membrane biosynthesis in somatic cells has been investigated by measuring the appearance and activity of specific enzymes, the incorporation of labeled precursors into membrane components and the effects of various inhibitory agents on this incorporation. In the case of male pronuclear envelope formation the contribution of membrane biosynthesis is not known with certainty. Analyses with fertilized *Arbacia* eggs demonstrate that the incorporation of labeled leucine into TCA-precipitable material can be inhibited with puromycin, with no apparent effect on pronuclear development. These observations suggest that *de novo* protein synthesis may not be directly involved in pronuclear envelope development.

Investigations employing isolated sperm nuclei incubated in cell-free extracts have added greatly to our fund of knowledge regarding the formation of the pronuclear envelope (Lohka and Masui, 1984b; Lohka, 1988; Vigers and Lohka, 1991; Longo *et al.*, 1994b; Collas *et al.*, 1995; Collas and Poccia, 1995a, b). Using this system, two models have been proposed for nuclear envelope assembly (Lohka, 1988).

1. The vesicular precursor model is based on observations similar to those described above, i.e. vesicles aggregate along the surface of the chromatin and fuse to form elongated cisternae, which in turn fuse to form a nuclear envelope with pores.
2. The prepore model is based on reconstitution experiments by Sheehan *et al.* (1988). Nuclear envelope assembly requires both the soluble and the particulate components of egg extracts (Lohka and Masui, 1984b). At ratios of 25:1 (soluble/particulate) complete nuclear envelopes do not form, but nuclear pores assemble along the periphery of the decondensed sperm chromatin with little or no membrane present. Sheehan *et al.* (1988) propose that components of the nuclear pores bind to chromatin to form prepores, and

membrane vesicles then bind to prepores, flatten and fuse to form a nuclear envelope.

The particulate portion of *Xenopus* egg extracts can be separated into two fractions, A and B, that have different properties and functions in nuclear envelope assembly (Vigers and Lohka, 1991). Vesicles in fraction B bind to chromatin in an ATP-independent manner and may be involved in pore formation. Vesicles in fraction A carry markers for the endoplasmic reticulum and represent a pool of cisternae that contribute to the expansion of the nuclear envelope. Parameters involved with nuclear envelope formation of sea urchin sperm nuclei incubated in egg cytosol have been analyzed (Collas and Poccia, 1995a, b). Decondensation in the absence of a nuclear envelope is only partial and only goes to completion with membrane and lamina assembly (Collas *et al.*, 1995). Binding of cytoplasmic vesicles to lipophilic substances at the apex and base of the sperm nucleus requires ATP (but not ATP hydrolysis) and is abolished by trypsin.

Assembly of the nuclear envelope is believed to follow that of the nuclear lamina, which in turn is regulated by the phosphorylation of its lamins (Gerace *et al.*, 1984). The chromatin surface has been shown to possess receptors for the inner membrane of the nuclear envelope (Chaudhary and Courvalin, 1993) and, although the receptors have not been identified, histone variants and nuclear scaffolding have been proposed as candidates (Newport, 1987). Type A lamins have also been proposed to mediate the attachment of chromatin with nuclear envelope precursor vesicles (Höger *et al.*, 1991; Glass and Gerace, 1990). The role that the nuclear lamins play in formation of the pronuclear envelope and the relationship of the nuclear matrix to DNA organization and replication in the male and female pronuclei have not been clearly defined (see Stricker *et al.*, 1989; Prather and Schatten, 1992). The germinal vesicle of *Xenopus* oocytes possesses lamin L_{III} which becomes soluble at the time of its breakdown (Stick and Hansen, 1985; Benavente *et al.*, 1985). Lamin L_{III} released from the germinal vesicle is believed to serve as a pool for the assembly of the nuclear lamina of the male and female pronuclei (Stick and Hausen, 1985). Lamins A/C and B are associated with the male and female pronuclei of mouse and sea urchin fertilized eggs (Stricker *et al.*, 1989; Schatten *et al.*, 1985a). Pronuclei of *Spisula* zygotes possess an A/C-type lamin referred to as lamin G (Maul and Schatten, 1986).

Experiments using cell-free extracts of *Xenopus* eggs have demonstrated that histone replacement is a prerequisite for pronuclear formation and DNA synthesis (reviewed in Perreault, 1992). DNA synthesis occurs only after nuclear assembly is completed (Naish *et al.*, 1987a, b; Leno and Laskey, 1991) and the integrity of the nuclear

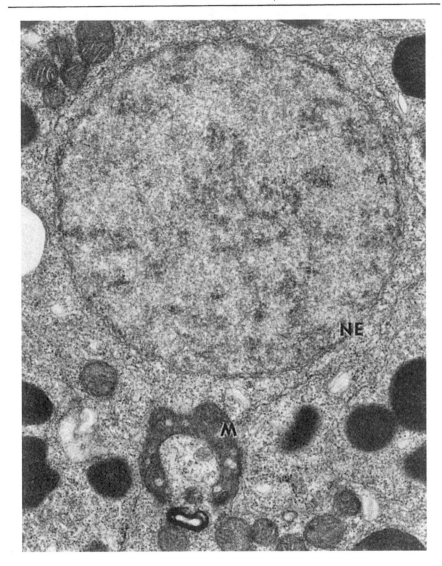

Fig. 10.7 Male pronucleus of a sea urchin (*Arbacia*) zygote. NE = nuclear envelope; M = sperm mitochondrion. See Longo, 1981b.

envelope is necessary for mechanisms that prevent reinitiation of DNA synthesis before mitosis (Blow and Laskey, 1988; Leno *et al.*, 1992).

With the formation of the pronuclear envelope, transformation of the sperm nucleus into a male pronucleus is essentially completed (Fig. 10.7). However, the male pronucleus continues to undergo morphogenetic changes which may include further enlargement, the continuation of chromatin dispersion and the acquisition of internuclear structures such as nucleoli, annulate lamellae and aggregations of tubular inclusions.

Aspects regulating
pronuclear development 11

11.1 DURATION OF PRONUCLEAR ASSEMBLY

The time required for male pronuclear formation varies (Longo, 1973, 1981b): in the sea urchin *Arbacia* it is about 8 min, in the surf clam, *Spisula*, 50–60 min, in the domestic fowl approximately 25 min and in mammals 2–4 hours. Although the bases for these time differences have not been determined, they most probably reflect a correlation between sperm nuclear transformation and meiotic maturation.

11.2 MALE AND FEMALE PRONUCLEAR DIFFERENCES

In sea urchins the male pronucleus is much smaller than the female. In other organisms, this size difference is less pronounced and may be reversed. Male pronuclei larger than the female are observed in rat and mouse zygotes; the opposite is the case in hamster and human fertilized eggs (Yanagimachi, 1994). If pronuclear migration is inhibited in sea urchin zygotes the male pronucleus increases in size, often attaining the dimensions of the female pronucleus. These observations suggest that factors responsible for the continued enlargement of the male pronucleus may be present in the zygote cytoplasm after the normal fertilization period, and the ultimate size of the male pronucleus may be related to time spent within the zygote cytoplasm. In mammalian zygotes, under similar circumstances, continued enlargement of the male pronucleus is not obvious.

With the exception of size in some cases and proximity to the first and/or second polar bodies, it is often difficult to distinguish the male from the female pronucleus. Differences in the contribution of the maternal and paternal genomes to the developing embryo have been demonstrated in the mouse (McGrath and Solter, 1984; Surani *et al.*, 1984, 1986). Chromosomes of both the male and female pronucleus are necessary for development to term: the paternal genome is crucial for development of extraembryonic tissues while the maternal is involved in embryogenesis during preimplantation and early postimplantation development. There is evidence demonstrating that these differences may be due to imprinting during oogenesis and spermatogenesis. What this implies is that differences between the parental chromosomes are heritable, they survive activation of the embryonic genome and are probably reprogrammed by epigenetic factors in the egg cytoplasm (Surani et al., 1986).

11.3 CELL CYCLE EVENTS

In eggs which are fertilized at an arrested stage of meiosis the incorporated sperm nucleus decondenses while the maternal chromatin is condensed and engaged in the completion of its meiotic divisions. How the zygote regulates these two very different nuclear activities has not been determined. One model proposed by Yanagimachi (1994) and depicted in Fig. 11.1 consists of essentially two stages.

The first is believed to be independent of egg activation and involves the decondensation of the sperm nucleus. The second, which is dependent on egg activation, involves the transformation of the decondensed sperm nucleus into a pronucleus.

Sperm decondensation, meiotic progression and formation of male and female pronuclei have been shown to occur in a coordinated

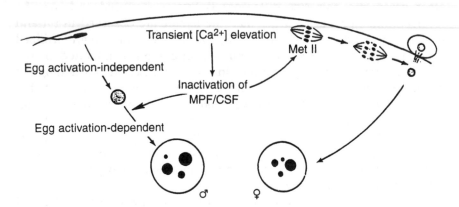

Fig. 11.1 Diagram illustrating that sperm nucleus decondensation is independent of egg activation, whereas the transformation of a decondensed sperm nucleus to a pronucleus is dependent on egg activation. Inactivation of MPF/CSF by a transient elevation of internal Ca^{2+} is important for egg activation. Reproduced with permission from Yanagimachi, 1994.

fashion in fertilized surf clam, starfish, hamster and human eggs (Perreault *et al.*, 1987; Wright and Longo, 1988; LaSalle and Testart, 1991; Longo *et al.*, 1991a; see Longo, 1990 for review). In each of these cases the sperm nucleus undergoes decondensation followed by a recondensation and then a second expansion concomitant with the formation of the male and female pronuclei (Fig. 11.2).

Observations with surf clam, hamster, human and starfish zygotes have demonstrated that these changes are closely coupled to processes attending meiotic maturation and female pronuclear development (Chen and Longo, 1983; Luttmer and Longo, 1988; Wright and Longo, 1988; LaSalle and Testart, 1991; Longo *et al.*, 1991a). Kinetic analysis of

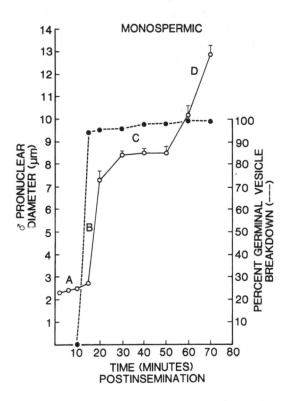

Fig. 11.2 Male pronuclear enlargement as determined by its change in diameter *versus* time in surf clam (*Spisula*) zygotes. Fertilized eggs were examined at timed intervals to determine the presence or absence of germinal vesicles (broken line) and the diameter of incorporated sperm nuclei (solid line). Expansion of the incorporated sperm chromatin is composed of four phases which are correlated with changes in the maternal chromatin: **(A)** slow, pregerminal vesicle breakdown; **(B)** rapid, germinal vesicle breakdown; **(C)** slow, meiotic divisions (polar body formation); and **(D)** rapid, completion of meiotic maturation/development of the female pronucleus. See Chen and Longo, 1983.

male pronuclear development in *Spisula* demonstrates that male pronuclear enlargement does not proceed at a constant rate but consists of four phases coordinated with major changes in the status of the maternal chromatin. The first phase is a short lag period prior to germinal vesicle breakdown in which the size of the sperm nucleus increases only slightly. This is followed by a rapid expansion of the sperm nucleus coordinated with germinal vesicle breakdown. With the development of the first meiotic spindle, sperm nuclear enlargement slows dramatically; this lasts until the completion of the meiotic divisions, when the developing male pronucleus undergoes a second rapid increase in size that correlates with female pronuclear development. A similar relationship of sperm chromatin expansion and meiotic maturation of the maternal chromatin is also seen in hamster zygotes (Fig. 11.3). In addition to demonstrating a coordination in activity between the maternally and paternally derived chromatin such an analysis has also shown quantitatively the effects of ploidy (polyspermy and polygyny) and metabolic and protein synthesis inhibitors during specific phases of fertilization (Table 11.1; Luttmer and Longo, 1988; Wright and Longo, 1988).

Working from such analyses a model has been proposed attempting to integrate possible relationships between factors regulating the morphogenesis of the maternal and paternal chromatin in eggs fertilized at an arrested stage of meiosis (Fig. 11.4).

As described above, observations from numerous laboratories have

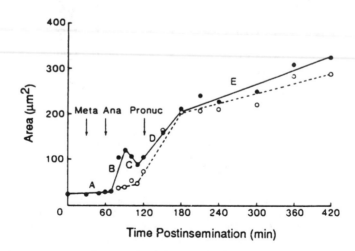

Fig. 11.3 Pronuclear enlargement rates of *in vitro* fertilized hamster eggs. Area measurements of expanding sperm nuclei (open circles) and maternal chromatin (closed circles). Meta = metaphase II; Ana = anaphase II; Pronuc = appearance of female pronucleus. See Wright and Longo, 1988.

Table 11.1 Effects of various agents on sperm nuclear enlargement in fertilized hamster eggs; PI = postinsemination (See Wright and Longo, 1988)

	Sperm nuclear area ($x \pm s.e.m.$)		
Treatment	90 min PI	110 min PI	3 h PI
Monospermy	120.9 ± 11.1	86.9 ± 4.3	210.6 ± 5.3
Polyspermy	126.9 ± 14.7	135.3 ± 9.4*	175.2 ± 13.0*
Polygyny	117.0 ± 13.7	75.9 ± 9.2	88.1 ± 6.2*
Colchicine	84.1 ± 11.0*	148.8 ± 11.7*	106.7 ± 12.2*
Puromycin	96.0 ± 14.4	123.6 ± 8.5*	153.5 ± 9.7*
Antimycin	129.3 ± 13.9	84.6 ± 12.9	76.3 ± 5.3*

* Statistically different compared to same time-point in controls.

demonstrated a 'factor(s)' or 'activity' to be present in eggs following germinal vesicle breakdown which acts on the newly incorporated sperm nucleus, inducing it to expand and presumably undergo changes whereby nucleoprotein composition becomes similar to that of the maternal chromatin. This activity, referred to here as F1, may be derived from substances originally located in the germinal vesicle (Masui and Clarke, 1979). In amphibians this factor has been identified and is nucleoplasmin (Ohsumi and Katagiri, 1991b; Philpott and Leno, 1992). This change gives rise to an expanded sperm nucleus, which is observed at the end of phase B. Also present in the egg is a second 'factor', MPF, which regulates the structure/function of the maternal chromatin and is responsible, at least in part, for the condensed state of

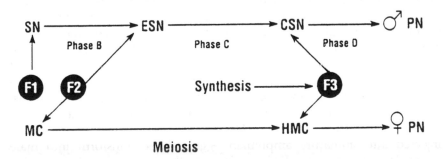

Fig. 11.4 Relationship and possible means of regulating male and female pronuclear development in eggs fertilized at an arrested stage of meiosis. (SN = sperm nucleus; ESN = expanded sperm nucleus; CSN = condensed sperm nucleus; PN = male and female pronuclei; MC = maternal chromatin; HMC = haploid maternal chromatin; F1–F3 = presumptive factors 1–3.) See Longo, 1990.

the maternal chromatin. It has been speculated that this factor, which is referred to here as F2, also acts on the transformed sperm nucleus such that the latter undergoes a condensation (phase C) as observed during polar body formation. A third 'factor' or 'activity' (F3), which may be synthesized in part during fertilization, is involved in transformations of condensed maternal and paternal chromatin such that they are able to undergo a second expansion characteristic of phase D. The coordinated effect of these factors synchronizes the activity of the maternal and paternal genomes with respect to events of the first mitosis and initiation of cell cycle processes characteristic of cleaving embryos (Murray and Hunt, 1993).

The importance of MPF as a regulator of chromatin structure/ function has been demonstrated in mouse zygotes (Borsuk and Manka, 1988; Hashimoto and Kishimoto, 1988). In fertilized mouse eggs treated with colcemid the meiotic spindle is destroyed and maternal chromatin remains condensed. In the meantime the incorporated sperm nucleus decondenses and then recondenses, and unless colcemid is removed both the maternal and paternal chromosomes do not form pronuclei (Borsuk and Manka, 1988). The disruption of spindle microtubules in this case is believed to maintain high levels of MPF (Hashimoto and Kishimoto, 1988). Recent investigations (Moos et al., 1995, 1996) indicate that elevated levels of MAP kinase activity in mouse eggs inhibit pronuclear formation, and that activation of MAP kinase subsequent to pronuclear formation results in precocious pronuclear envelope breakdown.

11.4 GERMINAL VESICLE BREAKDOWN

The eggs of many animals can be fertilized at an earlier stage than normal, e.g. amphibian and hamster eggs, which are normally inseminated at the second metaphase of meiosis, can be fertilized at meiotic prophase (germinal vesicle stage). In these cases, however, the incorporated sperm remains essentially unchanged (Dettlaff et al., 1964; Katagiri, 1974; Usui and Yanagimachi, 1976; Longo, 1978b; Hylander et al., 1981). It is only after germinal vesicle breakdown that the incorporated sperm nucleus undergoes transformations into a male pronucleus. Pronuclei that develop from sperm nuclei injected into mature amphibian eggs are capable of DNA synthesis, whereas sperm nuclei injected into immature amphibian eggs neither transform into male pronuclei nor synthesize DNA so long as the germinal vesicle remains intact (Skoblina, 1976; Moriya and Katagiri, 1976). The failure of sperm nuclei to synthesize DNA when injected into immature amphibian eggs may be due neither to a deficiency of DNA polymerase nor to an absence of deoxyribonucleotides but to additional factors which make the sperm DNA accessible for replication. Investigators have shown that factors required for morphogenesis of the sperm nucleus into a male pronucleus are in fact derived from the germinal vesicle (Katagiri and Moriya, 1976; Thadani, 1979; Balakier and Tarkowski, 1980; Lohka and Masui, 1983b; Yamada and Hirai, 1984). In the case of *Xenopus* eggs, one important germinal vesicle component involved with sperm chromatin decondensation is nucleoplasmin (Ohsumi and Katagiri, 1991b; Philpott *et al.*, 1991; Philpott and Leno, 1992). Whether or not a nucleoplasmin-like molecule is involved in sperm chromatin dispersion of other organisms has not been established. Furthermore, in some animals, e.g. sea urchins and mammals (Iwamatsu and Chang, 1972; Longo, 1978b), incorporated sperm nuclei exhibit greater differentiation with increasing oocyte maturation, which suggests that factors required for pronuclear development appear (or are activated) as the egg progresses through meiosis.

11.5 PROTEIN SYNTHESIS

Experiments with mammalian and some invertebrate eggs indicate that protein synthesis is essential for male pronuclear development after the initial decondensation of the sperm nucleus (Longo *et al.*, 1991b; Ding *et al.*, 1992). In the case of *Spisula* zygotes, the calcium ionophore A-23187 was able to override the effects of protein synthesis inhibition on pronuclear development, suggesting that both nascent proteins and Ca^{2+} signals are involved in regulating the status of the maternal and paternal chromatin during pronuclear development.

11.6 MALE PRONUCLEAR GROWTH FACTOR

A number of studies have indicated that transformation of the sperm nucleus into a male pronucleus may not be entirely dependent upon factors originating from the germinal vesicle. A male pronuclear growth factor is believed to be inactive or absent in *in vitro* matured rabbit, bovine and porcine oocytes (Thibault and Gérard, 1973; Motlik and Fulka, 1974; Trounson *et al.*, 1977; Mattioli *et al.*, 1988). The factor is formed in surrounding granulosa cells and then transported to the egg. Male pronuclear growth factor has not been characterized; its potential role in the development of the female pronucleus is unknown. It has been speculated that *in vitro* culture does not provide a 'natural' environment required for normal cytoplasm development. This suggestion is supported by studies demonstrating that marked qualitative changes occur in protein synthesis during maturation of oocytes *in vitro*. Based on these findings it has been suggested that proteins synthesized during the later stages of maturation are related directly to postmaturational events associated with fertilization and early development. Despite the evidence for a male pronuclear growth factor, other studies have shown that oocytes matured *in vitro* and then transferred to the oviducts of mated animals are able to be fertilized and undergo normal embryogenesis.

There is evidence to suggest that factors instrumental in the transformation of the sperm nucleus are present in limited quantities. For example, fully formed male pronuclei and undeveloped sperm nuclei may be found within the same cytoplasm of polyspermic and polygynic ova (Hunter, 1967a, b; Poccia *et al.*, 1978; Luttmer and Longo, 1988; Wright and Longo, 1988). In these instances, factors responsible for pronuclear development may have been exhausted or inactivated by developing pronuclei and, therefore, are absent or unable to influence remaining incorporated sperm nuclei.

11.7 FACTOR LONGEVITY

Studies have been conducted to determine the longevity of factors that are involved with the transformation of the sperm nucleus, i.e. for how long after insemination the egg cytoplasm is capable of supporting the development of a male pronucleus. For example, Borsuk and Tarkowski (1989) demonstrate that mouse eggs lose the ability to transform sperm nuclei into pronuclei between 3 and 5 hours after activation. Sperm nuclear changes have been examined in the sea urchin (*Arbacia*) and hamster following refertilization to establish the longevity of the factors regulating pronuclear development (Usui and Yanagimachi, 1976; Longo, 1984). It is possible to fuse sperm with *Arbacia* zygotes 20–30 min after insemination and well after pronuclear fusion (approximately 15 min postfertilization) while the zygote nucleus is undergoing DNA replication. Sperm incorporated into such zygotes undergo male pronuclear development, as observed in controls.

Observations of hamster and sea urchin zygotes reinseminated at later stages during fertilization and at different stages of cleavage indicate that regulators of sperm nuclear transformation may have a more general role, relating to factors involved with the cell cycle (Usui and Yanagimachi, 1976; Longo, 1983, 1984; see Murray and Hunt, 1993). Such an association is not difficult to envisage since many of the events of pronuclear development are similar to events of mitosis. Breakdown of the sperm nuclear envelope, chromatin dispersion and formation of the pronuclear envelope are structurally similar to events occurring at prophase and telophase of mitotically active cells. Furthermore, pronuclear development does not take place when sperm are incorporated into zygotes at the pronuclear stage, whereas sperm nuclear envelope breakdown and chromatin dispersion take place when sperm are incorporated into zygotes shortly before the first cleavage division, i.e. at prometaphase of mitosis.

11.8 FACTOR SPECIFICITY

Sperm incorporated into cultured somatic cells provide a system to determine whether factors necessary for the transformation of the sperm nucleus into a male pronucleus are present in cells other than eggs. Sperm phagocytosed by cultured cells may appear to undergo changes comparable to those observed at fertilization, but when closely examined, they are degenerative (Phillips *et al.*, 1976). There are instances where activation of the sperm nucleus (chromatin dispersion) appears to take place within somatic cells (VanMeel and Pearson, 1979). This is accompanied by a shift from the protamine to the histone type of basic protein associated with the transforming sperm nucleus. In some instances there is also the induction of RNA and DNA synthesis in the transformed sperm nucleus. These results suggest that factors may be present in somatic cells capable of inducing alterations in incorporated sperm nuclei similar to those occurring in fertilized eggs. Variations of this type of experiment include those fusing somatic cells with eggs (Czolowska *et al.*, 1984; Szöllösi *et al.*, 1986, 1988) and fusing fragments of immature and mature (activated and unactivated) with one another (Borsuk, 1991). Results of the former investigations indicate that factors regulating changes of incorporated sperm nuclei at fertilization are nonspecific. For example, thymocyte nuclei transform into pronuclei-like structures following fusion with activated mouse eggs. In the case of the latter experiments Borsuk (1991) has demonstrated that the cytoplasm of activated eggs and maturing oocytes contains different but complementary factors for transformation of the sperm nucleus into a male pronucleus.

Investigations have been carried out to determine whether factors responsible for sperm nuclear transformation in the eggs of one species are capable of interacting with the sperm nucleus of a different species to elicit pronuclear development. In those species studied, e.g. mussel (*Mytilus*) ♂ × sea urchin (*Arbacia*) ♀ or crosses between different species of mammal, changes in the sperm nucleus characteristic of pronuclear development do occur (Wu and Chang, 1973; Imai *et al.*, 1977; Barros and Herrera, 1977; Longo, 1977). These investigations indicate that factors involving the transformation of the sperm nucleus into a male pronucleus may not be species specific and are capable of interacting with sperm nuclei of evolutionary divergent organisms. The extent to which such interspecies combinations accomplish all the events of pronuclear formation has not been established. The male pronuclei of naturally fertilized *Mytilus* and *Arbacia* eggs differ when compared morphologically. Interestingly, in crosses of *Mytilus* ♂ × *Arbacia* ♀, the male pronuclei resemble those that develop from *Arbacia* sperm, suggesting that cytoplasmic factors determine the form of the pronucleus.

11.9 MICROINJECTED SPERM NUCLEI

Experiments in which sperm nuclei have been microinjected into eggs demonstrate that the structural integrity of the spermatozoon and the normal processes of gamete fusion and incorporation are not necessary prerequisites for pronuclear development (Skoblina, 1976; Moriya and Katagiri, 1976; Uehara and Yanagimachi, 1976). Moreover, frozen-thawed and freeze-dried human sperm injected into hamster eggs can develop into structures morphologically resembling male pronuclei and undergo DNA synthesis (Uehara and Yanagimachi, 1977; Katayose et al., 1992). The developmental potential of such zygotes could not be determined because of difficulties encountered in culturing hamster zygotes *in vitro* to a transferable stage. The nuclei of round spermatids have also been injected into hamster eggs and shown to develop into pronuclei that undergo DNA synthesis and become associated with the female pronucleus (Ogura and Yanagimachi, 1993). Interestingly, some eggs are not activated by the injection protocol; the spermatid nucleus decondenses and forms chromosomes 5–10 hours later.

11.10 INTRACELLULAR PH

The elevation of intracellular pH that occurs in the ova of some organisms at fertilization is necessary for egg activation and development of the male pronucleus (Chambers, 1976; Carron and Longo, 1980). In sea urchins, if alkalization of the egg cytoplasm is blocked, male pronuclear development is reversibly inhibited, which suggests that pronuclear formation is dependent upon cytoplasmic alkalization. Experiments with somatic cells also indicate that intracellular pH modulation influences nuclear structure.

Fate of incorporated sperm mitochondria, flagellum and perinuclear structures

12

In most animals, incorporation of the sperm nucleus is accompanied by the entry of the sperm mitochondria and components of the sperm flagellum, the axonemal complex (Longo, 1973; Fig. 12.1).

In the annelid worm *Nereis* (Wilson, 1925) and the Chinese hamster (Yanagimachi *et al.*, 1983) the sperm tail breaks off at its base and does not enter the egg. In ascidians, the spermatozoon, when it contacts the egg, loses its mitochondrion within 2 min of insemination. The mitochondria slide along the tail to its tip and are released (Lambert and Lambert, 1981). Hence, in this case the sperm mitochondria are not incorporated into the egg. This process is calcium-dependent and can be induced by egg water, elevated pH and low external sodium. The reaction is accompanied by a drop in extracellular pH suggestive of a release of protons from the sperm. The release of the mitochondria may be necessary for fertilization because of steric factors, so allowing the sperm to pass through the chorion, i.e. it may provide the push for sperm incorporation.

12.1 MITOCHONDRIA

Structural alterations of the sperm mitochondria have been observed in various mammalian species, including swelling and loss of cristae – processes believed to be indicative of degeneration. In the mouse sperm mitochondria begin to degenerate during the early stages of fertilization, whereas in the rat changes are not noted until cleavage. Sperm-derived mitochondria have been detected as late as the four-cell stage in rats (Shalgi *et al.*, 1994). Analysis of mitochondrial DNA of hybrid amphibian embryos derived from *Xenopus laevis* and *Xenopus mulleri*

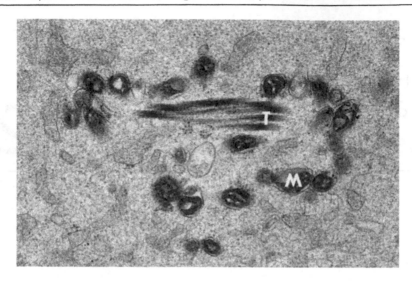

Fig. 12.1 Incorporated sperm mitochondria (M) and portion of the sperm tail (T) in a rabbit zygote. See Longo, 1976c.

indicate that mitochondria are maternally derived (Dawid and Blackler, 1972). This may not be the case in all organisms, as experiments with mussels indicate a paternal contribution to the pool of embryonic mitochondria (Hoeh *et al.*, 1991).

In the sea urchin *Arbacia* changes observed in sperm mitochondria during fertilization are not consistent. Some zygotes may exhibit a decrease in electron opacity, a loss of cristae and some swelling, while in others little or no change is apparent. Recognizable sperm mitochondria have been observed juxtaposed to the first mitotic apparatus approximately 60 min following insemination. Sperm mitochondria appear to decrease in size and become structurally similar to those maternally derived during the early stages of fertilization in the mussel *Mytilus*. As a result of these changes, they become indistinguishable from maternally-derived mitochondria in the surrounding cytoplasm.

12.2 FLAGELLUM

The incorporated sperm axoneme has been observed in mammalian zygotes, and has been followed by light and electron microscopy during cleavage. The external fibers and axonemal complex are observed in mouse, rat, rabbit, hamster and human zygotes (Zamboni, 1971; Shalgi *et al.*, 1994). In the rat, very little change is observed in the sperm axoneme up to the second division, following which it disappears (Szöllösi, 1965). The fibrous sheath of the rat sperm tail disappears prior to the first cleavage division. In the mouse, the axonemal complex often splays into multiple fibers following sperm incorporation and is not detected by the 16–32-cell stage (Simerly et al., 1993). The incorporated sperm axoneme of the sea urchin Arbacia appears to retain its structural integrity up to the first cleavage division; its fate at later stages of development is unknown (Longo, 1973). In the mussel Mytilus portions of the sperm axoneme are located adjacent to the sperm nucleus immediately after incorporation. Subsequently, the sperm axoneme is not observed, although long segments of microtubules, believed to be derived from this structure, may be found in association with the reorganizing sperm nucleus. Short segments of incorporated sperm axoneme have been observed in the surf clam, Spisula, and it is possible that little of this structure is taken into the egg upon insemination. In summary, what contribution components of the sperm tail make to the fertilized egg and developing embryo has not been established (see Karr, 1991).

12.3 PERINUCLEAR STRUCTURES

The incorporated postacrosomal complex of mouse zygotes has been observed up to the period of male pronuclear development (Stefanini et al., 1969). The fate of this structure has not been elucidated. Acid phosphatase activity has been demonstrated in the postacrosomal region of the mammalian spermatozoon, which may modify egg cytoplasmic components following sperm incorporation. Recent experiments with C. elegans demonstrate that a protein, Spe-11, found in sperm and not oocytes, localizes to the perinuclear region of sperm and functions directly during early embryogenesis (Browning and Strome, 1996). Spe-11 embryos (i.e. embryos derived from sperm of homozygous mutant animals) fail to complete meiosis, form a properly mitotic spindle and do not undergo cytokinesis. In contrast to the many known maternal factors required for embryogenesis, Spe-11 is the first paternally contributed factor to be identified and characterized.

Centrosome expression, sperm aster assembly and pronuclear migration

13

13.1 CENTROSOME

The centrosome of somatic cells serves as a microtubule-organizing center located at one pole of the nucleus. Most often it has an aster appearance; at its center is a cloud of some amorphous material that may or may not contain centrioles. Emanating from the amorphous material are fascicles of microtubules whose plus ends are directed away from the aster's center. Gamma tubulin is a distinctive component of the centrosome and is believed to nucleate microtubule growth (Joshi et al., 1992; Schatten, 1994).

Working largely from observations of the sea urchin egg, in which the oocyte's centrioles disappear some time during maturation, it was concluded that the egg possesses all the elements necessary for development except a division center or centrosome (Mazia, 1961). The spermatozoon, on the other hand, possesses a centrosome but lacks the necessary medium in which to function. Hence, it was inferred that the sperm supplies the division center normally responsible for the cleavage of the zygote. This was first outlined by Boveri (see Wilson, 1925), who postulated that the incorporation of sperm division center was a prerequisite for the organization of the cleavage spindle. Many investigations of egg–sperm systems support the idea that the sperm provides the centrosome at fertilization. There are some, however, that challenge the concept of the exclusive paternal derivation of the centrosome (Schatten, 1994).

1. Studies with mouse eggs indicate the participation of egg-derived centrosomes in the organization of the first cleavage spindle (Maro et al., 1985; Schatten et al., 1985b).

2. Centrosomes containing centrioles are produced in artificially activated eggs (Dirksen, 1961; see also Sorokin, 1968).
3. The presence of natural parthenogenesis indicates that in some species sperm-derived centrosomes are not utilized during egg activation (Schatten, 1994).

The origin, development and function of centrosomes in mouse fertilized eggs appears to differ from that observed in most other animals. Mouse sperm lack centrosomes, whereas the egg possesses 10–20 of these structures (Schatten et al., 1985b, 1986; Maro et al., 1985). Following sperm incorporation each of the egg centrosomes organizes an aster (Fig. 13.1) and the asters become associated with the developing male and female pronuclei during their movements to the center of the zygote.

This movement of the pronuclei is accompanied by the accumulation of mitochondria around the pronuclei (VanBlerkom, 1991). The centrosomes surround the pronuclei after the latter's migration to the center of the zygote and eventually become localized into two clusters that become the poles of the mitotic spindle. From experiments using drugs that disrupt microtubules, Schatten et al. (1985b, 1986) have concluded that the aster microtubules are needed for the migration and centration of the male and female pronuclei. Furthermore, the absence of a centrosome in the mouse spermatozoon, their presence in unfertilized eggs and their ability to give rise to asters in artificially activated eggs (Schatten et al., 1989) indicate that centrosomes of mouse embryos originate from the egg (Schatten, 1994). In other species there is strong evidence demonstrating that centrosomes brought into the egg with the sperm function as microtubule-organizing centers and ultimately give rise to an assembly of microtubules referred to as the sperm aster. Organisms whose eggs form sperm asters at fertilization are indicated in Table 13.1.

Investigations exploring the potential roles of centrosomal components in microtubule nucleation have been carried out in cell-free systems and demonstrate the importance of α-tubulin to microtubule assembly, as well as centrosome formation and activity (Schatten, 1994). Xenopus sperm, which possess the conserved centrosomal proteins, centrin and pericentrin, but not γ-tubulin, bind maternally derived γ-tubulin and nucleate microtubules (Doxsey et al., 1994; Félix et al., 1994; Stearns and Kirschner, 1994). Similar experiments have been carried out using human sperm which possess γ-tubulin (Schatten, 1994). However, in this case human sperm incubated in egg extract do not nucleate microtubule assembly unless first treated to reduce disulfide bonds.

The connection between the two events – disulfide bond reduction and microtubule assembly – has not been established. The development

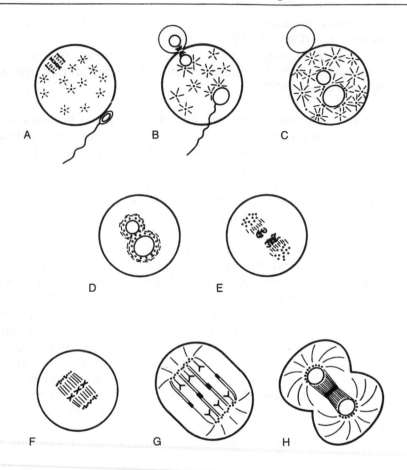

Fig. 13.1 Centrosome function in fertilized mouse eggs. Mouse sperm lack centrosomes and the unfertilized oocyte has approximately 16 cytoplasmic aggregates of centrosomal antigen as well as centrosomal bands at the meiotic spindle poles **(A)**. Each centrosomal focus organizes an aster and, after sperm incorporation, some foci along with their asters begin to associate with the developing male and female pronuclei **(B)**. When the pronuclei are closely apposed at the egg center, several foci are found in contact with the pronuclei and typically a pair reside between the adjacent pronuclei **(C)**. Towards the second half of the first cell cycle, the number of foci increases. At the end of interphase all the foci condense on the pronuclear surfaces and sheaths of microtubules circumscribe the adjacent pronuclei **(D)**. At prophase the centrosomes detach from the nuclear regions, appearing as two broad clusters **(E)** that aggregate into irregular bands at metaphase **(F)**; the first mitotic spindle is typically barrel-shaped, anastral and organized in the absence of centrioles. At anaphase and telophase the centrosomes widen somewhat **(G)**, and at cleavage the centrosomes appear on the poleward nuclear faces **(H)**. Triangular dots = centrosomal foci; lines = microtubules. Reproduced with permission from Schatten et al., 1986.

Table 13.1   Organisms whose eggs form sperm asters at fertilization

Organism	Reference
Ctenophores	Carre and Sardet, 1984; Houliston *et al.*, 1993
Nematodes	Albertson, 1984
Mollusks	Kuriyama *et al.*, 1986; Longo, 1983; Longo and Anderson, 1970b; Longo and Scarpa, 1991; Longo *et al.*, 1993
Echinoderms	Longo and Anderson, 1968; Schatten *et al.*, 1985b; Sluder *et al.*, 1989
Ascidians	Sawada and Schatten, 1989
Amphibians	Gerhardt and Keller, 1986; Klotz *et al.*, 1990
Fish	Lessman and Huver, 1981
Mammals	
Rabbit	Longo, 1976c; Yllera-Fernandez *et al.*, 1992
Sheep	LeGuen and Crozet, 1989; Crozet, 1990
Cow	Navara *et al.*, 1994
Human	Sathananthan *et al.*, 1991; Simerly *et al.*, 1995

of centrosomes, microtubule assembly and centriole duplication have been investigated in cytosol preparations of *Spisula* eggs (Palazzo *et al.*, 1992). Surf clam oocytes do not possess centrioles or asters (Kuriyama *et al.*, 1986). However, centrioles appear by 4 min of activation and duplicate by 15 min postactivation in concert with germinal vesicle breakdown and assembly of the meiotic spindle (Fig. 13.2). Mixtures of lysates from unactivated and activated *Spisula* oocytes are able to induce centriole formation, centriole duplication and astral microtubule assembly *in vitro*.

Based on their *in situ* observations, Kuriyama *et al.* (1986) have postulated that centrioles derived from the sperm and the egg of *Spisula* undergo specific behaviors as depicted in Fig. 13.2. Noteworthy aspects include the following.

1. Only centrioles of maternal origin are involved with meiotic maturation.
2. Sperm-derived centrioles do not become involved in aster formation until the completion of meiotic maturation. This dichotomy may not be due to the centrioles themselves but to pericentriolar material that is directly responsible for organizing microtubules (see Schatten, 1994).
3. The fate of the maternal centriole in this system, i.e. whether it disintegrates or is retained, has not been determined. This is an important

ACTIVATION

FERTILIZATION

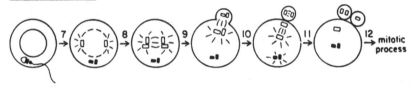

Fig. 13.2 Centriolar behavior in parthenogenetically activated and fertilized *Spisula* oocytes during maturation and cleavage. Open rectangle = maternal centriole; closed rectangle = paternal centriole. The numbers refer to steps of centriole development. Reproduced with permission from Kuriyama *et al.*, 1986.

question for, as pointed out by Sluder *et al.* (1989), only one parental centrosome can be utilized in development since the zygote undergoes DNA synthesis and centrosome duplication prior to first mitosis. An extra centrosome would potentially result in a quadra-polar rather than a bipolar spindle at the first cleavage division.

The inheritance of centrosomes has been studied in sea urchin and starfish eggs (Sluder and Rieder, 1985; Sluder *et al.*, 1989; 1993). Centro-somes inherited by the embryo have been shown to be derived from the sperm in sea urchins (Sluder and Rieder, 1985) and oysters (Longo *et al.*, 1993). Multiple sperm asters form in polyspermic oyster zygotes, which ultimately lead to the assembly of mitotic spindles having more than two poles. In contrast, polygynic zygotes form a single bipolar spindle on which all the maternally and paternally derived chromo-somes assemble. Fertilized sea urchin eggs incubated in colcemid do not undergo pronuclear fusion but the male and female pronuclei are able to undergo chromosome condensation and nuclear envelope breakdown. When the colcemid is inactivated in this situation only the paternally derived chromosomes become organized on a spindle. In addition to supporting the notion of the paternal inheritance of centro-somes, these observations also indicate the suppression of the female centrosome. How this happens in sea urchins has not been explored but experiments in fertilized starfish eggs indicate a loss of functional activity of the maternal centrosome (Sluder *et al.*, 1989, 1993). In other words, the reproductive capacity of the maternal centrosome is reduced between the first and the second meiotic division which correlates with the number of centrioles at each pole of the first and second meiotic spindles. Exactly how 'inactivation' of the maternally derived centrosome 'takes place' without affecting the paternally derived centrosome has not been determined. Results from experiments by Sluder *et al.* (1989) demonstrate that all maternal centrosomes are equivalent and that they are intrinsically different from the paternal. This intrinsic difference, in concert with changes in cytoplasmic conditions after meiosis, is believed to determine the selective loss of the maternal centrosome inherited from the meiotic II spindle.

13.2 SPERM ASTER

One of the earliest indications of sperm aster development is the dissociation of the centrioles from the sperm axoneme. In the mollusks *Mytilus* and *Spisula* the distal and proximal centrioles move into the cytoplasmic area in advance of the developing male pronucleus. In the sea urchin *Arbacia* only the distal centriole moves from the reorganizing sperm chromatin, while the proximal centriole is retained within an invagination of the male pronucleus, the centriolar fossa. Observations of rabbit zygotes did not disclose centrioles in the centrosphere of the developing sperm asters (Longo, 1976c; see also Szöllösi *et al.*, 1972; Magnuson and Epstein, 1984; Szöllösi and Ozil, 1991); However, in sheep and human zygotes, centrioles are associated with the developing sperm aster (LeGuen and Crozet, 1989; Sathananthan *et al.*, 1991).

Initially, large quantities of endoplasmic reticulum and microtubules aggregate in the pericentriolar region and form a dense matrix, referred to as the centrosphere (Longo, 1973, 1976c). The formation of the centrosphere and its subsequent growth tend to exclude components, such as mitochondria, yolk and lipid bodies, which become confined to the periphery of the aster. Radiating from the centrosphere are fascicles of microtubules interspersed among elements of endoplasmic reticulum. These radiating bundles of tubular-lamellar components are separated by clusters of yolk bodies, mitochondria and lipid droplets (Figs 13.3, 13.4).

During later stages of development, the sperm aster enlarges as a result of an accumulation of microtubules and endoplasmic reticulum. The morphology of the sperm aster is similar to that of the asters that comprise the mitotic and meiotic apparatus. However, they appear to differ in that the pericentriolar region of the meiotic and mitotic asters is frequently larger and composed of a dense matrix of fine-textured material. Changes in the sperm aster of sea urchin zygotes, leading to

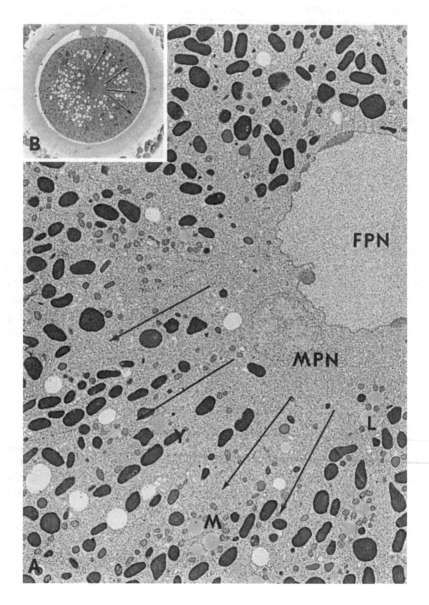

Fig. 13.3 **(A)** Portion of the sperm aster of a sea urchin (*Arbacia*) zygote showing radiating fascicles (arrows) containing microtubules and endoplasmic reticulum (not depicted because of the low magnification), which are separated by columns containing yolk (Y) and lipid (L) bodies and mitochondria (M). MPN and FPN = male and female pronuclei. **(B)** Sperm aster (arrows) of a rabbit zygote. See Longo, 1973.

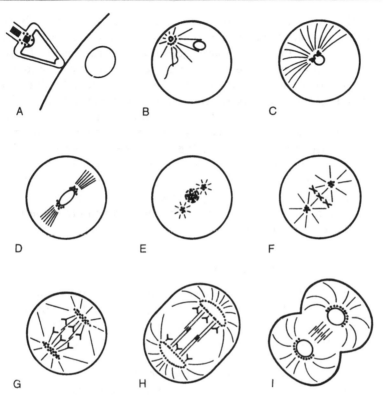

Fig. 13.4 Sperm aster development and centrosome function leading to the two-cell stage of sea urchins. The unfertilized egg lacks centrosomes, which are introduced along with the sperm centriole during incorporation **(A)**. As the microtubules of the sperm aster assemble, the centrosome spreads around the male pronucleus **(B)**. Following pronuclear migration the sperm aster resides at the junction between the pronuclei **(C)**, and separates around the time of pronuclear fusion. Centrosomes are found at opposing poles of the zygote nucleus at the streak stage **(D)**. At prophase, when the nuclear envelope breaks down, they are displaced into the cytoplasm and nucleate the formation of the bipolar mitotic apparatus **(E)**. They enlarge by metaphase but retain their spherical configuration **(F)**, and at anaphase they begin to flatten and spread with axes perpendicular to the mitotic axis **(G)**. At telophase **(H)**, the centrosomes have expanded into hemispheres as the astral microtubules disassemble within the asters and continue to elongate at the astral peripheries. Centrosomes are found on the poleward faces of the blastomere nuclei in cleaving eggs **(I)**. Reproduced with permission from Schatten *et al.*, 1986.

formation of the mitotic spindle and cleavage, have been reviewed (Paweletz and Mazia, 1989).

Formation of the sperm aster in the various organisms studied to date appears to differ temporally with respect to male pronuclear development. For example, in *Arbacia* morphogenesis of the sperm aster may begin during the early stages of chromatin dispersion but usually occurs with the formation of the male pronuclear envelope, about 6 min post-insemination. In the surf clam (*Spisula*), sperm aster formation commences following meiosis of the maternal chromatin and during formation of the male pronuclear envelope, approximately 50 min post-insemination (Longo and Anderson, 1970b; Kuriyama *et al.*, 1986; see also Longo *et al.*, 1993). This differs from the situation seen in starfish zygotes, where the sperm-derived centrosome organizes a sperm aster soon after sperm entry (Hirai *et al.*, 1982; Sluder *et al.*, 1989).

Sperm aster formation does not occur in inseminated *Arbacia* oocytes and apparently inseminated immature eggs of many species. This lack of development may be due to the absence of a cytoplasmic constituent(s) which acts in conjunction with the incorporated sperm components to initiate its morphogenesis.

13.3 PRONUCLEAR MIGRATION

Pronuclear migration involves those processes that are responsible for the movement of the male and female pronuclei from their site of formation to the region where they become associated. The end result is the juxtaposition of the pronuclei. In sea urchin zygotes male pronuclei migrate from 5–11 µm/min (Epel *et al.*, 1977; Schatten, 1981). The direction or path taken by the male pronucleus was the subject of investigations among early cytologists, who distinguished two components: (1) a penetration path, which is approximately vertical to the zygote's surface, and (2) a copulation path, which brings the pronuclei into close association (Wilson, 1925; Fig. 13.5).

Accompanying the male pronucleus during these movements is the sperm aster. How pronuclear movements are brought about has been the subject of considerable speculation.

1. Migration of the male pronucleus may be due to the elongation of sperm aster components; that the enlarging asters of dispermic eggs are pushed apart supports such a scheme.
2. Dense aggregations of microtubules in the cytoplasm between the male and female pronuclei have been observed in brine shrimp (*Artemia*) zygotes (Anteunis *et al.*, 1967).
3. Astral rays may adhere to peripheral structures and exert a tractive force that pulls the aster through the ooplasm (Wolf, 1978).
4. Sperm aster elements may establish a field of streaming cytoplasm which is ultimately responsible for moving the pronuclei together (Chambers, 1939; Hamaguchi *et al.*, 1986).

On the basis of fluorescent studies of cytoskeletal components of sea urchin zygotes Schatten (1984) suggested that the initial assembly of

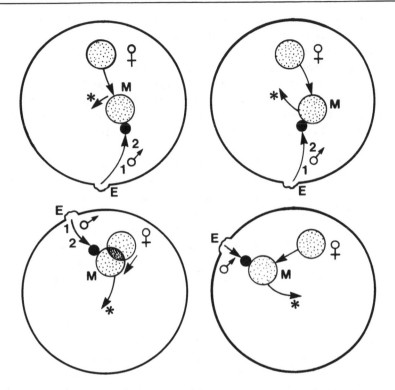

Fig. 13.5 Paths of the male and female pronuclei during their movements to each other and to the center of the zygote. E = entrance site of the spermatozoon; ♀ = original position of the female pronucleus; ♂ = path of the male pronucleus; M = meeting point of the pronuclei; ∗ = path taken by the zygote nucleus to become located within the center of the zygote. The track taken by the male pronucleus may be resolved into two components, a penetration path (1), nearly vertical to the egg surface and a copulation path (2) along which the male pronucleus moves towards the point of association with the female pronucleus.

microtubules on sperm centrioles pushes the male pronucleus to the center of the egg. This is believed to be a steric effect resulting from microtubular elongation. When sperm aster microtubules elongate to contact the surface of the female pronucleus, a swift migration of the female pronucleus to the male pronucleus occurs. During this movement, the sperm aster becomes asymmetrical. It is possible that disassembly of microtubules that interconnect the female and male pronuclei generate the force for movement of the female pronucleus. Following the migration of the female pronucleus to the sperm aster, the adjacent pronuclei move to the center of the zygote. This movement is believed to be due to an elongation of the sperm aster microtubules and concludes with the juxtaposition of the male and female pronuclei.

The importance of these movements is indicated by studies of *in vitro* fertilized human eggs, where failure of pronuclear apposition is not uncommon. Simerly *et al.* (1995) suggest that some cases of infertility may be due to defects in sperm aster formation and mechanisms that are responsible for bringing the parental genomes together.

The involvement of cytoskeletal components, specifically actin and microtubules, have been shown to be involved in pronuclear migration. Latrunculin and cytochalasin B, both inhibitors of microfilaments, block pronuclear migration in the mouse (Schatten and Schatten, 1986; Maro *et al.*, 1984, 1986), and indicate that actin is somehow involved in the apposition of the male and female pronuclei of this species. In oyster and sea urchin zygotes cytochalasin B has no effect on pronuclear migration (Schatten and Schatten, 1986; Longo *et al.*, 1993). In both sea urchins and mammals, the involvement of microtubules in pronuclear migration is indicated by studies in which chemical and physical methods known to destroy microtubular structures also prevent the movements and association of the pronuclei (Zimmerman and Zimmerman, 1967; Longo, 1976c; Schatten, 1984; Schatten *et al.*, 1985b; Maro *et al.*, 1986).

Aspects regulating the movement of the female pronucleus are unknown and it has not been clearly established whether or not they are related to mechanism(s) responsible for male pronuclear migration. In many zygotes, the female pronucleus is formed near where the pronuclei become associated and, therefore, it may undergo relatively little movement. In some species, the female pronucleus is associated with an aster-like structure which may function in the manner hypothesized for the sperm aster (Longo, 1973). Movement of the female pronucleus, however, can apparently occur in the absence of sperm aster. For example, artificially activated sea urchin eggs are distinguished by the migration of the female pronucleus to the center of the egg. In *Lytechinus* eggs there is a radially oriented population of microtubules which is not associated with the sperm aster (Mar, 1980). The microtubules originate in the egg cortex and elongate radially into the center of the egg. Elongation of the radially oriented population of microtubules is associated with the central migration of the male and female pronuclei in fertilized eggs or the female pronucleus in artificially activated ova (Harris, 1979; Harris *et al.*, 1980).

Association of the male and female pronuclei: the concluding events of fertilization

14

The end result of the migration of the male and female pronuclei is their juxtaposition. When adjacent, the pronuclei become associated in a characteristic fashion to establish the genome of the embryo and conclude the process of fertilization (Fig. 14.1).

Early studies of fertilization indicated that factors which govern the association of the males and female pronuclei may be closely but not absolutely correlated with the relation between the meiotic stage of the egg and the time it is normally inseminated. According to Wilson (1925), association of the male and female pronuclei may take essentially two forms.

1. Both the male and female pronuclei may give rise to a group of chromosomes for the ensuing cleavage division. In this form, there is an intermixing of the maternal and paternal chromosomes without fusion of the pronuclei. Eggs demonstrating this series of events are referred to as possessing the *Ascaris* type of fertilization (Wilson, 1925).
2. The pronuclei may fuse (pronuclear fusion) to produce a single zygote nucleus that later undergoes mitosis. Eggs exhibiting such a process are said to have the sea-urchin type of fertilization.

14.1 *ASCARIS* TYPE OF FERTILIZATION

Events associated with the *Ascaris* type of fertilization have been studied in the parasitic intestinal roundworm *Ascaris*, in the mollusks *Spisula*, *Mulinia*, *Crassostrea* and *Mytilus* and in various mammals, including humans (Zamboni, 1971; Longo, 1973; Wright *et al.*, 1990; Sathananthan and Trounson, 1985; Longo and Scarpa, 1991; Longo *et al.*, 1993). Eggs demonstrating this type of pronuclear association are

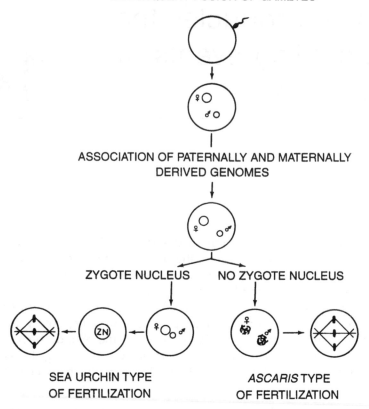

FERTILIZATION INTERACTION AND
SUBSEQUENT FUSION OF GAMETES

ASSOCIATION OF PATERNALLY AND MATERNALLY
DERIVED GENOMES

ZYGOTE NUCLEUS   NO ZYGOTE NUCLEUS

SEA URCHIN TYPE
OF FERTILIZATION

ASCARIS TYPE
OF FERTILIZATION

Fig. 14.1 Schematic representation of the principal ways in which the maternally and paternally derived genomes become associated during fertilization. ZN = zygote nucleus; ♂ = male pronucleus/chromosomes; ♀ = female pronucleus/ chromosomes. Reproduced with permission from Longo, 1973.

inseminated prior to the completion of meiosis, i.e. at meiotic prophase (germinal vesicle stage), metaphase I or metaphase II. Differences have been noted in each of the organisms studied and there is evidence that these variations may be related to the meiotic stage of the egg at insemination (Longo, 1973, 1990).

At the time of their association the male and female pronuclei of zygotes having the *Ascaris* type of fertilization are large spheroids (Figs 14.2, 14.3).

The pronuclei may become closely apposed and form nucleoplasmic projections that may interdigitate. Concomitantly, two asters become

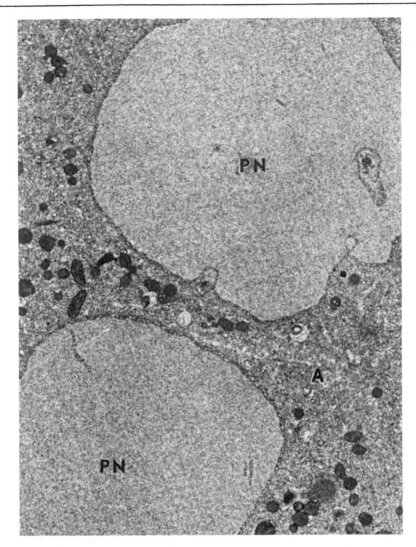

Fig. 14.2 Apposed male and female (PN) pronuclei associated with an aster (A) in a surf clam (*Spisula*) zygote. See Longo and Anderson, 1970b.

situated to either side of the associated pronuclei and establish what will become the poles of the mitotic spindle for first cleavage. During this period condensing chromosomes appear within the male and female pronuclei and the pronuclear envelopes break down (Fig. 14.4).

Ultimately, vesicles and elongated cisternae form that initially outline the condensing chromosomes but are eventually scattered within the

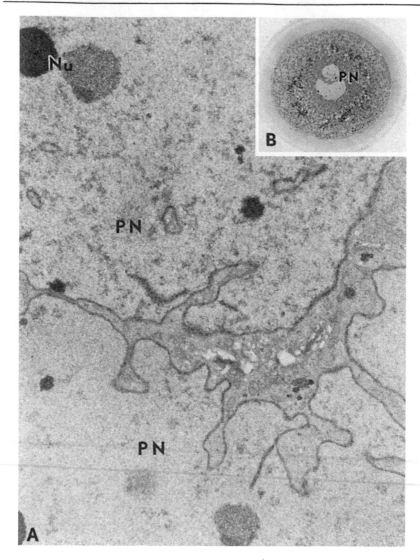

Fig. 14.3 (A) and **(B)**   Associated pronuclei (PN) in a rabbit zygote. Nu = nucleoli. See Longo and Anderson, 1969b.

cytoplasm. Following the disruption of the pronuclear envelopes, the chromosomal groups from each parent intermix on what will constitute the metaphase plate of the first mitotic apparatus. Microtubules become associated with the chromosomes and course to regions of the developing asters (Fig. 14.5).

During this process a zygote nucleus is not formed, i.e. the pronuclei

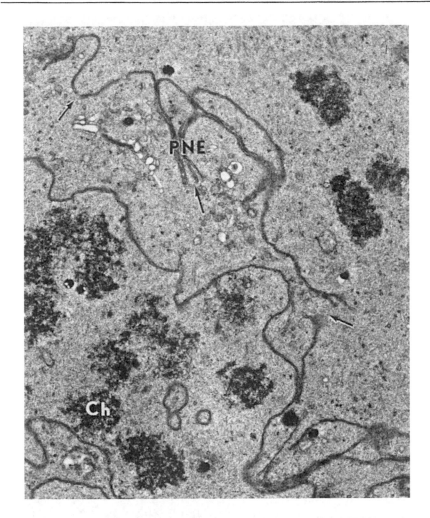

Fig. 14.4 Breakdown (arrows) of the pronuclear envelope (PNE) and the condensation of chromatin (Ch) in a rabbit zygote. See Longo and Anderson, 1969b.

do not fuse. The maternal and paternal genomes are associated within a single nucleus for the first time in the blastomeres of the two-cell stage (Fig. 14.6).

Many of the morphological aspects characteristic of the *Ascaris* type of fertilization are in fact early events normally associated with mitosis, i.e. prophase. Therefore, in these forms the later stages of fertilization also encompass early events of mitosis. Moreover, because eggs known to demonstrate the *Ascaris* type of fertilization are inseminated at an

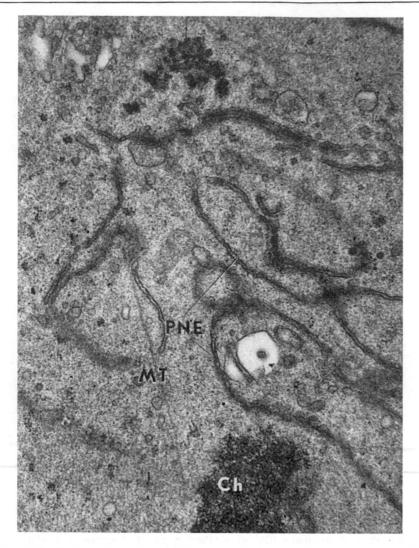

Fig. 14.5 Section through a rabbit zygote during an advanced stage of the breakdown of the male and female pronuclear envelopes (PNE). Ch = chromosomes associated with spindle microtubules (MT).

arrested stage of meiosis, fertilization in these forms involves events associated with three major processes:

1. meiotic maturation of the maternal chromatin;
2. the events of fertilization *per se*, e.g. sperm incorporation and pronuclear development;
3. the early stages of mitosis, i.e. prophase and metaphase.

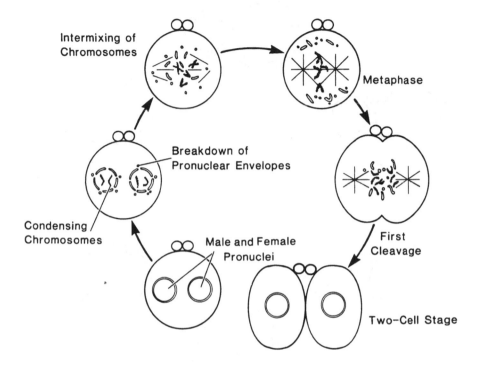

Fig. 14.6 Diagrammatic representation of pronuclear association in eggs undergoing the *Ascaris* type of fertilization.

14.2 SEA-URCHIN TYPE OF FERTILIZATION

The sea-urchin type of fertilization is observed in eggs that are inseminated after meiotic maturation. It involves a fusion of the nuclear envelopes of the male and female pronuclei (pronuclear fusion) and results in the formation of the zygote nucleus (Fig. 14.7).

This process has been studied in greatest detail in *Lytechinus* and in *Arbacia* and was first described ultrastructurally in plants (Jensen, 1964; Wilson, 1925; Longo, 1973). A specific site of contact and fusion does not exist along the periphery of the pronuclei and pronuclear fusion appears to occur randomly along the surfaces of both pronuclei. Following contact of the pronuclei, the outer membranes of both pronuclei are brought into apposition (Fig. 14.8). Subsequently, the inner membranes of the male and female pronuclear envelopes fuse

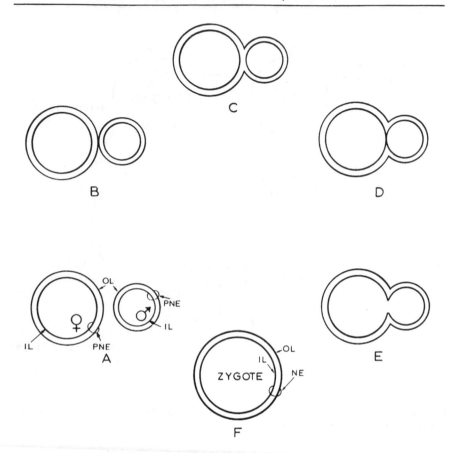

Fig. 14.7 Schematic representation of pronuclear association in eggs under-going the sea-urchin type of fertilization (pronuclear fusion). **(A)** Apposition of the (♂) male and (♀) female pronuclei. **(B)** Contact of the outer laminae (OL) of the male and female pronuclear envelopes (PNE). **(C)** Fusion of the outer laminae and apposition of the inner laminae (IL). **(D)** Contact of the inner laminae of the pronuclear envelopes. **(E)** Fusion of the inner laminae of the pronuclear envelopes. **(F)** Zygote nucleus with a continuous nuclear envelope (NE). Repro-duced with permission from Longo, 1973.

and form an internuclear bridge joining the former pronuclei (Fig. 14.9). How this process is regulated has not been established; however, fusion of somatic cell nuclei *in vitro* appears to be modulated by guanosine triphosphate (Paiement *et al.*, 1980).

Initially, the internuclear bridge connecting the former male and female pronuclei is small; it gradually increases in diameter, eventually yielding a spheroid zygote nucleus with a slight protuberance that

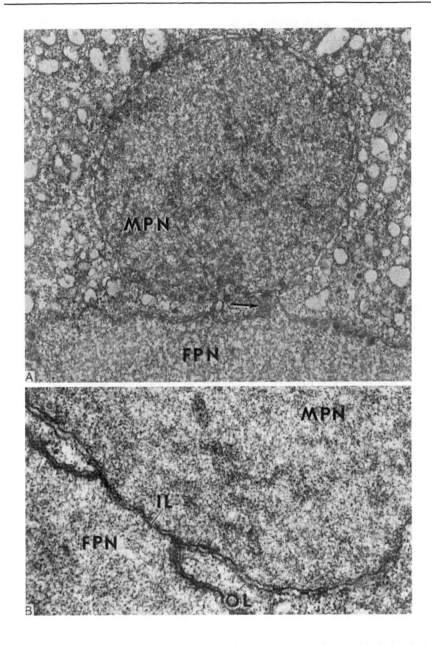

Fig. 14.8   (A) Initial stage in the fusion of the male (MPN) and female (FPN) pronuclei in the sea urchin *Arbacia* (arrow denotes site). See Longo and Anderson, 1968. **(B)** Male and female pronuclei (MPN and FPN) of the sea urchin *Arbacia* following fusion of the outer laminae (OL) of the male and female pronuclear envelope. The inner laminae (IL) of the pronuclear envelopes are parallel to one another. See Longo, 1973.

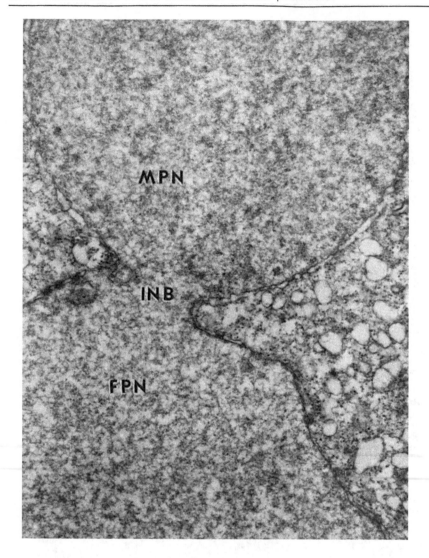

Fig. 14.9 Result of the fusion of the inner laminae of the male and female pronuclear envelopes in a sea urchin (*Arbacia*) zygote. An internuclear bridge (INB) connects the former male (MPN) and female (FPN) pronuclei. See Longo and Anderson, 1968.

represents the former male pronucleus (Fig. 14.10). The paternal chromatin is seen at the fusion site as an electron dense mass; eventually this material diffuses throughout the zygote nucleus and becomes indistinguishable from that of the female (Fig. 14.11).

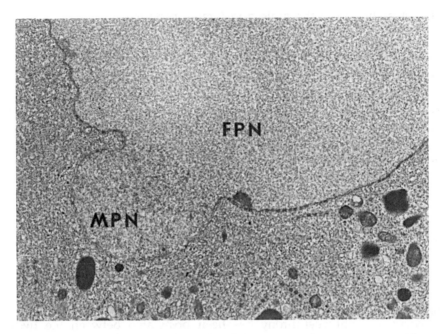

Fig. 14.10 Portion of a zygote nucleus of a fertilized sea urchin (*Arbacia*) egg containing a protuberance that was formerly the male pronucleus (MPN). FPN = former female pronucleus.

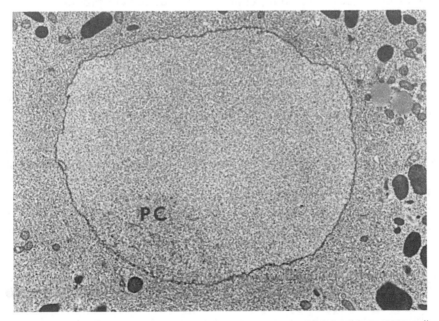

Fig. 14.11 Zygote nucleus of sea urchin *Arbacia* containing dense paternally derived chromatin (PC) at the region which marks the site of pronuclear fusion. Reproduced with permission from Longo and Anderson, 1968.

14.3 COMPARISON OF THE *ASCARIS* AND SEA-URCHIN TYPES OF FERTILIZATION

Differences between the sea-urchin type of fertilization and the *Ascaris* type of fertilization appear to be determined, at least in part, by the time interval between the entrance of the spermatozoon and the association of the pronuclei. By artificially prolonging this period the sea-urchin type of fertilization may take on the character of the *Ascaris* type (Wilson, 1925; Longo and Anderson, 1970c, d). For example, a lengthy delay of pronuclear fusion in *Arbacia* results in an enlargement of the male pronucleus and the acquisition of structural features normally associated with the female pronucleus. In some instances, the male pronucleus may undergo a similar series of events, observed in the mollusks *Spisula* and *Mytilus* and some mammals, i.e. vesiculation of the pronuclear envelopes and chromatin condensation.

In most instances morphogenesis of male and female pronuclei in eggs having the *Ascaris* type of fertilization appears to be synchronous. However, a slight asynchrony has been postulated between the two pronuclei regarding the onset of DNA replication, the acquisition of pronuclear envelopes and chromatin morphogenesis (see Ciemerych and Colowska, 1993). Hybrids made from fusing mouse zygotes or halves of zygotes containing either a male or female pronucleus with ovulated, metaphase II oocytes strongly indicate that organization of the maternal and paternal chromate in the pronucleate zygote is not identical. The basis for this difference has not been established.

In zygotes of humans there is a close association of the male and female pronuclei following their migration (Wright *et al.*, 1990; Zamboni *et al.*, 1966; Sathananthan and Trounson, 1985). Subsequently, the pronuclei undergo nuclear envelope breakdown and chromosome condensation, i.e. events characteristic of the *Ascaris* type of fertilization. Observations by Levron *et al.* (1995) indicate that uninucleate human zygotes from *in vitro* fertilization are relatively common and in the past have been considered the result of parthenogenetic activation or abnormal fertilization. They demonstrate, however, that some of these zygotes may actually come about as a result of early fusion, i.e. the coalescence of maternal and paternal chromosomes within the same pronucleus, or the fusion of male and female pronuclei. These observations are important, as they indicate that some single pronucleate zygotes may have the potential for further normal development and that human eggs may undergo both the sea-urchin and the *Ascaris* types of fertilization.

Subsequent to pronuclear fusion, the paternal chromatin is often seen at one pole of the zygote nucleus as a more condensed mass (Fig. 14.11). Eventually, the paternal chromatin loses this distinctive feature such that recognition of maternal *versus* paternal chromatin in the zygote nucleus becomes impossible. The actual association (intermixing?) of the two chromatin masses in this case has not been explored. A spatial separation of parental chromatin in hybrid embryos formed from hamster eggs and human sperm has been demonstrated (Brandriff and Gordon, 1992; Brandriff *et al.*, 1991). Separation of the two genomes may reflect subtle differences in the morphogenesis of the two pronuclei or molecular modifications of the maternal and paternal genomes related to imprinting (Monk *et al.*, 1987; Surani *et al.*, 1990).

A transient Ca^{2+} increase is occasionally observed just prior to breakdown of the pronuclear envelopes (Tombes *et al.*, 1992) that may be causal to processes characteristic of prophase nuclei (Whitaker and Patel, 1990). Although extracellular Ca^{2+} is not essential for pronuclear development, hamster pronucleate zygotes perish in its absence (Yanagimachi, 1982).

Wilson (1925) indicated that, because of the varying relation between the time of polar body formation and sperm incorporation, the sea-urchin and the *Ascaris* type of fertilization may represent extremes connected by a series of intermediate forms that have been described (Longo, 1973). Why zygotes exhibit a varied series of pronuclear events, modulated in a characteristic fashion, is an unexplained and interesting facet of fertilization. However, aside from the differences and similarities exhibited in fertilized eggs from various organisms, the fact of primary importance is the bringing together of two originally separated lines of heredity, thereby establishing the genetic composition of the embryo.

Manipulation and *in vitro* fertilization of human and other mammalian gametes

15

During the past 20 years there have been numerous developments in methodologies for *in vitro* fertilization (IVF) and the manipulation of human and other mammalian gametes. These developments are based on discoveries by investigators such as Pincus, Chang, Yanagimachi and others (see Chang, 1990 for references) who demonstrated the feasibility of IVF in mammals and did much to further our understanding of fertilization processes and early development.

Barriers to IVF in humans were dismantled in large part with the development of suitable media for culture of sperm and eggs, the initiation of the acrosome reaction *in vitro* and the maturation of fertilizable oocytes (Brinsden and Rainsbury, 1992a). The work of Steptoe and Edwards provided many of the methodological foundations for IVF in humans and the birth of Louise Brown in 1978 is recognized as an important milestone for the treatment of infertility in humans (Steptoe and Edwards, 1978). A number of treatises are available which provide comprehensive analyses of IVF, assisted reproductive procedures and manipulation of human and other mammalian gametes, including those edited by Jones *et al.* (1986), Wood and Trounson (1989), Dunbar and O'Rand (1991), Brinsden and Rainsbury (1992b), and Cohen *et al.* (1992a); see also Bavister *et al.* (1990).

As indicated by Bavister (1990), IVF in mammals presently serves and has the potential of serving a number of purposes including:

1. advancing the breeding of domestic species;
2. the breeding of animals with high intrinsic value, e.g. those bordering on extinction;
3. examination of human and other mammalian species' sperm chromosomal anomalies;

4. the treatment of certain forms of infertility, particularly those in humans.

IVF in domestic animals provides a means of improving domestic stocks and their numbers (see Toyoda and Naito, 1990; Dunbar and O'Rand, 1991). The eggs of many domestic species have been successfully fertilized *in vitro* and methods have been achieved for the culture of IVF bovine, ovine and porcine embryos and their successful implantation after transfer. Future impacts are expected with the adaptation of recombinant DNA technologies, i.e. the introduction of foreign genes into eggs of domestic species (Palmiter *et al.*, 1982; Rexroad and Pursel, 1988).

The application of IVF for species conservation has been advocated as a means of saving endangered species of animals, utilizing field- and zoo-based reproductive research (Wildt, 1990). There are problems in captive breeding of rare species involving behavioral incompatibilities or anomalies, the stress of captivity and limitations of the gene pool. IVF programs may be able to circumvent some of these difficulties and increase populations of endangered species.

Analyses of human sperm chromosome complements have been achieved by examining human sperm chromosomes after penetration of zona-free hamster eggs (Martin, 1990). This method of analysis can provide data on the frequency of chromosomal anomalies in a population, and give insights into the effects of aging, cryopreservation and other environmental factors on the frequency and type of abnormalities studied. This methodology may provide significant advances as prior research in this area has been inferential, relying heavily on studies of human spontaneous abortions and fetuses from prenatal diagnosis and newborns.

IVF has had its most significant application in the amelioration of certain types of infertility. Once techniques were realized to successfully achieve both the fertilization of human eggs *in vitro* and to transplant embryos derived from IVF, they rapidly became part of standard clinical practices with subtle differences from one laboratory to another. The standard methodology for human IVF is outlined in Fig. 15.1 and is discussed below. Additional methods, which in a sense grew out of these standard practices and involve greater manipulation of the gametes, are also described.

In general IVF consists of the following procedures: analysis of infertility and indicators for IVF, stimulation and monitoring for ovum production, oocyte collection, oocyte culture, semen collection and preparation for IVF, IVF, embryo culture and embryo transfer. Indications for IVF, gamete intrafallopian transfer (GIFT) or one of the varieties of assisted reproduction procedures (e.g. pronuclear stage

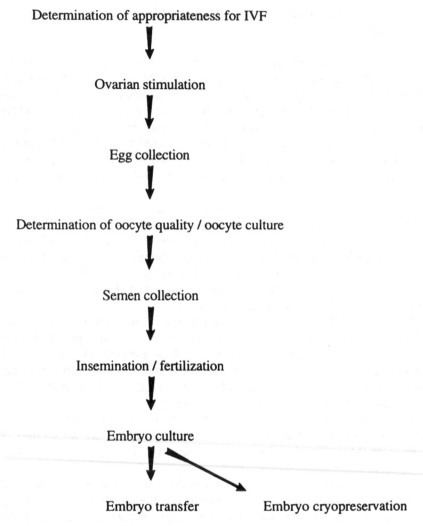

Determination of appropriateness for IVF

Ovarian stimulation

Egg collection

Determination of oocyte quality / oocyte culture

Semen collection

Insemination / fertilization

Embryo culture

Embryo transfer   Embryo cryopreservation

Fig. 15.1 Summary of some of the major events involving IVF and embryo transfer in humans.

tubal transfer (PROST); see Tan and Brinsden, 1992) are dependent on the accessibility of the ovaries and the health and patency of the fallopian tubes (Taylor and Kredentser, 1992). Specific aspects of infertility and categories of infertility that are considered include the following (Jones, 1986):

1. tubal diseases that have failed corrective therapy;
2. male infertility – although absolute criteria for seminal analysis that

characterize infertility remain poorly described; most infertile males have low sperm counts ($< 20 \times 10^6$/ml) and good motility appears to be essential for fertilization;

3. endometriosis is a critical factor and patients with this condition usually have prior endocrinological and/or surgical therapy without achieving pregnancy;
4. diethylstilbestrol exposure *in utero*, resulting in abnormalities of müllerian-duct-derived structures, e.g. the fallopian tubes and endometrial cavity may be deformed;
5. cervical factors – in some cases sperm are immobilized for unknown reasons;
6. immunological infertility – in some cases the patient has antisperm antibodies;
7. anovulation – the patient does not respond to standard methods of ovulation induction;
8. unknown causes – where there is no identifiable cause for infertility.

For the recruitment of eggs, the natural cycle has a number of disadvantages (Rosenwaks and Muasher, 1986). Only one egg is ovulated per cycle and this involves monitoring of the patient's hormone levels. Hence, the time and effort expended is considerable and usually too large for programs with a sizable number of patients. Thus, IVF programs rely on the use of stimulatory medications to recruit and develop multiple follicles. A variety of gonadotropin protocols have been used to recruit and retrieve multiple fertilizable oocytes for IVF (Rosenwaks and Muasher, 1986). Clomiphene citrate alone or in combination with human menopausal gonadotropin are the most commonly used stimulants for induction of multiple follicular development.

Methods for egg retrieval have been described (Acosta *et al.*, 1986; Leeton, 1989; Brinsden, 1992) and include laparoscopy and ultrasound-guided percutaneous follicular puncture. Laparoscopy takes place under anesthesia and concern for potential harmful effects of such agents on retrieved oocytes have been discussed, as there is little information on the effects of anesthetics on eggs after their retrieval (Acosta *et al.*, 1986). Routine laparoscopy is performed from the inferior margin of the umbilicus. Enlarged follicles are punctured, oocytes in cumulus are aspirated and aspirates are examined for oocyte–cumulus complexes. There are essentially three approaches that may be taken with respect to ultrasound-guided percutaneous follicular puncture: transvesical, perurethral or transvaginal. Using one of these approaches in combination with ultrasound scans, the physician guides a needle to an ovary and aspirates enlarged follicles as described for the laparoscopic method. Although follicular size and oocyte maturation are closely coupled, many of the eggs obtained by these methods may be immature. Consequently, oocytes must be cultured *in vitro* in order that they should complete meiotic maturation.

Following their retrieval, eggs in cumulus are evaluated morphologically (Veeck, 1986). Essentially three types of specimens are identified: mature; preovulatory eggs, i.e. immature oocytes; and degenerative/ atretic oocytes. Mature eggs and immature oocytes are cultured in culture medium for 3–30 hours to allow for cytoplasmic and nuclear maturation (Elder and Avery, 1992; Trounson, 1989a).

The volume of semen is measured after liquefication and sperm are analyzed for motility, morphology and concentration (number/ml). Sperm are washed in culture medium containing human fetal cord serum, which initiates the capacitation process (McDowell, 1986). The sperm concentration is then adjusted for insemination, which is usually in the neighborhood of 5×10^4/ml.

For insemination, the sperm are mixed with matured oocytes in culture dishes (Veeck and Maloney, 1986). Fertilization occurs some time during the ensuing 12–18 hours culture period, probably within the first 3 hours. Details of fertilization events are as described in previous chapters. Some 12–18 hours after mixing of the gametes the oocytes are transferred to fresh medium and examined for evidence of fertilization, i.e. the presence of male and female pronuclei. Determination of the number of polar bodies in the perivitelline space is often difficult; consequently it is not used as a consistent criterion for the determination of fertilization. The eggs are examined after the pronuclear stage 24 hours later (40 hours after mixing of the gametes) and at a time just prior to transfer (42–48 hours post mixing). Specimens are expected to be at the two- to eight-cell stage of development by 42–48 hours postinsemination.

Transfer of cleavage stage embryos has been described (Garcia, 1986) and involves the deposition of multiple cleavage stage embryos within a transfer catheter. The catheter is then introduced into the vagina and advanced into the uterine cavity until it reaches the fundus. The contents of the catheter are expelled. The catheter is removed, flushed with medium and both the catheter and the flushing are examined to establish that the embryos have been expelled into the uterus. Multiple embryo transfer increases the probability of establishing a pregnancy. However, it also gives rise to a greater risk of producing multiple pregnancy with associated obstetrical complications. Therefore, most programs transfer approximately three embryos.

The most inefficient step of the IVF–embryo transfer procedure is the embryo transfer itself. For during embryo transfer, and afterwards, all control of the embryo is lost. Furthermore, a multitude of things may happen to the embryos which may cause failure of implantation, including removal of the embryo with the catheter, lack of appropriate contact of the embryo with the endometrium and failure of appropriate stimulation of the corpus luteum from the embryo/implantation site.

Commonly employed variations of the IVF methodology include gamete intrafallopian transfer (GIFT) and pronuclear stage tubal transfer (PROST) (Jansen, 1989; Brinsden, 1992; Brinsden and Asch, 1992; Tan and Brinsden, 1992). These and similar procedures (see Tan and Brinsden, 1992) are often used to treat women whose fallopian tubes are normal. In GIFT the gametes are collected as described above and then transferred to the fallopian tubes. It is believed that because of the increased concentration of sperm and eggs placed in the fallopian tube and the general milieu of the female reproductive tract, there is an increased chance of successful conception. With PROST fertilized eggs at the pronuclear stage are transferred to the fallopian tubes. These methods represent a simplification of the IVF method and the results of GIFT and PROST are consistently as good or better than those of IVF (Trounson, 1989a). It is believed that this difference may be due to the reduced viability of embryos cultured *in vitro*.

In addition to IVF, GIFT and PROST, microsurgical methods have been devised to promote fertilization in humans and in other mammals (Cohen *et al.*, 1992a, b; Kimura and Yanagimachi, 1995). Three methods currently in use include zona drilling, subzonal insertion and direct sperm microinjection (intracytoplasmic sperm injection, ICSI). The three methods are related and are actually derivatives of one another. In zona drilling a slit is made, by micromanipulation in the zona pellucida, which allows ready access of sperm to the perivitelline space and the egg surface. For subzonal insertion, a spermatozoon is injected into the perivitelline space, i.e. just beneath the zona pellucida, so that the sperm may have ready access to and fuse with the egg plasma membrane. Direct sperm microinjection, as its name implies, is a method whereby a sperm carried in a microneedle is injected directly into the egg cytoplasm. All of the above methods have yielded fertilized ova and live births (Palermo *et al.*, 1992, 1993; Kimura and Yanagimachi, 1995).

The efficiency of IVF and other assisted reproductive methods of conception in humans has been discussed (Jones, 1986). As indicated, the efficiency of natural fertilization appears to vary from a low of 12% to a high of approximately 30%. In comparison, Jones (1986) makes the case that IVF has a higher efficiency, as 54% of patients became pregnant by their third attempt and 76% by their sixth. Consideration of how to access and express the results of pregnancy rates of IVF programs have been critically examined (Jones, 1986). Factors affecting the outcome of laboratory-assisted conception in the human have also been the source of extensive examination (see Van Blerkom, 1995).

Apparently viable embryos that are not transferred are cryopreserved for possible transfer at a later date (Trounson, 1989b; Sathananthan *et al.*, 1992; Mayer and Lanzendorf, 1986). Embryo survival rates range from 6–80%. Cryopreserved sperm may be used for IVF where there is difficulty in obtaining proper samples of semen at the time of egg recovery (Mayer and Lanzendorf, 1986). A number of laboratories have had success in the fertilization and cleavage of cryopreserved human eggs (Mayer and Lanzendorf, 1986; Chen, 1989). Advantages of cryopreserved eggs include their preservation for future fertilization, the availability of eggs in the event of ovarian failure and the elimination of the need for cycle synchronization between donor and recipient.

Studies have been carried out on parthenogenetically activated human eggs as an adjunct to IVF in an attempt to facilitate our understanding of the mechanics of activation and processes of early development (Johnson, 1992). In addition, recognition of the parthenogenetic state is necessary when and if it occurs spontaneously during early pregnancy loss. Human oocytes are parthenogenetically activated by acid Tyrode's solution and calcium ionophore, but at rates lower than those of mouse eggs under the same conditions. The reasons for the lower rates of activation are not clear. They may reflect intrinsic qualities of the gametes themselves or variability in the population available for study. That human eggs are insensitive to ethanol, cold temperature, hyaluronidase and pronase indicates an intrinsic resistance to activation. Artificially activated human eggs are capable of forming pronuclei and undergoing at least three rounds of cell division. The possibilities of further development are unknown.

References

Acosta, A. A., Chillik, C. and Lim, H. (1986) Laparoscopic harvest of eggs, in *In Vitro Fertilization – Norfolk* (eds H. W. Jones Jr, G. S. Jones, G. D. Hodgen and Z. Rosenwaks), Williams & Wilkins, Baltimore, MD, pp. 52–69.

Acott, T. S. and Hoskins, D. D. (1981) Bovine sperm forward motility protein: Binding to epididymal spermatozoa. *Biol. Reprod.*, **24**, 234–240.

Acott, T. S., Katz, D. F. and Hoskins, D. D. (1983) Movement characteristics of bovine epididymal spermatozoa: Effects of forward motility protein and epididymal maturation. *Biol. Reprod.*, **29**, 389–399.

Ahuja, K. K. (1982) Fertilization studies in the hamster: The role of cell-surface carbohydrates. *Exper. Cell Res.*, **140**, 353–362.

Aketa, K. (1973) Physiological studies on the sperm surface component responsible for sperm–egg bonding in sea urchin fertilization. *Exper. Cell Res.*, **80**, 439–441.

Aketa, K. and Ohta, T. (1977) When do sperm of the sea urchin, *Pseudocentrotus depressus*, undergo the acrosome reaction at fertilization? *Devel. Biol.*, **61**, 366–372.

Aketa, K. and Onitake, K. (1969) Effect on fertilization of antiserum against sperm-binding from homo- and heterologous sea urchin egg surfaces. *Exper. Cell Res.*, **56**, 84–86.

Aketa, K. and Tsuzuki, H. (1968) Sperm-binding capacity of the S-S reduced protein of the vitelline membrane of the sea urchin egg. *Exper. Cell Res.*, **50**, 675–676.

Albertini, D. F., Wickramasinghe, D., Messinger, S. *et al.* (1993) Nuclear and cytoplasmic changes during oocyte maturation, in *Preimplantation Embryo Development*, (ed. B. D. Bavister), Springer-Verlag, New York, pp. 3–21.

Albertson, D. G. (1984) Formation of the first cleavage spindle in nematode embryos. *Devel. Biol.*, **101**, 61–72.

Alexandre, H., vanCouwenberge, A. and Mulnard, J. (1989) Involvement of microtubules and microfilaments in the control of the nuclear movement during maturation of mouse oocyte. *Devel. Biol.*, **136**, 311–320.

Allen, R. D. (1953) Fertilization and artificial activation in the egg of the surf clam, *Spisula solidissima*. *Biol. Bull.*, **105**, 213–239.

Alliegro, M. C., Black, S. D. and McClay, D. R. (1992) Deployment of extracellular matrix proteins in sea urchin embryogenesis. *Microsc. Res. Tech.*, **22**, 2–10.

Almeida, E. A. C., Huovile, A. P. J., Sutherland, A. E. *et al.* (1995) Mouse egg integrin α6β1 functions as a sperm receptor. *Cell*, **81**, 1095–1104.

Almers, W. (1990) Exocytosis. *Annu. Rev. Physiol.*, **52**, 607–624.

Amano, T., Okita, Y. and Hoshi, M. (1992) Treatment of starfish sperm with egg jelly induces the degradation of histones. *Devel. Growth Differ.*, **34**, 99–106.

Anderson, E. (1968) Oocyte differentiation in the sea urchin, *Arbacia punctulata*, with particular reference to the cortical granules and their participation in the cortical reaction. *J. Cell Biol.*, **37**, 514–539.

Anderson, E. (1974) Comparative aspects of the ultrastructure of the female gamete. *Int. Rev. Cytol. Suppl.*, **4**, 1–70.

Anderson, W. A. and Eckberg, W. R. (1983) A cytological analysis of fertilization in *Chaetopterus pergamentaceus*. *Biol. Bull.*, **165**, 110–118.

Anstrom, J. A., Chin, J. E., Leaf, D. S. *et al.* (1988) Immunocytochemical evidence suggesting heterogeneity in the population of sea urchin cortical granules. *Devel. Biol.*, **125**, 1–7.

Anteunis, A., Fautrez-Firlefyn, N. and Fautrez, J. (1967) L'accolement des pronuclei de l'oeuf d'*Artemia salina*. *J. Ultrastruct. Res.*, **20**, 206–210.

Arezzo, F. (1989) Sea urchin sperm as a vector of foreign genetic information. *Cell Biol. Int. Rep.*, **13**, 391–404.

Austin, C. R. (1951) Observations on the penetration of sperm into the mammalian egg. *Austral. J. Sci. Res. (B)*, **4**, 581–596.

Austin, C. R. (1961) *The Mammalian Egg*, Blackwell Scientific Publications, Oxford.

Austin, C. R. (1965) *Fertilization*, Prentice Hall, Englewood Cliffs, NJ.

Austin, C. R. and Bishop, M. W. H. (1957) Fertilization in mammals. *Biol. Rev.*, **32**, 296–349.

Ayabe, T., Kopf, G. S. and Schultz, R. M. (1995) Regulation of mouse egg activation: Presence of ryanodine receptors and effects of micro-injected ryanodine and cyclic ADP ribose on uninseminated and inseminated eggs. *Development*, **121**, 2233–2244.

Azarnia, R. and Chambers, E. L. (1976) The role of divalent cations in activation of the sea urchin egg. I. Effect of fertilization on divalent cation content. *J. Exper. Zool.*, **198**, 65–78.

Baba, T., Azuma, S. Kashiwabara, and Toyoda, Y. (1994) Sperm from mice carrying a targeted mutation of the acrosin gene can penetrate the oocyte zona pellucida and effect fertilization. *J. Biol. Chem.*, **269**, 31845–31849.

Babcock, D. F., Singh, J. P. and Lardy, H. A. (1979) Alteration of membrane permeability to calcium ions during maturation of bovine spermatozoa. *Devel. Biol.*, **69**, 85–93.

Babcock, R. C., Mundy, C. N. and Whitehead, D. (1994) Sperm diffusion models and *in situ* confirmation of long-distance fertilization in the free-spawning asteroid *Acanthaster planci*. *Biol. Bull.*, **186**, 17–28.

Baccetti, B. and Afzelius, B. (1976) *The Biology of the Sperm Cell*, S. Karger, New York.

Baker, P. F. and Whitaker, M. J. (1978) Influence of ATP and calcium on the cortical reaction in sea urchin eggs. *Nature*, **276**, 513–515.

Baker, P. F. and Whitaker, M. J. (1979) Trifluoroperazine inhibits exocytosis in sea urchin eggs. *J. Physiol.*, **298**, 55.

Balakier, H. and Czolowska, R. (1977) Cytoplasmic control of nuclear maturation in mouse oocytes. *Exper. Cell Res.*, **110**, 466–469.

Balakier, H. and Tarkowski, A. K. (1980) The role of germinal vesicle karyoplasm in the development of male pronucleus in the mouse. *Exper. Cell Res.*, **128**, 79–85.

Barros, C. and Herrera, E. (1977) Ultrastructural observation of the incorporation of guinea pig spermatozoa into zona-free hamster oocytes. *J. Reprod. Fertil.*, **49**, 47–50.

Barros, C., Bedford, J. M., Franklin, L. E. and Austin, C. R. (1967) Membrane vesiculation as a feature of the mammalian acrosome reaction. *J. Cell Biol.*, **34**, C1–C5.

Barry, J. M. and Merriam, R. W. (1972) Swelling of hen erythrocyte nuclei in cytoplasm from *Xenopus* eggs. *Exper. Cell Res.*, **71**, 90–96.

Battaglia, D. and Gaddum-Rosse, P. (1986) The distribution of polymerized actin in the rat egg and its sensitivity to cytochalasin B during fertilization. *J. Exper. Zool.*, **237**, 97–105.

Bavister, B. D. (1990) Applications of IVF technology, in *Fertilization in Mammals* (eds B. D. Bavister, J. Cummins and E. R. S. Roldan), Serono Symposia USA, Norwell, MA, pp. 331–334.

Bavister, B. D., Cummins, J. and Roldan, E. R. S. (eds) (1990) *Fertilization in Mammals*, Serono Symposia USA, Norwell, MA.

Bedford, J. M. (1970) The saga of mammalian sperm from ejaculation to syngamy, in *Mammalian Reproduction*, (eds H. Gibian and E. J. Plotz), Springer-Verlag, New York, pp. 124–182.

Bedford, J. M. and Calvin, H. I. (1974) The occurrence and possible functional significance of -S-S-crosslinks in sperm heads, with particular reference to eutherian mammals. *J. Exper. Zool.*, **188**, 137–156.

Bedford, J. M. and Cooper, G. W. (1978) Membrane fusion events in the fertilization of vertebrate eggs, in *Cell Surface Reviews*, vol. 5 (eds G. Poste and G. L. Nicolson), Elsevier North-Holland, Amsterdam, pp. 65–125.

Bedford, J. M. and Hoskins, D. D. (1990) The mammalian spermatozoa: Morphology, biochemistry and physiology, in *Marshall's Physiology of Reproduction*, vol. 2, (ed. G. E. Lamming), Churchill Livingstone, Edinburgh, pp. 379–568.

Begg, D. A., Rebhun, L. I. and Hyatt, H. (1982) Structural organization of actin in the sea urchin egg cortex: Microvillar elongation in the absence of actin filament bundle formation. *J. Cell Biol.*, **93**, 24–32.

Bell, G. (1988) The origin and early evolution of germ cells as illustrated by the Volvocales, in *The Origin and Evolution of Sex*, (eds H. O. Halvorson and A. Monroy), Alan R. Liss, New York, pp. 221–256.

Bell, E. and Reeder, R. (1967) The effect of fertilization on protein synthesis in the egg of the surf-clam *Spisula solidissima*. *Biochim. Biophys. Acta*, **142**, 500–511.

Bellvé, A. R. and O'Brien, D. A. (1983) The mammalian spermatozoa: Structure and temporal assembly, in *Mechanism and Control of Animal Fertilization*, (ed. J. F. Hartman), Academic Press, New York, pp. 55–137.

Bellvé, A. R., Chandrika, R. C., Martinova, Y. and Barth, A. (1992) The perinuclear matrix as a structural element of the mouse sperm nucleus. *Devel. Biol.*, **47**, 451–465.

Bement, W. M. (1992) Signal transduction by calcium and protein kinase C during egg activation. *J. Exper. Zool.* **263**, 382–397.

Bement, W. M. and Capco, D. G. (1989) Activators of protein kinase C trigger cortical granule exocytosis, cortical contraction, and cleavage furrow formation in *Xenopus laevis* oocytes and eggs. *J. Cell Biol.*, **108**, 885–892.

Bement, W. M. and Capco, D. (1990) Protein kinase C acts downstream of calcium entry into the first mitotic interphase of *Xenopus laevis*. *Cell Regul.*, **1**, 315–326.

Bement, W. M., Gallicano, G. I. and Capco, D. G. (1992) Role of the cytoskeleton during early development. *Microsc. Res. Tech.*, **22**, 23–48.

Benavente, R. and Krohne, G. (1985) Change of karyoskeleton during spermatogenesis of *Xenopus*: Expression of lamin$_{IV}$, a nuclear lamina protein specific for the male germ line. *Proc. Natl Acad. Sci. USA*, **82**, 6176–6180.

Benavente, R., Krohne, G. and Franke, W. W. (1985) Cell type-specific expression of nuclear lamina proteins during development of *Xenopus laevis*. *Cell*, **41**, 177–190.

Bennett, J. and Mazia, D. (1981a) Interspecific fusion of sea urchin eggs. Surface events and cytoplasmic mixing. *Exper. Cell Res.*, **131**, 197–207.

Bennett, J. and Mazia, D. (1981b) Fusion of fertilized and unfertilized sea urchin eggs. Maintenance of cell surface integrity. *Exper. Cell Res.*, **134**, 494–498.

Bentley, J. K., Shimomura, H. and Garbers, D. L. (1986) Retention of a functional resact receptor in isolated sperm plasma membrane. *Cell*, **45**, 281–288.

Bentley, J. K., Khatra, A. S. and Garbers, D. L. (1988) Receptor-mediated activation of detergent-solubilized guanylate cyclase. *Biol. Reprod.*, **39**, 639–647.

Bernal, A., Torres, J., Reyes, A. and Rosado, A. (1980) Presence and regional distribution of sialyl transferase in the epididymis of the rat. *Biol. Reprod.*, **23**, 290–293.

Bernardini, G., Ferraguti, M. and Peres, A. (1986) The decrease of *Xenopus* egg membrane capacity during activation might be due to endocytosis. *Gamete Res.*, **14**, 123–127.

Berridge, M. J. (1993) Inositol trisphosphate and calcium signalling. *Nature*, **361**, 315–325.

Berridge, M. J. and Irvine, R. F. (1984) Inositol triphosphate, a novel second messenger in cellular signal transduction. *Nature*, **312**, 315–321.

Berrios, M. and Bedford, J. M. (1976) Oocyte maturation: Aberrant postfusion responses of the rabbit primary oocyte to penetrating spermatozoa. *J. Cell Sci.*, **39**, 1–12.

Bleil, J. D. and Wassarman, P. M. (1980a) Structure and function of the zona pellucida: Identification and characterization of the proteins of the mouse oocyte's zona pellucida. *Devel. Biol.*, **76**, 185–202.

Bleil, J. and Wassarman, P. (1980b) Mammalian sperm–egg interaction: Identification of a glyprotein in mouse zonae pellucidae possessing receptor activity for sperm. *Cell*, **20**, 873–882.

Bleil, J. and Wassarman, P. (1983) Sperm–egg interactions in the mouse: Sequence of events and induction of the acrosome reaction by a zona pellucida glyprotein. *Devel. Biol.*, **95**, 317–324.

Bleil, J. D. and Wassarman, P. M. (1986) Autoradiographic visualization of the mouse egg's sperm receptor bound to sperm. *J. Cell Biol.*, **102**, 1363–1371.

Bleil, J. D. and Wassarman, P. M. (1988) Galactose at the non-reducing terminus of O-linked oligosaccharides of mouse egg zona pellucida glycoprotein ZP3 is essential for the glycoprotein's sperm receptor activity. *Proc. Natl Acad. Sci. USA*, **85**, 6778–6782.

Bleil, J. D. and Wassarman, P. M. (1990) Identification of a ZP3-binding protein on acrosome-intact mouse sperm by photoaffinity crosslinking. *Proc. Natl Acad. Sci. USA*, **87**, 5563–5567.

Bleil, J. D., Beall, C. F. and Wassarman, P. M. (1981) Mammalian sperm–egg interaction: Fertilization of mouse eggs triggers modification of the major zona pellucida glycoprotein, ZP2. *Devel. Biol.*, **86**, 189–197.

Blobel, C. P., Wolfsberg, T. G., Turck, C. W. *et al.* (1992) A potential fusion peptide and an integrin ligand domain in a protein active in sperm–egg fusion. *Nature*, **356**, 248–252.

Bloch, D. P. (1969) A catalog of sperm histones. *Genetics Suppl.*, **61**, 93–111.

Bloch, D. P. and Hew, H. Y. C. (1960) Changes in nuclear histones during fertilization and early embryonic development in the pulmonate snail, *Helix aspersa*. *J. Biophys. Biochem. Cytol.*, **8**, 69–81.

Bloom, T. L., Szuts, E. Z. and Eckberg, W. R. (1988) Inositol trisphosphate, inositol phospholipid metabolism, and germinal vesicle breakdown in surf clam oocytes. *Devel. Biol.*, **129**, 532–540.

Blow, J. J. and Laskey, R. A. (1988) A role for the nuclear envelope in controlling DNA replication within the cell cycle. *Nature*, **332**, 546–548.

Bogart, J. P., Elinson, R. P. and Light, L. E. (1989) Temperature and sperm incorporation in polyploid salamanders. *Science*, **246**, 1032–1034.

Bonder, E. M., Fishkind, D. J., Cotran, N. M. and Begg, D. A. (1989) Cortical actin-membrane cytoskeleton of unfertilized sea urchin eggs: Analysis of the spatial organization and relationship of filamentous actin, nonfilamentous actin and egg spectrin. *Devel. Biol.*, **134**, 327–341.

Bonnell, B. S., Larabell, C. and Chandler, D. E. (1993) The sea urchin egg jelly coat is a three-dimensional fibrous network as seen by intermediate voltage electron microscopy and deep etching analysis. *Mol. Reprod. Devel.*, **35**, 18–188.

Bonnell, B. S., Keller, S. H., Vacquier, V. D. and Chandler, D. E. (1994) The sea urchin egg jelly coat consists of globular glycoproteins bound to a fibrous fucan superstructure. *Devel. Biol.*, **162**, 313–324.

Bonnell, B. S., Reinhart, D. and Chandler, D. E. (1996) *Xenopus laevis* egg jelly coats consist of small diffusible proteins bound to a complex system of structurally stable networks composed of high-molecular-weight glycoconjugates. *Devel. Biol.*, **174**, 32–42.

Borsuk, E. (1991) Anucleate fragments of parthenogenetic eggs and of maturing oocytes contain complementary factors required for development of a male pronucleus. *Mol. Reprod. Devel.*, **29**, 150–156.

Borsuk, E. and Manka, R. (1988) Behavior of sperm nuclei in intact and bisected metaphase II mouse oocytes fertilized in the presence of colcemid. *Gamete Res.*, **20**, 365–376.

Borsuk, E. and Tarkowski, A. K. (1989) Transformation of sperm nuclei into male pronuclei in nucleate and anucleate fragments of parthenogenetic mouse eggs. *Gamete Res.*, **24**, 471–481.

Brachet, J., Decroly, M. Ficq, A. and Quertier, J. (1963) Ribonucleic acid metabolism in unfertilized and fertilized sea-urchin eggs. *Biochim. Biophys. Acta*, **72**, 660–662.

Brackett, B. G., Baranska, W., Sawicki, W. and Koprowski, H. (1971) Uptake of heterologous genome by mammalian spermatozoa and its transfer to ova through fertilization. *Proc. Natl Acad. Sci. USA*, **68**, 353–357.

Bradley, M. P. and Forrester. I. T. (1980) A [Ca^{2+} + Mg^{2+}]-ATPase and active

Ca^{2+} transport in the plasma membranes isolated from ram sperm flagella. *Cell Calcium*, 1, 381–390.

Brandhorst, B. P. (1976) Two-dimensional gel patterns of protein synthesis before and after fertilization of sea urchin eggs. *Devel. Biol.*, 52, 310–317.

Brandis, J. W. and Raff, R. A. (1978) Translation of oogenetic mRNA in sea urchin eggs and early embryos. Demonstration of a change in translational efficiency following fertilization. *Devel. Biol.*, 67, 99–113.

Brandriff, B. F. and Gordon, L. A. (1992) Spatial distribution of sperm-derived chromatin in zygotes determined by fluorescence *in situ* hybridization. *Mutation Res.*, 296, 33–42.

Brandriff, B. F., Gordon, L. A., Segraves, R. and Pinkel, D. (1991) The male-derived genome after sperm–egg fusion: Spatial distribution of chromosomal DNA and paternal–maternal genomic association. *Chromosoma*, 100, 262–266.

Brinsden, P. R. (1992) Oocyte recovery and embryo transfer techniques for *in vitro* fertilization, in *A Textbook of In Vitro Fertilization and Assisted Reproduction*, (eds P. R. Brinsden and P. A. Rainsbury), Parthenon Publishing Group, Park Ridge, NJ, pp. 139–153.

Brinsden, P. R. and Asch, R. H. (1992) Gamete intra-fallopian transfer, in *A Textbook of In Vitro Fertilization and Assisted Reproduction*, (eds P. R. Brinsden and P. A. Rainsbury), Parthenon Publishing Group, Park Ridge, NJ, pp. 227–236.

Brinsden, P. R. and Rainsbury, P. A. (1992a) *A Textbook of In Vitro Fertilization and Assisted Reproduction*, Parthenon Publishing Group, Park Ridge, NJ.

Brinsden, P. R. and Rainsbury, P. A. (1992b) Introduction in, *A Textbook of In Vitro Fertilization and Assisted Reproduction*, (eds P. R. Brinsden and P. A. Rainsbury), Parthenon Publishing Group, Park Ridge, NJ, pp. 21–26.

Bronson, R. A. and Fusi, F. M. (1990) Evidence that an Arg-Gly-Asp adhesion sequence plays a role in mammalian fertilization. *Biol. Reprod.*, 43, 1019–1025.

Brown, C., von Glos, K. I. and Jones, R. (1983) Changes in plasma membrane glycoproteins of rat spermatozoa during maturation in the epididymis. *J. Cell Biol.*, 96, 256–264.

Brown, D. B., Blake, E. J., Wolgemuth, D. J., Gordon, K. and Ruddle, F. H. (1987) Chromatin decondensation and DNA synthesis in human sperm activated *in vitro* by using *Xenopus laevis* egg extracts. *J. Exper. Zool.*, 242, 215–231.

Browning, H. and Strome, S. (1996) A sperm-supplied factor required for embryogenesis in *C. elegans*. *Development*, 122, 391–404.

Bryan, J. (1970a) The isolation of a major structural element of the sea urchin fertilization membrane. *J. Cell Biol.*, 44, 635–644.

Bryan, J. (1970b) On the reconstitution of the crystalline components of the sea urchin fertilization membrane. *J. Cell Biol.*, 45, 606–614.

Bryan, J. and Kane, R. E. (1982) Actin gelatin in sea urchin egg extracts. *Methods Cell Biol.*, 25, 175–199.

Buck, W. R., Rakow, T. L. and Shen, S. S. (1992) Synergetic release of calcium in sea urchin eggs by caffeine and ryanodine. *Exper. Cell. Res.*, 202, 59–66.

Bullitt, E. S. A., DeRosier, D. J., Culuccio, L. M. and Tilney, L. G. (1988) Three dimensional reconstruction of an actin bundle. *J. Cell Biol.*, 107, 597–611.

Burks, D. J., Carballada, R., Moore, H. D. M. and Saling, P. M. (1995) Interaction of a tyrosine kinase from human sperm with the zona pellucida at fertilization. *Science*, **269**, 83–86.

Burnside, B., Kozak, C. and Kafatos, F. C. (1973) Tubulin determination by an isotope dilution-vinblastine precipitation method. The tubulin content of *Spisula* eggs and embryos. *J. Cell Biol.*, 59, 755–762.

Busa, W. B., Ferguson, J. E., Joseph, S. K. et al. (1985) Activation of frog (*Xenopus laevis*) egg by inositol trisphosphate. I. Characterization of Ca^{2+} release from intracellular stores. *J. Cell Biol.*, **101**, 677–683.

Calarco, P. G., Donahue, R. P. and Szöllösi, D. (1972) Germinal vesicle breakdown in the mouse oocyte. *J. Cell Sci.*, **10**, 369–385.

Calvin, H. I. and Bedford, J. M. (1971) Formation of disulphide bonds in the nucleus and accessory structures of mammalian spermatozoa during maturation in the epididymis. *J. Reprod. Fertil. Suppl.*, **13**, 65–75.

Cameron, L. A. and Poccia, D. L. (1994) *In vitro* development of the sea urchin male pronucleus. *Devel. Biol.*, **162**, 568–578.

Campanella, C. (1975) The site of spermatozoon entrance in the unfertilized egg of *Discoglossus pictus* (Anura): An electron microscope study. *Biol. Reprod.*, **12**, 439–447.

Campanella, C., Andreuccetti, P., Taddei, C. and Talevi, R. (1984) The modifications of cortical endoplasmic reticulum during *in vitro* maturation of *Xenopus laevis* oocytes and its involvement in cortical granule exocytosis. *J. Exper. Zool.*, **229**, 283–293.

Campanella, C., Talevi, R., Kline, D. and Nuccitelli, R. (1988) The cortical reaction in the egg of *Discoglossus pictus*: A study of the changes in the endoplasmic reticulum at activation. *Devel. Biol.*, **130**, 108–119.

Campisi, J. and Scandella, C. J. (1978) Fertilization-induced changes in membrane fluidity of sea urchin eggs. *Science*, **199**, 1336–1337.

Campisi, J. and Scandella, C. J. (1980a) Bulk membrane fluidity increases after fertilization of partial activation of sea urchin eggs. *J. Biol. Chem.*, **255**, 5411–5419.

Campisi. J. and Scandella, C. J. (1980b) Calcium-induced decrease in membrane fluidity of sea urchin egg cortex after fertilization. *Nature*, **286**, 185–186.

Carre, D. and Sardet, C. (1981) Sperm chemotaxis in Siphonophores. *Biol. Cell*, **40**, 119–128.

Carre, D. and Sardet, C. (1984) Fertilization and early development in *Beroe ovata*. *Devel. Biol.*, **105**, 188–195.

Carroll, E. J. and Epel, D. (1975) Isolation and biological activity of the proteases released by sea urchin eggs following fertilization. *Devel. Biol.*, **44**, 22–32.

Carroll, A. G. and Ozaki, H. (1979) Changes in the histones of the sea urchin *Strongylocentrotus purpuratus* at fertilization. *Cell Res.*, **119**, 307–315.

Carron, C. P. and Longo, F. J. (1980) Relation of intracellular pH and pronuclear development in the sea urchin, *Arbacia punctulata*. *Devel. Biol.*, **79**, 478–487.

Carron, C. P. and Longo, F. J. (1982) Relation of cytoplasmic alkanization to microvillar elongation and microfilament formation in the sea urchin egg. *Devel. Biol.*, **89**, 128–137.

Carron, C. P. and Longo, F. J. (1983) Filpin/sterol complexes in fertilized and unfertilized sea urchin membranes. *Devel. Biol.*, **99**, 482–488.

Carron, C. P. and Longo, F. J. (1984) Pinocytosis in fertilized sea urchin (*Arbacia punctulata*) eggs. *J. Exper. Zool.*, **231**, 413–422.

Cartaud, A., Boyer, J. and Ozon, R. (1984) Calcium sequestering activities of reticulum vesicles from *Xenopus laevis* oocytes. *Exper. Cell Res.*, 155, 565–574.

Chamberlin, M. E. and Dean, J. (1990) Human homolog of the mouse sperm receptor. *Proc. Natl Acad. Sci. USA*, 87, 6014–6018.

Chambers, E. L. (1939) The movement of the egg nucleus in relation to the sperm aster in the echinoderm egg. *J. Exper. Biol.*, 16, 409–424.

Chambers, E. L. (1976) Na is essential for activation of the inseminated sea urchin egg. *J. Exper. Zool.*, 197, 149–154.

Chambers, E. L. (1980) Fertilization and cleavage of eggs of the sea urchin *Lytechinus variegatus* in Ca^{2+}-free sea water. *Eur. J. Cell Biol.*, 22, 476.

Chambers, E. L. (1989) Fertilization in voltage-changed sea urchin eggs, in *Mechanisms of Egg Activation*, (eds R. Nuccitelli, G. N. Cherr and W. H. Clark Jr), Plenum Press, New York, pp. 1–18.

Chambers, E. L. and Hinkley, R. E. (1979) Non-propagated cortical reactions induced by the divalent ionophore A-23187 in eggs of the sea urchin, *Lytechinus variegatus*. *Exper. Cell Res.*, 124, 441–446.

Chambers, E. L. and McCulloh, D. H. (1990) Excitation, activation and sperm entry in voltage-clamped sea urchin eggs. *J. Reprod. Fertil. Suppl.*, 42, 117–132.

Chambers, E. L., Pressman, B. C. and Rose, B. (1974) The activation of sea urchin eggs by the divalent ionophores A-23187 and X-537A. *Biochem. Biophys. Res.*, 60, 126–132.

Chandler, D. E. and Heuser, J. (1979) Membrane fusion during secretion: Cortical granule exocytosis in sea urchin eggs as studied by quick-freezing and freeze-fracture. *J. Cell Biol.*, 83, 91–108.

Chandler, D. E. and Heuser, J. (1980) The vitelline layer of the sea urchin egg and its modification during fertilization. *J. Cell Biol.*, 84, 618–632.

Chandler, D. E. and Heuser, J. (1981) Postfertilization growth of microvilli in the sea urchin egg: New views from eggs that have been quick-frozen, freeze-fractured and deeply etched. *Devel. Biol.*, 92, 393–400.

Chang, M. C. (1951) The fertilizing capacity of spermatozoa deposited into the fallopian tubes. *Nature*, 168, 697–698.

Chang, M. C. (1990) Reminiscence on the study of animal reproduction and association with reproductive biologists, in *Fertilization in Mammals*, (eds B. D. Bavister, J. Cummins and E. R. S. Roldan), Serono Symposia USA, Norwell, MA, pp. 383–400.

Channing, C. P., Anderson, L. D. Hoover, D. J. *et al.* (1982) The role of non-steroidal regulators in control of oocyte and follicular maturation. *Rec. Prog. Horm. Res.*, 38, 331–408.

Charbonneau, M. and Grey, R. D. (1984) The onset of activation responsiveness during maturation coincides with the formation of the cortical endoplasmic reticulum in oocytes of *Xenopus laevis*. *Devel. Biol.*, 102, 90–97.

Chaudhary, N. and Courvalin, J.-C. (1993) Stepwise reassembly of the nuclear envelope at the end of mitosis. *J. Cell Biol.*, 122, 295–306.

Chen, C. (1989) Oocyte freezing, in *Clinical In Vitro Fertilization*, (eds C. Wood and A. Trounson), Springer-Verlag, New York, pp. 113–126.

Chen, D. Y. and Longo, F. J. (1983) Sperm nuclear dispersion coordinate with meiotic maturation in fertilized *Spisula solidissma* eggs. *Devel. Biol.*, 99, 217–224.

Chen, H. Y., Brinster, R. L. and Merz, E. A. (1980) Changes in protein synthesis following fertilization of the mouse ovum. *J. Exper. Zool.*, 212, 355–360.

Chen, H. Y., Trumbauer, M. E., Ebert, K. M. *et al.* (1986) Developmental change in the response of mouse eggs to injected genes, in *Molecular Developmental Biology*, (ed. L. Bogorad), Alan R. Liss, New York, pp. 149–159.

Cherr, G. N., Drobnis, E. Z. and Katz, D. F. (1988) Localization of cortical granule constituents before and after exocytosis in the hamster egg. *J. Exper. Zool.*, **246**, 81–93.

Cherr, G. N., Meyers, S. A., Yudin, A. I. *et al.* (1996) The PH-20 protein in cynomolgus macaque spermatozoa: Identification of two different forms exhibiting hyaluronidase activity. *Devel. Biol.*, **175**, 142–153.

Chiba, K., Kado, R. T. and Jaffe, L. A. (1990) Development of calcium release mechanisms during starfish oocyte maturation. *Devel. Biol.*, **146**, 300–306.

Chiba, K., Longo, F. J., Kontani, K., Katada, T. and Hoshi, M. (1995) A periodic network of G proteins βγ subunit coexisting with cytokeratin filament in starfish oocytes. *Devel. Biol.*, **169**, 415–420.

Cicirelli, M. F., Robinson, K. R. and Smith, L. D. (1983) Internal pH of *Xenopus* oocytes: A study of the mechanism and role of pH changes during meiotic maturation. *Devel. Biol.*, 100, 133–146.

Ciemerych, M. A. and Czolowska, R. (1993) Differential chromatin condensation of female and male pronuclei in mouse zygotes. *Mol. Reprod. Devel.*, **34**, 73–80.

Citkowitz, E. (1971) The hyaline layer: Its isolation and role in echinoderm development. *Devel. Biol.*, **24**, 348–362.

Clapham, D. E. (1995) Calcium signaling. *Cell*, **80**, 259–268.

Clapper, D. L. and Lee, H. C. (1985) Inositol trisphosphate induces calcium release from non-mitochondrial stores in sea urchin egg homogenates. *J. Biol. Chem.*, **260**, 13947–13954.

Clapper, D. L., Walseth, T. F., Dargie, P. L. and Lee, H. C. (1987) Pyridine nucleotide metabolites stimulate calcium release from sea urchin egg microsomes desensitized to inositol trisphosphate. *J. Biol. Chem.*, **262**, 9561–9568.

Clark, W. H. Jr and Griffin, F. J. (1988) The morphology and physiology of the acrosome reaction in the sperm of the decapod, *Sicyonia ingentis*. *Develop. Growth Differ.*, **30**, 451–462.

Clark, W. H. Jr, Lynn, J. W., Yudin, A. I. and Persyn, H. O. (1980) Morphology of the cortical reaction in the eggs of *Penaeus aztecus*. *Biol. Bull.*, **158**, 175–186.

Clarke, H. J. and Masui, Y. (1983) The induction of reversible and irreversible chromosome decondensation by protein synthesis inhibition during meiotic maturation of mouse oocytes. *Devel. Biol.*, **97**, 219–301.

Clegg, E. D. (1983) Mechanisms of mammalian sperm capacitation, in *Mechanism and Control of Animal Fertilization*, (ed. J. F. Hartmann), Academic Press, New York, pp. 177–212.

Clegg, K. B. and Denny, P. C. (1974) Synthesis of rabbit globin in a cell-free protein synthesis system utilizing sea urchin egg and zygote ribosomes. *Devel. Biol.*, **37**, 263–272.

Clegg, K. B. and Pikó, L. (1982) RNA synthesis and cytoplasmic polyadenylation in the one-cell mouse embryo. *Nature*, **295**, 342–345.

Cohen, J., Malter, H. E., Talansky, B. E. and Grito, J. (1992a) *Micromanipulation of Human Gametes and Embryos*, Raven Press, New York.

Cohen, J., Malter, H. E. and Talansky, B. E. (1992b) Microsurgical fertilization, in *A Textbook of In Vitro Fertilization and Assisted Reproduction*, (eds P. R.

Brinsden and P. A. Rainsbury), Parthenon Publishing Group, Park Ridge, NJ, pp. 205–226.

Coll, J. C., Bowden, B. F. and Clayton, M. N. (1990) Chemistry and coral reproduction. *Chem. Britain*, **August**, 761–763.

Collas, P. and Poccia, D. (1995a) Formation of the sea urchin male pronucleus *in vitro*: Membrane-independent chromatin decondensation and nuclear envelope-dependent nuclear swelling. *Mol. Reprod. Devel.*, **42**, 106–113.

Collas, P. and Poccia, D. L. (1995b) Lipophilic organizing structures of sperm nuclei target membrane vesicle binding and are incorporated into the nuclear envelope. *Devel. Biol.*, **169**, 123–125.

Collas, P., Pinto-Correia, C. and Poccia, D. (1995) Lamin dynamics during sea urchin male pronucleus formation *in vitro*. *Exper. Cell Res.*, **219**, 687–698.

Collier, J. R. (1979) Nucleic acid chemistry in the *Ilyanassa* embryo. *Am. Zool.*, **16**, 483–500.

Collins, F. (1976) A re-evaluation of the fertilizin hypothesis of sperm agglutination and the description of a novel form of sperm adhesion. *Devel. Biol.*, **49**, 381–394.

Colonna, R. and Tatone, C. (1993) Protein kinase C-dependent and independent events in mouse egg activation. *Zygote*, **1**, 243–256.

Colwin, L. H. and Colwin, A. L. (1967) Membrane fusion in relation to sperm–egg association, in *Fertilization*, vol. 1, (eds C. B. Metz and A. Monroy), Academic Press, New York, pp. 295–367.

Conrad, G. W. and Williams. D. C. (1974a) Polar lobe formation and cytokinesis in fertilized eggs of *Ilyanassa obsoleta*. I. Ultrastructure and effects of cytochalasin B and colchicine. *Devel. Biol.*, **36**, 363–378.

Conrad, G. W. and Williams, D. C. (1974b) Polar lobe formation and cytokinesis in fertilized eggs of *Ilyanassa obsoleta*. II. Large bleb formation caused by high concentrations of exogenous calcium ions. *Devel. Biol.*, **37**, 280–294.

Cook, S. P., Brokaw, C. J., Muller, C. H. and Babcock, D. F. (1994) Sperm chemotaxis: Egg peptides control cytosolic calcium to regulate flagellar responses. *Devel. Biol.*, **165**, 10–19.

Craig, S. P. and Piatigorsky, J. (1971) Protein synthesis and development in the absence of cytoplasmic RNA synthesis in non-nucleate egg fragments and embryos of sea urchins: Effect of ethidium bromide. *Devel. Biol.*, **24**, 213–232.

Cross, N. L. and Elinson, R. P. (1980) A fast block to polyspermy in frogs mediated by changes in the membrane potential. *Devel. Biol.*, **75**, 187–198.

Crossley, I., Whalley, T. and Whitaker, M. J. (1991) Guanosine 5´-thiotriphosphate may stimulate phosphoinositide messenger production in sea urchin eggs by a different route than the fertilizing sperm. *Cell Regul.*, **2**, 121–133.

Crozet, N. (1990) Behavior of the sperm centriole during sheep oocyte fertilization. *Eur. J. Cell Biol.*, **53**, 321–332.

Cummins, J. M. and Yanagimachi, R. (1982) Sperm–egg ratios and the site of the acrosome reaction during *in vitro* fertilization in the hamster. *Gamete Res.*, **5**, 239–256.

Czolowska, R., Modlinski, J. A. and Tarkowski, A. K. (1984) Behavior of thymocyte nuclei in non-activated and activated mouse oocytes. *J. Cell Sci.*, **69**, 19–34.

Daar, I., Yew, N. and VandeWoude, G. F. (1993) Inhibition of *mos*-induced oocyte maturation by protein kinase A. *J. Cell Biol.*, **120**, 1197–1202.

Dacheux, J. L., Dacheux, F. and Paquignon, M. (1989) Changes in sperm surface membrane and luminal protein fluid content during epididymal transit in the boar. *Biol. Reprod.*, **40**, 633–651.

Daentl, D. L. and Epstein, C. J. (1971) Developmental interrelationships of uridine uptake, nucleotide formation and incorporation into RNA by early mammalian embryos. *Devel. Biol.*, **24**, 428–442.

Dale, B. (1988) Primary and secondary messengers in the activation of ascidian eggs. *Exper. Cell Res.*, **177**, 205–211.

Dale, B. and Monroy, A. (1981) How is polyspermy prevented? *Gamete Res.*, **4**, 151–169.

Dale, B., DeFelice, L. J. and Ehrenstein, G. (1985) Injection of a soluble sperm extract into sea urchin eggs triggers the cortical reaction. *Experientia*, **41**, 1068–1070.

Dale, B., Tosti, E. and Iaccarino, M. (1995) Is the plasma membrane of the human oocyte reorganized following fertilization and early cleavage? *Zygote*, **3**, 31–36.

Dan, J. C. (1967) Acrosome reaction and lysins, in *Fertilization*, (eds C. Metz and A. Monroy), Academic Press, New York, pp. 237–293.

Dan K. and Nakajima, T. (1965) On the morphology of the mitotic apparatus isolated from echinoderm eggs. *Embryologia*, **3**, 187–200.

Dangott, L. J., Jordan, J. E., Bellet, R. A. and Garbers, D. L. (1989) Cloning of the mRNA for the protein that crosslinks to the egg peptide speract. *Proc. Natl Acad. Sci. USA*, **86**, 2128–2132.

Danilchik, M. V. and Hille, M. B. (1981) Sea urchin egg and embryo ribosomes: Differences in translational activity in a cell-free system. *Devel. Biol.*, **84**, 291–298.

Danilchik, M. V., Yablonka-Reuveniz, Z., Moon, R. T. *et al.* (1986) Separate ribosomal pools in sea urchin embryos: Ammonia activates a movement between pools. *Biochemistry*, **25**, 3696–3702.

Darszon, A. Guerrero, A., Liévano, A. *et al.* (1988) Ionic channels in sea urchin sperm physiology. *N. Physiol. Sci.*, **3**, 181–185.

Das, N. K., Micou-Eastwood, J. and Alfert, M. (1975) Cytochemical studies on the protamine-type protein transition in sperm nuclei after fertilization and the early embryonic histones of *Urechis caupo*. *Devel. Biol.*, **43**, 333–339.

Davidson, E. H. (1976) *Gene Activity in Early Development*, Academic Press, New York.

Dawid, I. B. and Blackler, A. M. (1972) Maternal and cytoplasmic inheritance of mitochondrial DNA in *Xenopus*. *Devel. Biol.*, **29**, 152–161.

Decker, S. J. and Kinsey, W. H. (1983) Characterization of cortical secretory vesicles from the sea urchin egg. *Devel. Biol.*, **96**, 37–45.

Deguchi, R. and Osanai, K. (1994a) Repetitive intracellular Ca^{2+} increases at fertilization and the role of Ca^{2+} in meiosis reinitiation from the first metaphase in oocytes of marine bivalves. *Devel. Biol.*, **163**, 162–174.

Deguchi, R. and Osanai, K. (1994b) Meiosis reinitiation from first prophase is dependent on the levels of intracellular Ca^{2+} and pH in oocytes of the bivalves, *Mactra chinensis* and *Limaria hakodatensis*. *Devel. Biol.*, **166**, 587–599.

Deguchi, R. and Osanai, K. (1995) Serotonin-induced meiosis reinitiation from first prophase and from first metaphase in oocytes of the marine bivalve *Hiatella flaccida*: Respective changes in intracellular Ca^{2+} and pH. *Devel. Biol.*, **171**, 483–496.

DeLeon, C. V., Cox, K. H., Angerer, L. M. and Angerer, R. C. (1983) Most early variant histone mRNA is contained in the pronucleus of sea urchin eggs. *Devel. Biol.*, **100**, 197–206.

Denny, P. C. and Tyler, A. (1964) Activation of protein biosynthesis in nonnucleate fragments of sea urchin eggs. *Biochem. Biophys. Res. Commun.*, **14**, 245–249.

De Petrocellis, B. and Rossi, M. (1976) Enzymes of DNA biosynthesis in developing sea urchins: Changes in ribonucleotide reductase, thymidine and thymidylate kinase activities. *Devel. Biol.*, **48**, 250–257.

DeSantis, R., Jammuno, G. and Rosati, F. (1980) A study of the chorion and the follicle cells in relation to sperm–egg interaction in the ascidian *Ciona intestinalis*. *Devel. Biol.*, **74**, 490–499.

DeSantis, R., Shirakawa, H., Nakada, K. *et al.* (1992) Evidence that metalloendoproteases are involved in gamete fusion of *Ciona intestinalis*, Ascida. *Devel. Biol.*, **153**, 165–171.

Dessev, G. and Goldman, R. (1988) Meiotic breakdown of nuclear envelope in oocytes of *Spisula solidissima* involves phosphorylation and release of nuclear lamin. *Devel. Biol.*, **130**, 543–550.

Dessev, G., Palazzo, R., Rebhun, L. and Goldman, R. (1989) Disassembly of the nuclear envelope of *Spisula* oocytes in a cell-free system. *Devel. Biol.*, **131**, 496–504.

Dessev, G., Iovcheva-Dessev, C., Bischoff, J. R. *et al.* (1991) A complex containing p34^{cdc2} and cyclin B phosphorylates the nuclear lamin and disassembles nuclei of clam oocytes *in vitro*. *J. Cell Biol.*, **112**, 523–533.

Detering, N. K., Decker, G. L., Schmell, E. D. and Lennarz, W. L. (1977) Isolation and characterization of plasma membrane-associated cortical granules from sea urchin eggs. *J. Cell Biol.*, **75**, 899–914.

Detlaff, T. A., Nikitina, L. A. and Stroeva, O. G. (1964) The role of the germinal vesicle in oocyte maturation in anurans as revealed by the removal and transplantation of nuclei. *J. Embryol. Exper. Morphol.*, **12**, 851–873.

Dewel, W. C. and Clark, W. H. Jr (1974) A fine structural investigation of surface specializations and the cortical reaction in eggs of the cnidarian *Bunodosoma cavernata*. *J. Cell Biol.*, **60**, 78–91.

Ding, J., Moor, R. M. and Foxcroft, G. R. (1992) Effects of protein synthesis on maturation, sperm penetration and pronuclear development in porcine oocytes. *Mol. Reprod. Devel.*, **33**, 59–66.

Dirksen, E. R. (1961) The presence of centrioles in artificially activated sea urchin eggs. *J. Biophys. Biochem. Cytol.*, **11**, 244–252.

Dohmen, M. R. and Van Der Mey, J. C. (1977) Local surface differentiations at the vegetal pole of the eggs of *Nassarius reticulatus, Buccinum undatum* and *Crepidula fornicata* (Gastropoda, Prosobranchia). *Devel. Biol.*, **61**, 104–113.

Donovan, M. and Hart, N. H. (1982) Uptake of ferritin by the mosaic egg surface of *Brachydanio*. *J. Exper. Zool.*, **223**, 299–304.

Dorée, M. and Guerrier, P. (1975) Site of action of 1-methyladenine in inducing oocyte maturation in starfish. *Exper. Cell Res.*, **91**, 296–300.

Downs, S. M. (1995) Control of the resumption of meiotic maturation in mammalian oocytes, in *Cambridge Reviews in Human Reproduction, Vol. 2: Gametes – The Oocyte*, (eds J. L. Yovich, J. L. Simpson and T. Chard), Cambridge University Press, Cambridge.

Doxsey, S. J., Stein, P., Evans, L. *et al.* (1994) Pericentrin, a highly conserved centrosome protein involved in microtubule organization. *Cell*, **76**, 639–650.

Drobnis, E. Z. and Overstreet, J. W. (1992) 1. Natural history of mammalian spermatozoa in the female reproductive tract, in *Oxford Review of Reproductive Biology*, vol. 14, (ed. S. R. Milligan), Oxford University Press, New York, pp. 1–45.

Dubé, F. (1988) The relationship between early ionic events, the pattern of protein synthesis, and oocyte activation in the surf clam, *Spisula solidissima*. *Devel. Biol.*, **126**, 233–241.

Dubé, F. and Coutu, L. (1990) The sodium requirement for activation of surf clam oocytes. *Cell Biol. Int. Rep.*, **14**, 463–471.

Dubé, F. T., Schmidt, T., Johnson, C. H. and Epel, D. (1985) The hierarchy of requirements for an elevated intracellular pH during early development of sea urchin embryos. *Cell*, **40**, 657–666.

Dubé, F., Dufresne, L., Coutu, L. and Clotteau, G. (1991) Protein phosphorylation during activation of surf clam oocytes. *Devel. Biol.* **146**, 473–482.

Ducibella, T. and Buetow, J. (1994) Competence to undergo normal, fertilization-induced cortical activation develops after metaphase I of meiosis in mouse oocytes. *Devel. Biol.*, **165**, 95–104.

Ducibella, T., Anderson, E., Albertini, D. F. *et al.* (1988a) Quantitative studies of changes in cortical granule number and distribution in the mouse oocyte during meiotic maturation. *Devel. Biol.*, **130**, 184–197.

Ducibella, T., Rangarajan, S. and Anderson, E. (1988b) The development of mouse oocyte cortical reaction competence is accompanied by major changes in cortical vesicles and not cortical granule depth. *Devel. Biol.*, **130**, 789–792.

Ducibella, T., Kurasawa, S., Rangarajan, S. *et al.* (1990) Precocious loss of cortical granules during mouse oocyte meiotic maturation and correlation with an egg-induced modification of the zona pellucida. *Devel. Biol.*, **137**, 46–55.

Dunbar, B. S. and O'Rand, M. G. (1991) *A Comparative Overview of Mammalian Fertilization*, Plenum Press, New York.

Eckberg, W. R. (1988) Intracellular signal transduction and amplification mechanisms in the regulation of oocyte maturation. *Biol. Bull.*, **174**, 95–108.

Eckberg, W. R. and Miller, A. L. (1995) Propagated and non-propagated calcium transients during egg activation in the annelid, *Chaetopterus*. *Devel. Biol.*, **172**, 654–664.

Eckberg, W. R., Stutz, E. Z. and Carroll, A. G. (1987) Protein kinase C activity, protein phosphorylation and germinal vesicle breakdown in *Spisula* oocytes. *Devel. Biol.*, **124**, 57–64.

Ecker, R. E. and Smith, L. D. (1971) The nature and fate of *Rana pipiens* proteins synthesized during maturation and early cleavage. *Devel. Biol.*, **24**, 559–576.

Ecklund, P. S. and Levine, L. (1975) Mouse sperm basic nuclear protein, electrophoretic characterization and fate after fertilization. *J. Cell Biol.*, **66**, 251–262.

Eisen, A., Kiehart, D. P., Wieland, S. J. and Reynolds, G. T. (1984) Temporal sequence and spatial distribution of early events of fertilization in single sea urchin eggs. *J. Cell Biol.*, **99**, 1647–1654.

Elder, K. T. and Avery, S. M. (1992) Routine gamete handling: oocyte collection and embryo culture, in *A Textbook of In Vitro Fertilization and Assisted Reproduction*, (eds P. R. Brinsden and P. A. Rainsbury), Parthenon Publishing Group, Park Ridge, NJ, pp. 155–170.

Elinson, R. P. (1980) The amphibian egg cortex in fertilization and early development, in *The Cell Surface: Mediator of Developmental Processes*. (eds S. Subtelny and N. K. Wessels), Academic Press, New York, pp. 217–234.

Endo, Y., Lee, M. A. and Kopf, G. S. (1987) Evidence for the role of a guanine nucleotide-binding regulatory protein in the zona pellucida-induced mouse sperm acrosome reaction. *Devel. Biol.*, **119**, 210–226.

Endo, Y., Lee, M. A. and Kopf, G. S. (1988) Characterization of an islet activating protein sensitive site in mouse sperm that is involved in the zona pellucida-induced acrosome reaction. *Devel. Biol.*, **129**, 12–24.

Eng, L. A. and Metz, C. B. (1980) Sperm head decondensation by a high molecular weight fraction of sea urchin egg homogenates. *J. Exper. Zool.*, **212**, 159–167.

Epel, D. (1967) Protein synthesis in sea urchin eggs: A 'late' response to fertilization. *Proc. Natl Acad. Sci. USA*, **57**, 899–906.

Epel, D. (1969) Does ADP regulate respiration following fertilization of sea urchin eggs? *Exper. Cell Res.*, **58**, 312–319.

Epel, D. (1978) Mechanisms of activation of sperm and egg during fertilization of sea urchin gametes. *Curr. Topics Devel. Biol.*, **12**, 185–246.

Epel, D. (1980) Experimental analysis of the role of intracellular calcium in the activation of the sea urchin egg at fertilization, in *The Cell Surface: Mediator of Developmental Processes*, (eds S. Subtelny and N. K. Wessels), Academic Press, New York, pp. 169–185.

Epel, D. (1989) Arousal of activity in sea urchin eggs at fertilization, in *The Cell Biology of Fertilization*, (eds H. Schatten and G. Schatten), Academic Press, New York, pp. 361–385.

Epel, D., Steinhardt, R. A., Humphreys, T. and Mazia, D. (1974) An analysis of the partial metabolic depression of sea urchin eggs by ammonia: The existence of independent pathways. *Devel. Biol.*, **40**, 245–255.

Epel, D. Cross, N. L. and Epel, N. (1977) Flagellar motility is not involved in the incorporation of the sperm into the egg at fertilization. *Devel. Growth Diff.*, **19**, 15–21.

Epel, D., Patton, C., Wallace, R. W. and Cheung, W. Y. (1981) Calmodulin activates NAD kinase of sea urchin eggs: An early event of fertilization. *Cell*, **23**, 543–549.

Eppig, J. J. and Downs, S. M. (1984) Chemical signals that regulate mammalian oocyte maturation. *Biol. Reprod.*, **30**, 1–11.

Epstein, C. J. and Daentl, D. L. (1971) Precursor, pools and RNA synthesis in preimplantation mouse embryos. *Devel. Biol.*, **26**, 517–524.

Epstein, C. J., Daentl, D. L., Smith, S. A. and Kwok, L. W. (1971) Guanine metabolism in preimplantation mouse embryos. *Biol. Reprod.*, **5**, 308–313.

Evans, J. P., Schultz, R. M. and Kopf, G. S. (1995) Identification and localization of integrin subunits in oocytes and eggs of the mouse. *Mol. Reprod. Devel.*, **40**, 211–220.

Ezzell, R. M. and Szego, C. M. (1979) Luteinizing hormone-accelerated redistribution of lysosome-like organelles preceding dissolution of the nuclear envelope in rat oocytes maturing *in vitro*. *J. Cell Biol.*, **82**, 264–277.

Fankhauser, G. (1948) The organization of the amphibian egg during fertilization and cleavage. *Ann. NY Acad. Sci.*, **49**, 684–708.

Fansler, B. and Loeb, L. A. (1969) Sea urchin nuclear DNA polymerase. II. Changing localization during early development. *Exper. Cell Res.*, **57**, 305–310.

Fansler, B. and Loeb, L. A. (1972) Sea urchin nuclear DNA polymerase. IV. Reversible association of DNA polymerase with nuclei during the cell cycle. *Exper. Cell Res.*, **75**, 433–441.

Fawcett, D. W. and Hollenberg, R. D. (1963) Changes in the acrosome of guinea pig spermatozoa during passage through the epididymis. *Zeitschr. Zellforsch*, **60**, 276–292.

Félix, M.-A., Antony, C., Wright, M. and Maro, B. (1994) Centrosome assembly *in vitro*: Roles of γ-tubulin recruitment in *Xenopus* sperm aster formation. *J. Cell Biol.*, **124**, 19–31.

Ferguson, J. E. and Shen, S. S. (1984) Evidence of phospholipase A2 in the sea urchin egg: Its possible involvement in the cortical granule reaction. *Gamete Res.*, **9**, 329–338.

Feuchter, F. A., Vernon, R. B. and Eddy, E. M. (1981) Analysis of the sperm surface with monoclonal antibodies: Topographically restricted antigens appearing in the epididymis. *Biol. Reprod.*, **24**, 1099–1110.

Firtel, R. A. and Monroy, A. (1970) Polysomes and RNA synthesis during early development of the surf clam *Spisula solidissima*. *Devel. Biol.*, **21**, 87–104.

Fisher, P. A. (1987) Disassembly and reassembly of nuclei in cell-free systems. *Cell*, **48**, 175–176.

Fisher, G. W. and Rebhun, L. I. (1983) Sea urchin egg cortical granule exocytosis is followed by a burst of membrane retrieval via uptake into coated vesicles. *Devel. Biol.*, **99**, 456–472.

Fissore, R. A. and Robl, J. M. (1993) Sperm, inositol trisphosphate, and thimerosal-induced intracellular Ca^{2+} elevations in rabbit eggs. *Devel. Biol.*, **159**, 122–130.

Florman, H. M. (1994) Sequential focal and global elevation of sperm intracellular Ca^{2+} are initiated by the zona pellucida during acrosomal exocytosis. *Devel. Biol.*, **165**, 152–164.

Florman, H. M. and Storey, B. (1982) Mouse gamete interactions: The zona pellucida is the site of the acrosome reaction leading to fertilization *in vitro*. *Devel. Biol.*, **91**, 121–130.

Florman, H. M. and Wassarman, P. M. (1985) O-linked oligosaccharides of mouse egg ZP3 account for its sperm receptor activity. *Cell*, **41**, 313–324.

Florman, H. M., Bechtol, K. B. and Wassarman, P. M. (1984) Enzymatic dissection of the functions of the mouse egg's receptor for sperm. *Devel. Biol.*, **106**, 243–255.

Florman, H. M., Corron, M. E., Kim, T. D.-H. and Babcock, D. F. (1992) Activation of voltage-dependent calcium channels of mammalian sperm is required for zona pellucida-induced acrosomal exocytosis. *Devel. Biol.*, **152**, 304–314.

Focarelli, R. and Rosati, F. (1995) The 220-kDa vitelline coat glycoprotein mediates sperm binding in the polarized egg of *Unio elongatulus* through O-linked oligosaccharides. *Devel. Biol.*, **171**, 606–614.

Foerder, C. A. and Shapiro, B. M. (1977) Release of ovoperoxidase from sea urchin eggs hardens the fertilization membrane with tryosine crosslinks. *Proc. Natl Acad. Sci. USA*, **74**, 4214–4218.

Foerder, D. A., Klebanoff, S. J. and Shapiro, B. M. (1978) Hydrogen peroxide production, chemiluminescence and the respiratory burst of fertilization: Interrelated events in early sea urchin development. *Proc. Natl Acad. Sci. USA*, **75**, 3183–3187.

Foltz, K. R. (1994) The sea urchin egg receptor for sperm. *Semin. Devel. Biol.,* **5,** 243–253.

Foltz, K. R. and Lennarz, W. J. (1990) Purification and characterization of an extracellular fragment of the sea urchin egg receptor for sperm, *J. Cell Biol.,* **11,** 2951–2959.

Foltz, K. R. and Lennarz, W. J. (1992) Identification of the sea urchin receptor for sperm using an antiserum raised against its extracellular domain. *J. Cell Biol.,* **116,** 647–658.

Foltz, K. R. and Lennarz, W. J. (1993) The molecular basis of sea urchin gamete interactions at the egg plasma membrane. *Devel. Biol.,* **158,** 46–61.

Foltz, K. R. and Shilling, F. M. (1993) Receptor-mediated signal transduction and egg activation. *Zygote,* **1,** 276–279.

Foltz, K. R., Partin, J. S. and Lennarz, W. J. (1993) Sea urchin egg receptors for sperm: Sequence similarity of binding domain and hsp 70. *Science,* **259,** 1421–1425.

Franke, W. W. (1974) Structure, biochemistry and functions of the nuclear envelope. *Int. Rev. Cytol. Suppl.,* **4,** 72–236.

Freeman, G. and Miller, R. L. (1982) Hydrozoan eggs can only be fertilized at the site of polar body formation. *Devel. Biol.,* **94,** 142–152.

Friend, D. S. (1980) Freeze-fracture alterations in guinea pig sperm membranes preceding gamete fusion, in *Membrane–Membrane Interactions,* (ed. N. B. Gilula), Raven Press, New York, pp. 153–165.

Friend, D. S., Orci, L., Perrelet, A. and Yanagimachi, R. (1977) Membrane particle changes attending the acrosome reaction in guinea pig spermatozoa. *J. Cell Biol.,* **74,** 561–577.

Frye, L. D. and Edidin, M. (1970) The rapid intermixing of cell surface antigens after formation of mouse–human heterokayons. *J. Cell Sci.,* **7,** 319–335.

Fujiwara, A., Taguchi, K. and Yasumasu, I. (1990) Fertilization membrane formation in sea urchin eggs induced by drugs known to cause Ca^{2+} release from isolated sarcoplasmic reticulum. *Devel. Growth Differ.,* **32,** 303–314.

Fujiwara, T., Nakada, K., Shirakawa, H. and Miyazaki, S. (1993) Development of inositol trisphosphate-induced calcium release mechanism during maturation of hamster oocyte. *Devel. Biol.,* **156,** 69–79.

Fulka, J. Jr, Jung, T. and Moor, R. M. (1992) The fall of biological maturation promoting factor (MPF) and histone H1 kinase activity during anaphase and telophase in mouse oocytes. *Mol. Reprod. Devel.,* **32,** 378–382.

Gabel, C. A., Eddy, E. M. and Shapiro, B. M. (1979) After fertilization, sperm surface components remain as a patch in sea urchin and mouse embryos. *Cell,* **18,** 207–215.

Gaddum-Rosse, P. and Blandau, R. J. (1972) Comparative studies on the proteolysis of fixed gelatin membranes by mammalian sperm acrosomes. *Am. J. Anat.,* **134,** 133–144.

Galione, A., Lee, H. C. and Busa, W. B. (1991) Ca^{2+}-induced Ca^{2+} release in sea urchin egg homogenates: Modulation by cyclic ADP-ribose. *Science,* **253,** 1143–1146.

Galione, A., McDougall, A., Busa, W. B. *et al.* (1993) Redundant mechanisms of calcium-induced calcium release underlying calcium waves during fertilization of sea urchin eggs. *Science,* **261,** 348–352.

Gallicano, G. I., Schwarz, S. M., McGaughey, R. W. and Capco, D. G. (1993)

Protein kinase C, a pivotal regulator of hamster egg activation, functions after elevation of intracellular free calcium. *Devel. Biol.*, **156**, 94–106.

Garanga, S. and Redi, C. A. (1988) Chromatin topology during the transformation of the mouse sperm nucleus into pronucleus *in vivo*. *J. Exper. Zool.*, **246**, 187–193.

Garbers, D. L. and Hardman, J. G. (1976) Effects of egg factors on cyclic nucleotide metabolism in sea urchin sperm. *J. Cyclic Nucleotide Res.*, **2**, 59–70.

Garbers, D. L. and Kopf, G. S. (1980) The regulation of spermatozoa by calcium and cyclic nucleotides. *Adv. Cyclic Nucleotide Res.*, **13**, 251–306.

Garbers, D. L., First, N. L. and Lardy, H. A. (1973a) The stimulation of bovine epididymal sperm metabolism by cyclic nucleotide phosphodiesterase inhibitors. *Biol. Reprod.*, **8**, 589–598.

Garbers, D. L., First, N. L., Gorman, S. K. and Lardy, H. A. (1973b) The effects of cyclic nucleotide phosphodiesterase inhibitors on ejaculated porcine spermatozoan metabolism. *Biol. Reprod.*, **8**, 599–606.

Garcia, J. E. (1986) Conceptus transfer, in *In Vitro Fertilization – Norfolk*, (eds H. W. Jones Jr, G. S. Jones, G. D. Hodgen and Z. Rosenwaks), Williams & Wilkins, Baltimore, MD, pp. 215–220.

Gard, D. L. (1992) Microtubule organization during maturation of *Xenopus* oocytes: Assembly and rotation of the meiotic spindles. *Devel. Biol.*, **151**, 516–530.

Gard, D. L., Cha, B.-J. and Roeder, A. D. (1995) F-actin is required for spindle anchoring and rotation in *Xenopus* oocytes: a re-examination of the effects of cytochalasin B on oocyte maturation. *Zygote*, **3**, 17–26.

Gardiner, D. M. and Grey, R. D. (1983) Membrane junctions in *Xenopus* eggs: Their distribution suggests a role in calcium regulation. *J. Cell. Biol.*, **96**, 1159–1163.

Garner, D. L., Easton, M. P., Munson, M. E. and Doane, M. A. (1975) Immunofluorescent localization of bovine acrosin. *J. Exper. Zool.*, **191**, 127–131.

Gaunt, S. J. (1983) Spreading of a sperm surface antigen within the plasma membrane of the egg after fertilization in the rat. *J. Embryol. Exper. Morphol.*, **75**, 259–270.

Gavin, A. C., Cavadore, J.-C. and Schorderet-Slatkine, S. (1994) Histone H1 kinase activity, germinal vesicle breakdown and M-phase entry in mouse oocytes. *J. Cell Sci.*, **107**, 275–283.

Gerace, L. and Burke, B. (1988) Functional organization of the nuclear envelope. *Ann. Rev. Cell Biol.*, **4**, 335–374.

Gerace, L., Comeau, C. and Benson, M. (1984) Organization and modification of nuclear lamina structure. *J. Cell Sci. Suppl.*, **1**, 137–160.

Gerhart, J. C. and Keller, R. (1986) Region specific cell activities in amphibian gastrulation. *Ann. Rev. Cell Biol.*, **2**, 201–229.

Gerton, G. L. and Hedrick, J. L. (1986) The vitelline envelope to fertilization envelope conversion in eggs of *Xenopus laevis*. *Devel. Biol.*, **116**, 1–7.

Gianaroli, L. Tosti, E., Magli, C. *et al.* (1994) Fertilization current in the human oocyte. *Mol. Reprod. Devel.*, **38**, 209–214.

Gilbert, S. F. (1991) *Developmental Biology*, Sinauer Assoc., Sunderland, MA.

Gilkey, J. C., Jaffe, L. F., Ridgeway, E. G. and Reynolds, G. T. (1978) A free calcium wave traverses the activating egg of the medaka, *Oryzias latipes*. *J. Cell Biol.*, **76**, 448–466.

Gillot, I., Payan, P., Girard, J.-P. and Sardet, C. (1990) Calcium in sea urchin egg during fertilization. *Int. J. Devel. Biol.*, 34, 117–125.

Gillot, I., Ciapa, B., Payan, P. and Sardet, C. (1991) The calcium content of cortical granules and the loss of calcium from sea urchin eggs at fertilization. *Devel. Biol.*, 146, 396–405.

Ginsberg, A. S. (1987) Egg cortical reaction during fertilization and its role in block to polyspermy. *Sov. Sci. Rev. F. Physiol. Gen. Biol.* 1, 307–375.

Girard, J., Payan, P. and Sardet, C. (1982) Changes in intracellular cations following fertilization of sea urchin eggs. Na^+/H^+ and Na^+/K^+ exchanges. *Exper. Cell Res.*, 142, 215–221.

Giudice, G. (1973) *Developmental Biology of the Sea Urchin Embryo*, Academic Press, New York.

Glabe, C. G. (1985a) Interaction of the sperm adhesive protein, bindin, with phospholipid vesicles. I. Specific association of bindin with gel-phase phospholipid vesicles. *J. Cell Biol.*, 100, 794–799.

Glabe, C. G. (1985b) Interaction of the sperm adhesive protein, bindin, with phospholipid vesicles. II. Bindin induces the fusion of mixed phase vesicles that contain phosphatidylcholine and phosphatidyl serine *in vitro*. *J. Cell Biol.*, 100, 800–806.

Glabe, C. G. and Clark, D. (1991) The sequence of the *Arbacia punctulata* bindin cDNA and implication for the structural basis of species-specific sperm adhesion and fertilization. *Devel. Biol.*, 143, 282–288.

Glabe, C. G. and Lennarz, W. J. (1981) Isolation and partial characterization of a high molecular weight egg surface glyco-conjugate implicated in sperm adhesion. *J. Supramol. Struct.*, 15, 387–394.

Glabe, C. G. and Vacquier, V. D. (1978) Egg surface glycoprotein receptor for sea urchin sperm bindin. *Proc. Natl Acad. Sci. USA*, 75, 881–885.

Glass, J. R. and Gerace, L. (1990) Lamins A and C bind and assemble at the surface of mitotic chromosomes. *J. Cell Biol.*, 111, 1047–1057.

Glassner, M., Jones, J-L., Kligman, I. *et al.* (1991) Immunocytochemical and biochemical characterization of guanine nucleotide-binding regulatory proteins in mammalian spermatozoa. *Devel. Biol.*, 146, 438–450.

Golbus, M. S., Calarco, P. G. and Epstein, C. J. (1973) The effects of inhibitors of RNA synthesis (α-amanitin and actinomycin D) on preimplantation mouse embryogenesis. *J. Exper. Zool.*, 186, 207–216.

Goldstein, J. L., Anderson, R. G. W. and Brown, M. S. (1979) Coated pits, coated vesicles and receptor mediated endocytosis. *Nature*, 279, 679–685.

Gong, X., Dubois, D. H., Miller, D. J. and Shur, B. D. (1995) Activation of a G protein complex by aggregation of β-1,4,-galactosyltransferase on the surface of sperm. *Science*, 269, 1718–1721.

Gordon, M., Dandekar, P. V. and Eager, P. R. (1978) Identification of phosphatases on the membranes of guinea pig sperm. *Anat. Rec.*, 191, 123–134.

Goudeau, H. and Goudeau, M. (1989) Electrical responses to fertilization and spontaneous activation in decapod crustacean eggs: Characteristics and role, in *Mechanisms of Egg Activation*, (eds R. Nuccitelli, G. N. Cherr and W. H. Clark Jr), Plenum Press, New York, pp. 61–88.

Goudeau, H., Goudeau, M. and Guibourt, N. (1992) The fertilization potential and associated membrane potential oscillations during the resumption of meiosis in the egg of the ascidian *Phallusia mammillata*. *Devel. Biol.*, 153, 227–241.

Gould, M. and Stephano, J. L. (1989) How do sperm activate eggs in *Urechis* (as well as in polychaetes and molluscs)?, in *Mechanisms of Egg Activation*, (eds R. Nuccitelli, G. N. Cherr and W. H. Clark Jr), Plenum Press, New York, pp. 201–214.

Gould, M. C. and Stephano, J. L. (1991) Peptides from sperm acrosome protein that initiate egg development. *Devel. Biol.*, **146**, 509–518.

Gould, M., Stephano, J. L. and Holland, L. Z. (1986) Isolation of protein from *Urechis* sperm acrosomal granules that binds sperm to eggs and initiates development. *Devel. Biol.*, **117**, 306–318.

Gould-Somero, M., Holland, L. and Paul, M. (1977) Cytochalasin B inhibits sperm penetration into eggs of *Urechis caupo* (Echiura). *Devel. Biol.*, **58**, 11–22.

Graham, C. F. (1966) The regulation of DNA synthesis and mitosis in multi-nucleate frog eggs. *J. Cell Sci.*, **1**, 363–374.

Grainger, J. L., Winkler, M. M., Shen, S. S. and Steinhardt, R. A. (1979) Intracellular pH controls protein synthesis rate in the sea urchin egg and early embryo. *Devel. Biol.*, **68**, 396–406.

Grandin, N. and Charbonneau, M. (1989) Intracellular pH and the increase in protein synthesis accompanying activation of *Xenopus* eggs. *Biol. Cell.*, **63**, 321–330.

Gratwohl, E. K., Kellenberger, E., Lorand, L. and Noll, H. (1991) Storage, ultra-structural targeting and function of toposomes and hyaline in sea urchin embryogenesis. *Mech. Devel.*, **33**, 127–138.

Green, G. R. and Poccia, D. L. (1985) Phosphorylation of sea urchin sperm H1 and H2B histones precedes chromatin decondensation and H1 exchange during pronuclear formation. *Devel. Biol.*, **108**, 235–245.

Green, J. D. and Summers, R. G. (1980) Ultrastructural demonstration of trypsin-like protease in acrosomes of sea urchin sperm. *Science*, **209**, 398–400.

Gross, P. R. (1964) The immediacy of genomic control during early development. *J. Exper. Zool.*, **157**, 21–38.

Gross, P. R. and Cousineau, G. H. (1964) Macromolecule synthesis and the influence of actinomycin on early development. *Exper. Cell Res.*, **33**, 368–395.

Gross, P. R., Kraemer, K. and Malkin, L. I. (1965) Base composition of RNA synthesized during cleavage of the sea urchin embryo. *Biochem. Biophys. Res. Commun.*, **18**, 569–575.

Gross, K. W., Jacobs-Lorena, M., Baglioni, C. and Gross, P. R. (1973) Cell-free translation of material messenger RNA from sea urchin eggs. *Proc. Natl Acad. Sci. USA*, **70**, 2614–2618.

Groudine, M. and Conkin, K. F. (1985) Chromatin structure and *de novo* methylation of sperm DNA: Implications for activation of the paternal genome. *Science*, **228**, 1061–1068.

Gualtieri, R., Campanella, C. and Andreuccetti, P. (1992) Cytochemical Ca^{2+} distribution in activated *Discoglossus pictus* eggs: A gradient in the predetermined site of fertilization. *Devel. Growth Diff.*, **34**, 127–135.

Guerrier, P., Brassart, M., David, C. and Moreau, M. (1986) Sequential control of meiosis reinitiation by pH and Ca^{2+} in oocytes of the Prosobranch Mollusk, *Patella vulgata*. *Devel. Biol.*, **114**, 315–324.

Gulyas, B. J. (1980) Cortical granules of mammalian eggs. *Int. Rev. Cytol.*, **63**, 357–392.

Gulyas, B. J. and Schmell, E. D. (1980) Ovoperoxidase activity in ionophore treated mouse eggs. I. Electron microscopic localization. *Gamete Res.*, **3**, 167–177.

Gundersen, G. G., Gabel, C. A. and Shapiro, B. M. (1982) An intermediate state of fertilization involved in internalization of sperm components. *Devel. Biol.*, **93**, 59–72.

Guo, X., Cooper, K., Hershberger, W. K. and Chew, K. K. (1992a) Genetic consequences of blocking polar body I with cytochalasin B in fertilized eggs of the Pacific oyster, *Crassostrea gigas:* I. Ploidy of resultant embryo. *Biol. Bull.*, **183**, 381–386.

Guo, X., Hershberger, W. K., Cooper, K. and Chew, K. K. (1992b) Genetic consequences of blocking polar body I with cytochalasin B in fertilized eggs of the Pacific oyster, *Crassostrea gigas:* II. Segregation of chromosomes. *Biol. Bull.*, **183**, 387–393.

Gurdon, J. B. (1967) On the origin and persistence of a cytoplasmic state inducing nuclear DNA synthesis in frogs' eggs. *Proc. Natl Acad. Sci. USA*, **58**, 545–552.

Gurdon, J. B. (1976) Injected nuclei in frog eggs: Fate, enlargement and chromatin dispersal. *J. Embryol. Exper. Morphol.*, **36**, 523–540.

Gurdon, J. B. and Woodland, H. R. (1968) The cytoplasmic control of nuclear activity in animal development. *Biol. Rev.*, **43**, 233–267.

Gurdon, J. B., Birnstiel, M. L. and Speight, V. A. (1969) The replication of purified DNA introduced into living egg cytoplasm. *Biochim. Biophys. Acta*, **174**, 614–628.

Gwatkin, R. B. L. (1977) *Fertilization Mechanisms in Man and Mammals*, Plenum Press, New York.

Haccard, O., Sarcevic, B., Lewellyn, A. *et al.* (1993) Induction of metaphase arrest in cleaving *Xenopus* embryos by MAP kinase. *Science*, **262**, 1262–1265.

Hafner, M., Petzelt, C., Nobiling, R. *et al.* (1988) Wave of free calcium at fertilization in the sea urchin egg visualized with fura-2. *Cell Motil. Cytoskel.*, **9**, 271–277.

Hall, H. G. (1978) Hardening of the sea urchin envelope by peroxidase catalyzed phenolic coupling of tryosines. *Cell*, **15**, 343–355.

Hamaguchi, M. S. and Hiramoto, Y. (1978) Protoplasmic movement during polar body formation in starfish oocytes. *Exper. Cell Res.*, **112**, 55–62.

Hamaguchi, M. S., Hamaguchi, Y. and Hiramoto, Y. (1986) Microinjected polystyrene beads move along astral rays in sand dollar eggs. *Devel. Growth Differ.*, **28**, 461–470.

Hamilton, D. W. (1977) The epididymis, in *Frontiers in Reproduction and Fertility Control*, (eds R. O. Greep and M. A. Koblinsky), MIT Press, Cambridge, MA, pp. 411–426.

Hamilton, D. W. (1980) UDP-galactose: N-acetylglucosamine galactosyltranferase in fluids from rat testis and epididymis. *Biol. Reprod.*, **23**, 377–385.

Hamilton, D. W., Wenstrom, J. C. and Baker, J. B. (1986) Membrane glycoproteins from spermatozoa: partial characterization of an integral Mr = ca. 24,000 molecule from the rat spermatozoa that is glycosylated during epididymal maturation. *Biol. Reprod.*, **34**, 925–936.

Hampl, A. and Eppig, J. J. (1995) Translational regulation of the gradual increase in histone H1 kinase activity in maturing mouse oocyte. *Mol. Reprod. Devel.*, **40**, 9–15.

Hansbrough, J. R. and Garbers, D. L. (1981a) Speract. Purification and characterization of a peptide associated with eggs that activates spermatozoa. *J. Biol. Chem.*, **256**, 1447–1452.

Hansbrough, J. R. and Garbers, D. L. (1981b) Sodium-dependent activation of sea urchin spermatozoa by speract and monensin. *J. Biol. Chem.*, **256**, 2235–2241.

Hardy, D. M., Harumi, T. and Garbers, D. L. (1994) Sea urchin sperm receptors for egg peptides. *Semin. Devel. Biol.*, **5**, 217–224.

Harris, P. (1979) A spiral cortical fiber system in fertilized sea urchin eggs. *Devel. Biol.*, **68**, 525–532.

Harris, P., Osborn, M. and Weber, K. (1980) A spiral array of microtubules in the fertilized sea urchin egg cortex examined by indirect immunofluorescence and electron microscopy. *Exper. Cell Res.*, **126**, 227–236.

Hart, N. H. and Donovan, M. E. (1983) Fine structure of the chorion and site of sperm entry into the egg of *Brachydanio*. *J. Exper. Zool.*, **227**, 277–296.

Hart, N. H., Becker, K. A. and Wolenski, J. S. (1992) The sperm entry site during fertilization of the zebrafish egg: Localization of actin. *Mol. Reprod. Devel.* **32**, 217–228.

Hartmann, J. F., Gwatkin, R. B. and Hutchison, C. F. (1972) Early contact interactions between mammalian gametes *in vitro*: Evidence that the vitellus influences adherence between sperm and zona pellucida. *Proc. Natl Acad. Sci. USA*, **69**, 2767–2769.

Harvey, E. B. (1956) *The American Arbacia and Other Sea Urchins*, Princeton University Press, Princeton, NJ.

Hashimoto, H. and Kishimoto, T. (1988) Regulation of meiotic metaphase by a cytoplasmic maturation-promoting factor during mouse oocyte maturation. *Devel. Biol.*, **126**, 242–252.

Heald, R. and McKeon, F. (1990) Mutations of phosphorylation sites in lamin A that prevent nuclear lamina disassembly in mitosis. *Cell*, **61**, 579–589.

Hecht, N. B. (1974) A DNA polymerase isolated from bovine spermatozoa. *J. Reprod. Fertil.*, **41**, 345–354.

Hecht, N. B. and Williams, J. L. (1979) Nuclear and mitochondrial DNA-dependent RNA polymerases in bovine spermatozoa. *J. Reprod. Fertil.*, **57**, 157–165.

Henson, J. H., Begg, D. A., Beaulieu, S. M. *et al.* (1989) A calsequestrin-like protein in the endoplasmic reticulum of the sea urchin: Localization and dynamics in the egg and first cell cycle embryo. *J. Cell Biol.*, **109**, 149–161.

Herlands, L. and Maul, G. G. (1994) Characterization of a major nucleoplasmin-like germinal veside protein which is rapidly phosphorylated before germinal vesicle breakdown in *Spisula solidissima*. *Devel. Biol.*, **161**, 530–537.

Hille, M. B. (1974) Inhibitor of protein synthesis isolated from ribosomes of unfertilized eggs and embryos of sea urchins. *Nature*, **249**, 556–558.

Hille, M. B. and Albers, A. A. (1979) Efficiency of protein synthesis after fertilization of sea urchin eggs. *Nature*, **278**, 469–471.

Hille, M. B., Danilchik, M. V., Colin, A. M. and Moon, R. T. (1985) Translational control in echinoid eggs and early embryos, in *The Cellular and Molecular Biology of Invertebrate Development*, (eds R. H. Sawyer and R. M. Showman), University of South Carolina Press, pp. 91–124.

Hinkley, R. E., Wright, B. D. and Lynn, J. W. (1986) Rapid visual detection of sperm–egg fusion using the DNA-specific fluorochrome Hoechst 33342. *Devel. Biol.*, **118**, 148–154.

Hirai, S., Nagahama, Y., Kishimoto, T. and Kanatani, H. (1982) Cytoplasmic maturity revealed by the structural changes in incorporated spermatozoan during the course of starfish, *Asterina pectinifera* oocyte maturation. *Devel. Growth Differ.*, **23**, 465–478.

Hiramoto, Y. (1962) Microinjection of the live spermatozoa into sea urchin eggs. *Exper. Cell Res.*, **27**, 416–426.

Hoeh, W. R., Blakley, K. H. and Brown, W. M. (1991) Heteroplasmy suggests limited biparental inheritance of *Mytilus* mitochondrial DNA. *Science*, **257**, 1488–1490.

Hoffer, A. P., Shalev, M. and Frisch, D. H. (1981) Ultrastructure and maturational changes in spermatozoa in the epididymis of the pigtail monkey, *Macaca nemestrina*. *J. Androl.*, **2**, 140–146.

Hofmann, A. and Glabe, C. (1994) Bindin, a multifunctional sperm ligand and the evolution of new species. *Semin. Devel. Biol.*, **5**, 233–242.

Höger, T. H., Krohne, G. and Kleinschmidt, J. A. (1991) Interaction of *Xenopus* lamins A and LII with chromatin *in vitro* mediated by a sequence element in the carboxyterminal domain. *Exper. Cell Res.*, **197**, 280–289.

Holland, N. D. (1979) Electron microscopic study of the cortical reaction of an ophiuroid echinoderm. *Tissue Cell*, **11**, 445–455.

Holland, L. Z. and Holland, N. D. (1992) Early development in the lancelet (=Amphioxus) *Branchiostoma floridae* from sperm entry through pronuclear fusion: Presence of vegetal pole plasm and lack of conspicuous ooplasmic segregation. *Biol. Bull.*, **182**, 77–96.

Holmberg, S. R. M. and Johnson, M. H. (1979) Amino acid transport in the unfertilized and fertilized mouse egg. *J. Reprod. Fertil.*, **56**, 223–231.

Homa, S. T. (1995) Calcium and meiotic maturation of the mammalian oocyte. *Mol. Reprod. Devel.*, **40**, 122–134.

Hoodbhoy, T. and Talbot, P. (1994) Mammalian cortical granules: Contents, fate and function. *Mol. Reprod. Devel.*, **39**, 439–448.

Hoshi, M. (1985) Lysins, in *Biology of Fertilization*, vol. 2, (eds C. B. Metz and A. Monroy), Academic Press, New York, pp. 431–462.

Hoshi, M., Numakunai, T. and Sawada, H. (1981) Evidence for participation of sperm proteinases in fertilization of the solitary ascidian, *Halocythia roretzi*: Effects of protease inhibitors. *Devel. Biol.*, **86**, 117–121.

Hoshi, J., Amano, T., Okita, Y. *et al.* (1990) Egg signals for triggering the acrosome reaction in starfish spermatozoa. *J. Reprod. Fertil. (Suppl.)*, **42**, 23–31.

Hoshi, M., Takizawa, S. and Hirohashi, N. (1994) Glycosidases, proteases and ascidian fertilization. *Semin. Devel. Biol.*, **5**, 201–208.

Hoskins, D. D., Brandt. H. and Acott, T. S. (1978) Initiation of sperm motility in the mammalian epididymis. *Fed. Proc.*, **37**, 2534–2542.

Hoskins, D. D., Johnson, D., Brandt, H. and Acott, T. S. (1979) Evidence for a role for a forward motility protein in the epididymal development of sperm motility, in *The Spermatozoon*, (eds D. W. Fawcett and J. M. Bedford), Urban & Schwarzenberg, Baltimore, MD, pp. 43–53.

Hosoya, H., Mabuchi, I. and Sakai, H. (1982) Actin modulating protein in the sea urchin egg. I. Analysis of G-actin-binding proteins by DNase I-affinity chromatography and purification of a 17 000 molecular weight component. *J. Biochem.*, **92**, 1853–1862.

Houk, M. S. and Epel, D. (1974) Protein synthesis during hormonally induced meiotic maturation and fertilization in starfish oocytes. *Devel. Biol.*, **40**, 298–310.

Houliston, E. and Elinson, R. P. (1991) Evidence for the involvement of microtubules, ER, and kinesin in the cortical rotation of fertilized frog eggs. *J. Cell Biol.*, **114**, 1017–1028.

Houliston, E., Carre, D., Johnston, J. A. and Sardet, C. (1993) Axis establishment and microtubule-mediated waves prior to first cleavage in *Beroe ovata*. *Development*, **117**, 75–87.

House, C. R. (1994) Confocal ratio-imaging of intracellular pH in unfertilized mouse oocytes. *Zygote*, **2**, 37–45.

Howlett, S. K. (1985) A set of proteins showing cell cycle dependent modification in the early mouse embryo. *Cell*, **45**, 387–396.

Howlett, S. K. and Bolton, V. N. (1985) Sequence and regulation of morphological and molecular events during the first cell cycle of mouse embryogenesis. *J. Embryol. Exper. Morphol.*, **87**, 175–206.

Huez, G., Marbaix, G., Hubert, E. *et al.* (1974) Role of the polyadenylate segment in the translation of globin messenger RNA in *Xenopus* oocytes. *Proc. Natl Acad. Sci. USA*, **71**, 3143–3146.

Humphreys, W. J. (1967) The fine structure of cortical granules in eggs and gastrulae of *Mytilus edulis*. *J. Ultrastruct. Res.*, **17**, 314–326.

Humphreys, T. (1969) Efficiency of translation of messenger RNA before and after fertilization of sea urchin eggs. *Devel. Biol.*, **20**, 435–458.

Humphreys, T. (1971) Measurements of messenger RNA entering polysomes upon fertilization of sea urchin eggs. *Devel. Biol.*, **26**, 201–208.

Hunter, R. H. F. (1967a) Polyspermic fertilization in pigs during the luteal phase of the estrous cycle. *J. Exper. Zool.*, **165**, 451–460.

Hunter, R. H. F. (1967b) Sperm–egg interactions in the pig: Monospermy, extensive polyspermy and the formation of chromatin aggregates. *J. Anat.*, **122**, 43–59.

Hylander, B. L. and Summers, R. G. (1981) The effect of local anesthetics and ammonia on cortical granule-plasma membrane attachment in the sea urchin egg. *Devel. Biol.*, **86**, 1–11.

Hylander, B. L. and Summers, R. G. (1982a) An ultrastructure immunocytochemical localization of hyalin in the sea urchin egg. *Devel. Biol.*, **93**, 368–380.

Hylander, B. L. and Summers, R. G. (1982b) Observations on the role of the cortical reaction in surface changes at fertilization. *Cell Diff.*, **11**, 267–270.

Hylander, B. L., Anstrom, J. and Summers, R. G. (1981) Premature sperm incorporated into the primary oocyte of the polychaete *Pectinaris*: Male pronuclear formation and oocyte maturation. *Devel. Biol.*, **82**, 382–387.

Hyne, R. V. and Garbers, D. L. (1979) Calcium-dependent increase in adenosine 35'-monophosphate and induction of the acrosome reaction in guinea pig spermatozoa. *Proc. Natl Acad. Sci. USA*, **76**, 5699–5703.

Ikadai, H. and Hoshi, M. (1981) Biochemical studies on the acrosome reaction of the starfish, *Asterias amurensis*: Purification and characterization of the acrosome reaction-inducing substance. *Devel. Growth Differ.*, **23**, 81–88.

Imai, H., Niwa, K. and Iritani, A. (1977) Penetration *in vitro* of zona-free hamster eggs by ejaculated boar spermatozoa. *J. Reprod.*, **51**, 495–497.

Infante, J. P. and Huszagh, V. A. (1985) Synthesis of highly unsaturated phosphatidylcholine in the development of sperm motility: A role for epididymal glycerol-3-phosphoryl-choline. *Mol. Cell Biochem.*, **69**, 3–9.

Infante, J. P. and Nemer, M. (1968) Heterogeneous ribonucleoprotein particles in the cytoplasm of sea urchin embryos. *J. Mol. Biol.*, **32**, 543–565.

Iwamatsu, T. and Chang, M. C. (1972) Sperm penetration *in vitro* of mouse oocyte at various times during maturation. *J. Reprod. Fertil.*, **31**, 237–247.

Iwao, Y., Sakamoto, N., Takahara, K. *et al.* (1993) The egg nucleus regulates the behavior of sperm nuclei as well as cycling of MPF in physiological polyspermic newt eggs. *Devel. Biol.*, **160**, 15–27.

Iwasa, K. H., Ehrenstein, G., DeFelice, L. and Russel, J. T. (1990) High concentrations of inositol 1,4,5-trisphosphate in sea urchin sperm. *Biochem. Biophys. Res. Commun.*, **172**, 932–938.

Jaffe, L. A. (1976) Fast block to polyspermy in sea urchin eggs is electrically mediated. *Nature*, **261**, 68–71.

Jaffe, L. A. and Terasaki, M. (1993) Structural changes of the endoplasmic reticulum of sea urchin eggs during fertilization. *Devel. Biol.*, **156**, 566–573.

Jaffe, L. A., Hagiwara, S. and Kado, R. T. (1978) The time course of cortical vesicle fusion in sea urchin eggs observed as membrane capacitance changes. *Devel. Biol.*, **67**, 243–248.

Jaffe, L. A., Sharp, A. P. and Wolf, D. P. (1983) Absence of an electrical polyspermy block in the mouse. *Devel. Biol.*, **96**, 317–323.

Jaffe, L. F. (1983) Sources of calcium in egg activation. A review and hypothesis. *Devel. Biol.*, **99**, 265–276.

Jansen, R. P. S. (1989) Gamete intra-fallopian transfer, in *Clinical In Vitro Fertilization*, (eds C. Wood and A. Trounson), Springer-Verlag, New York, pp. 63–80.

Jeffrey, W. R. (1984) Pattern formation by ooplasmic segregation in ascidian eggs. *Biol. Bull.*, **166**, 277–298.

Jenkins, N. A., Kaumeyer, J. F., Young, E. M. and Raff, R. A. (1978) A test for masked message: The template activity of messenger ribonucleoprotein particles isolated from sea urchin eggs. *Devel. Biol.*, **63**, 279–298.

Jensen, W. A. (1964) Observations on the fusion of nuclei in plants. *J. Cell Biol.*, **23**, 669–672.

Johnson, M. H. (1992) The parthenogenetic activation and development of human oocytes, in *A Textbook of In Vitro Fertilization and Assisted Reproduction*, (eds P. R. Brinsden and P. A. Rainsbury), Parthenon Publishing Group, Park Ridge, NJ, pp. 361–367.

Johnson, C. H. and Epel, D. (1982) Starfish oocyte maturation and fertilization: Intracellular pH is not involved in activation. *Devel. Biol.*, **92**, 461–469.

Johnson, R. T. and Rao, P. N. (1971) Nucleo-cytoplasmic interactions in the achievement of nuclear synchrony in DNA synthesis and mitosis in multinucleate cells. *Biol. Rev.*, **46**, 97–155.

Johnson, J. D., Epel, D. and Paul, M. (1976) Intracellular pH and activation of sea urchin eggs after fertilization. *Nature*, **262**, 661–664.

Johnston, R. N. and Paul, M. (1977) Calcium influx following fertilization of *Urechis caupo* eggs. *Devel. Biol.*, **57**, 364–374.

Jones, H. W. Jr (1986) Overall results from *in vitro* fertilization, in *In Vitro Fertilization – Norfolk*, (eds H. W. Jones Jr, G. S. Jones, G. D. Hodgen and Z. Rosenwaks), Williams & Wilkins, Baltimore, MD, pp. 288–294.

Jones, H. W. Jr (1989) Indications for *in vitro* fertilization, in *In Vitro Fertilization – Norfolk*, (eds H. W. Jones Jr, G. S. Jones, G. D. Hodgen and Z. Rosenwaks), Williams & Wilkins, Baltimore, MD, pp. 1–7.

Jones, H. W. Jr, Jones, G. S., Hodgen, G. D. and Rosenwaks, Z. (1986) *In Vitro Fertilization – Norfolk*, Williams & Wilkins, Baltimore, MD.

Joshi, H. C., Palacios, M. J., McNamara, L. and Cleveland, D. W. (1992) γ-Tubulin is a centrosomal protein required for cell cycle-dependent microtubule nucleation. *Nature*, **356**, 80–83.

Kadam, A. L., Kadam, P. A. and Koide, S. S. (1990) Calcium requirement for 5-hydroxytryptamine-induced maturation of *Spisula* oocytes. *Invert. Reprod. Devel.*, **18**, 165–168.

Kanatani, H., Sharai, H., Nakanishi, K. and Kurokawa, T. (1969) Isolation and identification of meiosis inducing substance in starfish *Asterias amurensis*. *Nature*, **221**, 273–274.

Kane, R. E. (1973) Hyaline release during normal sea urchin development and its replacement after removal at fertilization. *Exper. Cell Res.*, **81**, 301–311.

Karp, G. C. (1973) Autoradiographic patterns of [³H]-uridine incorporation during the development of the mollusc, *Acmaea scutum. J. Embryol. Exper. Morphol.*, **29**, 15–25.

Karr, T. L. (1991) Intracellular sperm/egg interactions in *Drosophilia*: A three-dimensional structural analysis of a paternal product in the developing egg. *Mech. Devel.*, **34**, 101–111.

Katagiri, C. (1974) A high frequency of fertilization in premature and mature coelomic toad eggs after enzymic removal of vitelline membrane. *J. Embryol. Exper. Morphol.*, **31**, 573–587.

Katagiri, C. and Moriya, M. (1976) Spermatozoon response to the toad egg matured after removal of germinal vesicle. *Devel. Biol.*, **50**, 235–241.

Katayose, H., Matsuda, J. and Yanagimachi, R. (1992) The ability of dehydrated hamster and human sperm nuclei to develop into pronuclei. *Biol. Reprod.*, **47**, 277–284.

Kato, K. H., Washitani-Nemoto, S., Hino, A. and Nemoto, S. (1990) Ultrastructural studies on the behavior of centrioles during meiosis of starfish oocytes. *Devel. Growth Diff.*, **32**, 41–49.

Katsura, S. and Tominaga, A. (1974) Peroxidatic activity of catalase in the cortical granules of sea urchin eggs. *Devel. Biol.*, **40**, 292–297.

Kawahara, H. and Yokosawa, H. (1994) Intracellular calcium mobilization regulates the activity of 26S proteasome during the metaphase-anaphase transition in the ascidian meiotic cell cycle. *Devel. Biol.* **166**, 623–633.

Kay, E. S. and Shapiro, B. M. (1985) The formation of the fertilization membrane of the sea urchin egg. *Biol. Fertil.*, **3**, 45–80.

Kay, E. S. and Shapiro, B. M. (1987) Ovoperoxidase localization effects dityrosine crosslinking of specific polypeptides after assembly of the sea urchin fertilization envelope. *Devel. Biol.*, **121**, 325–334.

Keller, S. H. and Vacquier, V. D. (1994) Isolation of acrosome reaction inducing glycoproteins from sea urchin egg jelly. *Devel. Biol.*, **162**, 304–312.

Kidder, G. M. (1976) RNA synthesis and the ribosomal cistrons in early molluscan development. *Am. Zool.*, **16**, 501–520.

Kimura, Y. and Yanagimachi, R. (1995) Intracytoplasmic sperm injection in the mouse. *Biol. Reprod.*, **52**, 709–720.

Kinsey, W. H. (1995) Protein tyrosine kinase activity during egg activation is important for morphogenesis at gastrulation in the sea urchin embryo. *Devel. Biol.*, **172**, 704–707.

Kinsey, W. H. (1996) Biphase activation of *fyn* kinase upon fertilization of the sea urchin egg. *Devel. Biol.*, **174**, 281–287.

Kinsey, W. H., Rubin, J. A. and Lennarz, W. J. (1980) Studies on the specificity of sperm binding in echinoderm fertilization. *Devel. Biol.*, **74**, 245–250.

Kirschner, M., Gerhart, J. C., Hara, K. and Ubbels, G. A. (1980) Initiation of the cell cycle and establishment of bilateral symmetry in *Xenopus* eggs, in *The Cell Surface: Mediator of Developmental Processes*, (eds S. Subtelny and N. K. Wessels), Academic Press, New York, pp. 187–215.

Kishimoto, T. (1988) Regulation of metaphase by a maturation-promoting factor. *Devel. Growth Differ.*, **30**, 105–115.

Kishimoto, T., Hirai, S. and Kanatani, H. (1981) Role of germinal vesicle material in producing maturation-promoting factor in starfish oocyte. *Devel. Biol.*, **81**, 177–181.

Kishimoto, T., Yamazaki, K., Kato, Y., Koide, S. S. and Kanatani, H. (1984) Induction of starfish oocyte maturation by maturation-promoting factor of mouse and surf clam oocytes. *J. Exper. Zool.*, **231**, 293–295.

Kline, D. (1988) Calcium dependent events at fertilization of the frog egg: Injection of a calcium buffer blocks ion channel opening, exocytosis and formation of pronuclei. *Devel. Biol.*, **126**, 346–361.

Kline, D. and Kline, J. T. (1992) Repetitive calcium transients and the role of calcium in exocytosis and cell cycle activation in the mouse egg. *Devel. Biol.*, **149**, 80–89.

Kline, D. and Stewart-Savage, J. (1984) The timing of cortical granule fusion, contents dispersal and endocytosis during fertilization of the hamster egg: An electrophysiological and histochemical study. *Devel. Biol.*, **162**, 277–287.

Kline, D., Simoncini, L., Mandel, G. *et al.* (1988) Fertilization events induced by neurotransmitters after injection of mRNA in *Xenopus* eggs. *Science*, **241**, 464–467.

Kline, D., Kado, R. T., Kopf, G. S. and Jaffe, L. A. (1990) Receptors, G-proteins, and activation of the amphibian egg, in *Mechanisms of Fertilization*, (ed. B. Dale), NATO ASI Series, vol. A45, Springer-Verlag, Berlin, pp. 529–541.

Kline, D., Kopf, G. S., Muncy, L. F. and Jaffe, L. A. (1991) Evidence for the involvement of a pertussis toxin-insensitive G-protein in egg activation by the frog, *Xenopus laevis*. *Devel. Biol.*, **143**, 218–229.

Kline, J. T. and Kline, D. (1994) Regulation of intracellular calcium in the mouse egg: Evidence for inositol trisphosphate-induced calcium release, but not calcium-induced calcium release. *Biol. Reprod.*, **50**, 193–203.

Klotz, C., Dabauvalle, M.-C., Paintrand, M. *et al.* (1990) Parthenogenesis in *Xenopus* eggs requires centrosomal integrity. *J. Cell Biol.*, **110**, 405–415.

Kobrinsky, E., Ondrias, K. and Marks, A. R. (1995) Expressed ryanodine receptor can substitute for the inositol 1,4,5-trisphosphate receptor in *Xenopus laevis* oocytes during progesterone-induced maturation. *Devel. Biol.*, **172**, 531–540.

Koch, R. A., Johnson, J. S. and Lambert, C. C. (1993) Structure of the ascidian vitelline coat and its role in fertilization. *J. Reprod. Devel.*, **39** (Suppl.), 35–36.

Koehler, J. K. (1978) The mammalian sperm surface: Studies with specific labeling techniques. *Int. Rev. Cytol.*, **54**, 73–108.

Koehler, J. K. and Gaddum-Rosse, P. (1975) Media induced alterations of the membrane associated particles of the guinea pig sperm tail. *J. Ultrastuct. Res.*, **51**, 106–118.

Koehler, J. K., Wurschmidt, U. and Larsen, M. P. (1983) Nuclear and chromatin structure in rat spermatozoa. *Gamete Res.*, **8**, 357–370.

Kopecny, V. and Pavlok, A. (1984) Association of newly synthetized ^{3}H-arginine-labeled proteins with chromatin structure at fertilization. *Gamete Res.*, **9**, 399–408.

Kopf, G. S. and Garbers, D. L. (1980) Calcium and fucose-sulfate rich polymer regulate sperm cyclic nucleotide metabolism and the acrosome reaction. *Biol. Reprod.*, **22**, 1118–1126.

Kopf, G. S. and Gerton, G. L. (1991) The mammalian sperm acrosome and the acrosome reaction, in *Elements of Mammalian Fertilization*, vol. 1, (ed. P. M. Wassarman), CRC Press, Boca Raton, FL, pp. 153–203.

Kosower, N. S., Katayose, H. and Yanagimachi, R. (1992) Thiol-disulfide status and acridine orange fluorescence of mammalian sperm nuclei. *J. Androl.*, **13**, 342–348.

Krohne, G. K. and Franke, W. W. (1980) Immunological identification and localization of the predominant nuclear protein of the amphibian oocyte nucleus. *Proc. Natl Acad. Sci. USA*, **77**, 1034–1038.

Kubiak, J. Z., Weber, M., dePennart, H. *et al.* (1993) The metaphase II arrest in mouse oocytes is controlled through microtubule-dependent destruction of cyclin B in the presence of CSF. *EMBO J*, **12**, 3773–3778.

Kume, S., Muto, A., Aruga, J. *et al.* (1993) The *Xenopus* IP3 receptor: Structure, function and localization in oocytes and eggs. *Cell*, **73**, 555–570.

Kunkle, M., Longo, F. J. and Magun, B. E. (1978a) Nuclear protein changes in the maternally and paternally derived chromatin at fertilization. *J. Exper. Zool.*, **203**, 371–380.

Kunkle, M., Magun, B. E. and Longo, F. J. (1978b) Analysis of isolated sea urchin nuclei incubated in egg cytosol. *J. Exper. Zool.*, **203**, 381–390.

Kupitz, Y. and Atlas, D. (1993) A putative ATP-activated Na$^+$ channel involved in sperm-induced fertilization. *Science*, **261**, 484–486.

Kuriyama, R., Borisy, G. G. and Masui, Y. (1986) Microtubule cycles in oocytes of the surf clam, *Spisula solidissima*: An immunofluorescence study. *Devel. Biol.*, **114**, 115–160.

Kyozuka, K. and Osanai, K. (1994a) Functions of the egg envelope of *Mytilus edulis* during fertilization. *Bull. Mar. Biol. Stat. Asamushi.* **19**, 79–92.

Kyozuka, K. and Osanai, K. (1994b) Cytochalasin B does not block sperm penetration into denuded starfish oocytes. *Zygote*, **2**, 103–109.

Labbe, J. C., Picard, A., Peaucellier, G. *et al.* (1989) Purification of MPF from starfish: Identification as the H1 histone kinase p34^{cdc2} and a possible mechanism for its periodic activation. *Cell*, **57**, 253–263.

Lambert, C. C. and Lambert, G. (1981) The ascidian sperm reaction: Ca^{2+} uptake in relation to H$^+$ efflux. *Devel. Biol.*, **88**, 312–317.

Lanzendorf, S. E., Slusser, J., Malloney, M. K. *et al.* (1988) A preclinical evaluation of pronuclear formation by microinjection of human spermatozoa into human oocytes. *Fertil. Steril.*, **49**, 835–842.

Larabell, C. A. and Chandler, D. E. (1988a) The extracellular matrix of *Xenopus laevis* eggs: A quick-freeze, deep-etch analysis of its modification at fertilization. *J. Cell Biol.*, **107**, 731–741.

Larabell, C. A. and Chandler, D. E. (1988b) *In vitro* formation of the 'S' layer, a unique component of the fertilization envelope in *Xenopus laevis* eggs. *Devel. Biol.*, **130**, 356–364.

Larabell, C. A. and Chandler, D. E. (1991) Fertilization induced changes in the vitelline layer of echinoderm and amphibian eggs: self-assembly of an extracellular matrix. *J. Elect. Microsc. Tech.*, **17**, 294–318.

Larabell, C. A. and Nuccitelli, R. (1992) Inositol lipid hydrolysis contributes to the Ca^{2+} wave in the activating egg of *Xenopus laevis*. *Devel. Biol.*, **153**, 347–355.

Lasalle, B. and Testart, J. (1991) Sequential transformation of human sperm nucleus in human egg. *J. Reprod. Fertil.*, **91**, 393–402.

Laskey, R. A. and Gurdon, J. B. (1973) Induction of polyoma DNA synthesis by injection into frog egg cytoplasm. *Eur. J. Biochem.*, **37**, 467–471.

Laskey, R. A., Honda, B. M., Mills, A. D. and Finch, J. T. (1978) Nucleosomes are assembled by an acidic protein which binds histones and transfers them to DNA. *Nature*, **275**, 416–420.

Lathan, K. E., Garrels, J. I., Chan, C. and Solter, D. (1991) Quantitative analysis of protein synthesis in mouse embryos. I. Extensive reprogramming at the one- and two-cell stages. *Development*, **112**, 921–932.

Lavitrano, M., Camaioni, A., Fazio, V. M. *et al.* (1989) Sperm cells as vectors for introducing foreign DNA into eggs: Genetic transformation of mice. *Cell*, **57**, 717–723.

Lechleiter, J. D. and Clapham, D. E. (1992) Molecular mechanisms of intracellular calcium excitability in *X. laevis* oocytes. *Cell*, **69**, 283–294.

Lechleiter, J. D., Girard, S., Peralta, E. and Clapham, D. E. (1991) Spiral calcium wave propagation and annihilation in *Xenopus laevis* oocytes. *Science*, **252**, 123–126.

Lee, H. C. (1991) Specific binding of cyclic ADP-ribose to calcium-storing microsomes from sea urchin eggs. *J. Biol. Chem.*, **266**, 2276–2281.

Lee, H. C. and Steinhardt, R. A. (1981) pH changes associated with meiotic maturation in oocytes of *Xenopus laevis*. *Devel. Biol.*, **85**, 358–369.

Lee, H. C., Johnson, C. and Epel, D. (1983) Changes in internal pH associated with initiation of motility and acrosome reaction of sea urchin sperm. *Devel. Biol.*, **95**, 31–45.

Lee, H. C., Walseth, T. F., Bratt, G. T. *et al.* (1989) Structural demonstration of a cyclic metabolite of NAD^+ with intracellular Ca^{2+}-mobilizing activity. *J. Biol. Chem.*, **264**, 1608–1615.

Lee, H. C., Aarhus, R. and Walseth, T. F. (1993) Calcium mobilization by dual receptors during fertilization of sea urchin eggs. *Science*, **261**, 352–355.

Leeton, J. (1989) Oocyte pick-up, in *Clinical In Vitro Fertilization*, (eds C. Wood and A. Trounson), Springer-Verlag, New York, pp. 23–31.

Le Guen, P. and Crozet, N. (1989) Microtubule and centrosome distribution during sheep fertilization. *Eur. J. Cell Biol.*, **48**, 239–249.

Le Lannou, D., Colleu, D., Boujard, D. and Segalin, J. (1985) Stabilization of nuclear chromatin in human spermatozoa during capacitation *in vitro*, in *Human In Vitro Fertilization*, (eds J. Testan and R. Frydman), Elsevier, Amsterdam, pp. 149–152.

Leno, G. H. and Laskey, R. A. (1991) The nuclear membrane determines the timing of DNA replication in *Xenopus* egg extracts. *J. Cell Biol.*, **112**, 557–566.

Leno, G. H., Downes, C. S. and Laskey, R. A. (1992) The nuclear membrane prevents replication of human G2 nuclei but not G1 nuclei in *Xenopus* egg extract. *Cell*, **69**, 151–158.

Lessman, C. A. and Huver, C. W. (1981) Quantification of fertilization induced gamete changes and sperm entry without egg activation in a teleost egg. *Devel. Biol.*, **84**, 218–224.

Levron, J., Munné, S., Willadsen, S. *et al.* (1995) Male and female genomes associated in a single pronucleus in human zygotes. *Biol. Reprod.*, **52**, 653–657.

Leyton, L. and Saling, P. (1989a) 95kd sperm proteins bind ZP3 and serve as tyrosine kinase substrates in response to zona binding. *Cell*, **57**, 1123–1130.

Leyton, L. and Saling, P. (1989b) Evidence that aggregation of mouse sperm receptors by ZP3 triggers the acrosome reaction. *J. Cell Biol.*, **108**, 2163–2168.

Leyton, L., LeGuen, P., Bunch, D. and Saling, P. M. (1992) Regulation of mouse gamete interaction by a sperm tyrosine kinase. *Proc. Natl Acad. Sci. USA*, **89**, 11692–11695.

Lillie, F. R. (1919) *Problems of Fertilization*, University of Chicago Press, Chicago, IL.

Lindsay, L. L., Hertzler, P. L. and Clark, W. H. Jr (1992) Extracellular Mg^{2+} induces an intracellular Ca^{2+} wave during oocyte activation in the marine shrimp *Sicyonia ingentis*. *Devel. Biol.*, **152**, 94–102.

Loeb, L. A. and Fansler, B. (1970) Sea urchin DNA polymerase. III. Intracellular migration of DNA polymerase in early developing sea urchin embryos. *Biochim. Biophys. Acta*, **217**, 50–55.

Loeb, L. A., Fansler, B., Williams, R. and Mazia, D. (1969) Sea urchin DNA polymerase. I. Localization in nuclei during rapid DNA synthesis. *Exper. Cell Res.*, **57**, 298–304.

Lohka, M. J. (1988) The reconstitution of nuclear envelopes in cell-free extracts. *Cell Biol. Internat. Rep.*, **12**, 833–848.

Lohka, M. J. and Maller, J. L. (1985) Induction of nuclear envelope breakdown, chromosome condensation and spindle formation in cell-free extracts. *J. Cell Biol.*, **101**, 518–526.

Lohka, M. J. and Maller, J. L. (1987) Regulation of nuclear formation and breakdown in cell-free extracts of amphibian eggs, in *Molecular Regulation of Nuclear Events in Mitosis and Meiosis*, (eds R. A. Schlegel, M. S. Halleck and P. N. Rao), Academic Press, New York, pp. 67–109.

Lohka, M. J. and Masui, Y. (1983a) Formation *in vitro* of sperm pronuclei and mitotic chromosomes induced by amphibian ooplasmic components. *Science*, **220**, 719–721.

Lohka, M. J. and Masui, Y. (1983b) The germinal vesicle material required for sperm pronuclear formation is located in the soluble fraction of egg cytoplasm. *Exper. Cell Res.*, **148**, 481–491.

Lohka, M. J. and Masui, Y. (1984a) Effects of Ca^{2+} ions on the formation of metaphase chromosomes and sperm pronuclei in cell-free preparations from unactivated *Rana pipiens* eggs. *Devel. Biol.*, **103**, 434–442.

Lohka, M. J. and Masui, Y. (1984b) Roles of cytoplasmic particles in nuclear envelope assembly and sperm pronuclear formation in cell-free preparations from amphibian eggs. *J. Cell Biol.*, **98**, 1222–1230.

Lohka, M. J., Hayes, M. K. and Maller, J. L. (1988) Purification of maturation-promoting factor, an intracellular regulator of early mitotic events. *Proc. Natl Acad. Sci. USA*, **85**, 3009–3013.

Longo, F. J. (1972) The effects of cytochalasin B on the events of fertilization in the surf clam, *Spisula solidissima*. I. Polar body formation. *J. Exper. Zool.*, **182**, 321–344.

Longo, F. J. (1973) Fertilization: A comparative ultrastructural review. *Biol. Reprod.*, **9**, 149–215.

Longo, F. J. (1976a) Ultrastructural aspects of fertilization in spiralian eggs. *Amer. Zool.*, **16**, 375–394.

Longo, F. J. (1976b) Derivation of the membrane comprising the male pronuclear envelope in inseminated sea urchin eggs. *Devel. Biol.*, **49**, 347–368.

Longo, F. J. (1976c) Sperm aster in rabbit zygotes: Its structure and function. *J. Cell Biol.*, **69**, 539–547.

Longo, F. J. (1977) An ultrastructural study of cross-fertilization (*Arbacia* ♀ × *Mytilus* ♂). *J. Cell Biol.*, **73**, 14–26.

Longo, F. J. (1978a) Effects of cytochalasin B on sperm–egg interactions. *Devel. Biol.*, **67**, 249–265.

Longo, F. J. (1978b) Insemination of immature sea urchin (*Arbacia punctulata*) eggs. *Devel. Biol.*, **62**, 271–291.

Longo, F. J. (1980) Organization of microfilaments in sea urchin *Arbacia punctulata* eggs at fertilization: Effects of cytochalasin B. *Devel. Biol.*, **74**, 422–433.

Longo, F. J. (1981a) Morphological features of the surface of the sea urchin (*Arbacia punctulata*) egg: Oolemma-cortical granule association. *Devel. Biol.*, **84**, 173–182.

Longo, F. J. (1981b) Regulation of pronuclear development, in *Bioregulators of Reproduction*, (eds G. Jagiello and C. Vogel), Academic Press, New York, pp. 529–557.

Longo, F. J. (1982) Integration of sperm and egg plasma membrane components at fertilization. *Devel. Biol.*, **89**, 409–416.

Longo, F. J. (1983) Meiotic maturation and fertilization, in *Biology of Mollusca*, vol. 3, (eds K. M. Wilbur, N. H. Verdonk, J. A. M van den Biggelaar and A. S. Tompa), Academic Press, New York, pp. 49–89.

Longo, F. J. (1984) Transformation of sperm nuclei incorporated into sea urchin (*Arbacia punctulata*) embryos at different stages of the cell cycle. *Devel. Biol.*, **103**, 168–181.

Longo, F. J. (1986) Surface changes at fertilization: Integration of sea urchin (*Arbacia punctulata*) sperm and oocyte plasma membranes. *Devel. Biol.*, **116**, 143–159.

Longo, F. J. (1988) Reorganization of the egg surface at fertilization. *Int. Rev. Cytol.*, **113**, 233–269.

Longo, F. J. (1989a) Egg cortical architecture, in *The Cell Biology of Fertilization*, (eds H. Schatten and G. Schatten), Academic Press, New York, pp 108–138.

Longo, F. J. (1989b) Incorporation and dispersal of sperm surface antigens in plasma membranes of inseminated sea urchin (*Arbacia punctulata*) eggs and oocytes. *Devel. Biol.*, **131**, 37–43.

Longo, F. J. (1989c) Incorporation and fate of specific sperm plasma membrane components following insemination as revealed by ultrastructural immunocytochemistry. *Gamete Res.*, **23**, 215–228.

Longo, F. J. (1990) Dynamics of sperm nuclear transformation at fertilization, in *Fertilization in Mammals*, (eds B. D. Bavister, J. Cummins and E. R. S. Roldan), Serono Symposia, Norwell, MA, pp. 297–307.

Longo, F. J. and Anderson, E. (1968) The fine structure of pronuclear development and fusion in the sea urchin, *Arbacia punctulata*. *J. Cell Biol.*, **39**, 339–368.

Longo, F. J. and Anderson, E. (1969a) Cytological aspects of fertilization in the lamellibranch, *Mytilus edulis*. I. Polar body formation and development of the female pronucleus. *J. Exper. Zool.*, **172**, 69–96.

Longo, F. J. and Anderson, E. (1969b) Cytological events leading to the formation of the two-cell stage in the rabbit: Association of the maternally and paternally derived genomes. *J. Ultrastruct. Res.*, **29**, 86–118.

Longo, F. J. and Anderson, E. (1970a) An ultrastructural analysis of fertilization in the surf clam, *Spisula solidissima*. I. Polar body formation and development of the female pronucleus. *J. Ultrastruct. Res.*, **33**, 495–514.

Longo, F. J. and Anderson, E. (1970b) An ultrastructural analysis of fertilization in the surf clam, *Spisula solidissima*. II. Development of the male pronucleus and the association of the maternally and paternally derived chromosomes. *J. Ultrastruct. Res.*, **33**, 515–527.

Longo, F. J. and Anderson, E. (1970c) A cytological study of the relation of the cortical reaction to subsequent events of fertilization in urethane-treated eggs of the sea urchin, *Arbacia punctulata*. *J. Cell Biol.*, **47**, 646–665.

Longo, F. J. and Anderson, E. (1970d) The effects of nicotine on fertilization in the sea urchin, *Arbacia punctulata*. *J. Cell Biol.*, **46**, 308–325.

Longo, F. J. and Anderson, E. (1974) Gametogenesis, in *Concepts of Development*, (eds J. Lash and J. R. Whittaker), Sinauer Assoc., Stamford, CN, pp. 3–47.

Longo, F. J. and Chen, D. Y. (1985) Development of cortical polarity in mouse eggs: Involvement of the meiotic apparatus. *Devel. Biol.*, **107**, 382–394.

Longo, F. J. and Kunkle, M. (1977) Synthesis of RNA by male pronuclei of fertilized sea urchin eggs. *J. Exper. Zool.*, **201**, 431–438.

Longo, F. J. and Kunkle, M. (1978) Transformation of sperm nuclei upon insemination. *Curr. Top. Devel. Biol.*, **12**, 149–184.

Longo, F. J. and Plunkett, W. (1973) The onset of DNA synthesis and its relation to morphogenetic events of the pronuclei in activated eggs of the sea urchin, *Arbacia punctulata*. *Devel. Biol.*, **30**, 56–67.

Longo, F. J. and Scarpa, J. (1991) Expansion of the sperm nucleus and association of the maternal and paternal genomes in fertilized *Mulina lateralis* eggs. *Biol. Bull.*, **180**, 56–64.

Longo, F. J., Lynn, J. W., McCulloh, D. H. and Chambers, E. L. (1986) Correlative ultrastructural and electrophysiological studies of sperm–egg interactions of the sea urchin, *Lytechinus variegatus*. *Devel. Biol.*, **118**, 155–166.

Longo, F. J., Krohne, G. and Franke, W. W. (1987) Basic proteins of the perinuclear theca of mammalian spermatozoa and spermatids: A novel class of cytoskeletal elements. *J. Cell Biol.*, **105**, 1105–1120.

Longo, F. J., Cook, S. and Mathews, L. (1991a) Pronucleus formation in starfish eggs inseminated at different stages of meiotic maturation: Correlation of sperm nuclear transformations and activity of the maternal chromatin. *Devel. Biol.*, **147**, 62–72.

Longo, F. J., Cook, S., Mathews, L. and Wright, S. J. (1991b) Nascent protein requirement for completion of meiotic maturation and pronuclear development: Examination of fertilized and A-23187-activated surf clam, *Spisula solidissima* eggs. *Devel. Biol.*, **148**, 75–86.

Longo, F. J., Mathews, L. and Hedgecock, D. (1993) Morphogenesis of maternal and paternal genomes in fertilized oyster eggs, *Crassostrea gigas:* Effects of cytochalasin B at different periods during meiotic maturation. *Biol. Bull.*, **185**, 197–214.

Longo, F. J., Cook, S., McCulloh, D. H. *et al.* (1994a) Stages leading to and following fusion of sperm and egg plasma membranes. *Zygote*, **2**, 317–331.

Longo, F. J., Mathews, L. and Palazzo, R. E. (1994b) Sperm nuclear transformations in cytoplasmic extracts from surf clam, *Spisula solidissima* oocytes. *Devel. Biol.*, **162**, 245–258.

Longo, F. J., Ushiyama, A., Chiba, K. and Hoshi, M. (1995) Ultrastructural localization of acrosomal reaction-inducing substance (ARIS) on sperm of the starfish, *Asterias amurensis*. *Mol. Reprod. Devel.*, **41**, 91–99.

Lopata, A., Sathananthan, A. H., McBain, J. C. *et al.* (1980) The ultrastructure of the preovulatory human egg fertilized *in vitro*. *Fertil. Steril.*, **33**, 12–20.

Lopez, L. C. and Shur, B. D. (1987) Redistribution of mouse sperm surface galactosyltransferase after the acrosome reaction. *J. Cell Biol.*, **105**, 1663–1670.

Lopez, A., Miraglia, S. J. and Glabe, C. G. (1993) Structure/function analysis of the sea urchin sperm adhesive protein bindin. *Devel. Biol.*, **156**, 24–33.

Lopo, A. C. (1983) Sperm–egg interactions in invertebrates, in *Mechanism and Control of Animal Fertilization*, (ed. J. F. Hartmann), Academic Press, New York, pp. 269–324.

Lorca, T., Labbe, J. C., Devault, A. *et al.* (1992) Dephosphorylation of cdc2 on threonine 161 is required for cdc2 kinase inactivation and normal anaphase. *EMBO J.*, **11**, 2381–2390.

Lorca, T., Cruzalegui, F. H., Fesquet, D. *et al.* (1993) Calmodulin-dependent protein kinase II mediates inactivation of MPF and CSF upon fertilization of *Xenopus* eggs. *Nature*, **366**, 270–273.

Luchtel, D. L. (1976) An ultrastructural study of the egg and early cleavage stages of *Lymnaea stagnalis*, a pulmonate mollusc. *Am. Zool.*, **16**, 405–419.

Luthardt, F. W. and Donahue, R. P. (1973) Pronuclear DNA synthesis in mouse eggs: An autoradiographic study. *Exper. Cell Res.*, **82**, 143–151.

Luttmer, S. and Longo, F. J. (1985) Ultrastructure and morphometric observations of cortical endoplasmic reticulum in *Arbacia*, *Spisula* and mouse eggs. *Devel. Growth Differ.*, **27**, 349–359.

Luttmer, S. J. and Longo, F. J. (1987) Rates of male pronuclear enlargement in sea urchin zygotes. *J. Exper. Zool.*, **243**, 289–298.

Luttmer, S. J. and Longo, F. J. (1988) Sperm nuclear transformations consist of enlargement and condensation coordinate with stage of meiotic maturation in fertilized *Spisula solidissima* oocytes. *Devel. Biol.*, **128**, 86–96.

Mabuchi, I. (1973) ATPase in the cortical layer of sea urchin egg, its properties and interaction with cortex protein. *Biochem. Biophys. Acta.*, **297**, 317–332.

Mabuchi, I. (1981) Purification from starfish eggs of a protein that depolymerizes actin. *J. Biochem.*, **89**, 1341–1344.

Mabuchi, I., Hamaguchi, Y., Kobayashi, T. *et al.* (1985) Alpha-actinin from sea urchin eggs: Biochemical properties, interaction with actin and distribution in the cell during fertilization and cleavage. *J. Cell Biol.*, **100**, 375–583.

McCulloh, D. H. and Chambers, E. L. (1986a) When does the sperm fuse with the egg? *J. Gen. Physiol.*, **88**, 38a–39a.

McCulloh, D. H. and Chambers, E. L. (1986b) Fusion and 'unfusion' of sperm and egg are voltage dependent in the sea urchin, *Lytechinus variegatus*. *J. Cell Biol.*, **103**, 236a.

McCulloh, D. H. and Chambers, E. L. (1991) A localized zone of increased conductance progresses over the surface of the sea urchin egg during fertilization. *J. Gen. Physiol.*, **97**, 579–604.

McCulloh, D. H. and Chambers, E. L. (1992) Fusion of membranes during ferti-lization: Increases of sea urchin eggs' membrane capacitance and membrane conductance at the site of contact with the sperm. *J. Gen. Physiol.*, **99**, 137–175.

McCulloh, D. H., Rexroad, C. E. and Levitan, H. (1983) Insemination of rabbit eggs is associated with slow depolarization and repetitive diphasic membrane potentials. *Devel. Biol.*, **95**, 372–377.

McDougall, A. and Sardet, C. (1995) Function and characteristics of repetitive calcium waves associated with meiosis. *Curr. Biol.*, **5**, 318–328.

McDowell, J. S. (1986) Preparation of spermatozoa for insemination *in vitro*, in *In Vitro Fertilization – Norfolk*, (eds H. W. Jones Jr, G. S. Jones, G. D. Hodgen and Z. Rosenwaks), Williams & Wilkins, Baltimore, MD, pp. 162–167.

McGrath, J. and Solter, D. (1984) Completion of mouse embryogenesis requires both the maternal and paternal genomes. *Cell*, **37**, 179–183.

McIntosh, J. R. (1983) The centrosome as an organizer of the cytoskeleton, in *Spatial Organization of Eukaryotic Cells*, (ed. J. R. McIntosh), Alan R. Liss, New York, pp. 115–142.

McLean, K. (1976) Some aspects of RNA synthesis in oyster development. *Am. Zool.*, **16**, 521–528.

McLean, K. W. and Whiteley, A. H. (1974) RNA synthesis during the early development of the Pacific oyster, *Crassostrea gigas*. *Exper. Cell Res.*, **87**, 132–138.

McPherson, S. and Longo, F. J. (1993) Chromatin structure–function alterations during mammalian spermatogenesis: DNA nicking and repair in elongating spermatids. *Eur. J. Histochem.*, **37**, 109–128.

McPherson, S. M., McPherson, P. S., Mathews, L. *et al.* (1992) Cortical localiza-tion of a calcium release channel in sea urchin eggs. *J. Cell Biol.*, **116**, 1111–1121.

Magnuson, T. and Epstein, C. J. (1984) Oligosyndactly: A lethal mutation in the mouse that results in mitosis arrest very early in development. *Cell*, **38**, 823–833.

Maleszewski, M. (1992) Behavior of sperm nuclei incorporated into parthenoge-netic mouse eggs prior to the first cleavage division. *Mol. Reprod. Devel.*, **33**, 215–221.

Maller, C. (1973) The participation of the embryonic genome during early cleavage in the rabbit. *Devel. Biol.*, **32**, 453–459.

Maller, J. L. (1985) Regulation of amphibian oocyte maturation. *Cell Diff.* **16**, 211–221.

Maller, J. L. (1994) Biochemistry of cell cycle checkpoints at the G2/M and metaphase/anaphase transitions. *Semin. Devel. Biol.*, **5**, 183–190.

Maller, J. L. and Krebs, E. G. (1977) Progesterone-stimulated meiotic cell division in *Xenopus* oocytes: Induction by regulatory subunit and inhibition by catalytic subunit of adenosine 3´:5´-monophosphate-dependent protein kinase. *J. Biol. Chem.*, **252**, 1712–1718.

Maller, J. L. and Krebs, E. G. (1980) Regulation of oocyte maturation. *Curr. Topics Cell Reg.*, **16**, 271–311.

Manes, C. (1973) The participation of the embryonic genome during early cleavage in the rabbit. *Devel. Biol.*, **32**, 453–459.

Mar, H. (1980) Radial cortical fibers and pronuclear migration in fertilized and artificially activated eggs of *Lytechinus pictus*. *Devel. Biol.*, **78**, 1–13.

Margulis, L., Sagan, D. and Olendzerski, L. (1985) What is Sex?, in *The Origin and Evolution of Sex*, (eds H. O. Halvorson and A. Monroy), Alan R. Liss, New York, pp. 69–85.

Maro, B., Johnson, M. H., Pickering, S. J. and Flach, G. (1984) Changes in actin distribution during fertilization of the mouse egg. *J. Embryol. Exper. Morphol.*, **81**, 211–237.

Maro, B. Howlett, S. K. and Webb, M. (1985) Non-spindle microtubule organizing centers in metaphase II-arrested mouse oocytes. *J. Cell Biol.*, **101**, 1665–1672.

Maro, B., Johnson, M. H., Webb, M. and Flach, G. (1986) Mechanism of polar body formation in the mouse oocyte an interaction between the chromosomes, the cytoskeleton and the plasma membrane. *J. Embryol. Exper. Morphol.*, **92**, 11–32.

Maro, B., Kubiak, J. Z., Verlhac, M.-H. and Winston, N. J. (1994) Interplay between the cell cycle control machinery and the microtubule network in mouse oocytes. *Semin. Devel. Biol.*, **5**, 191–198.

Martin, R. H. (1990) Analysis of human sperm chromosome complements, in *Fertilization in Mammals*, (eds B. D. Bavister, J. Cummins and E. R. S. Roldan), Serono Symposia USA, Norwell, MA, pp. 365–372.

Masui, Y. (1985) Meiotic arrest in animal oocytes, in *Biology of Fertilization*, (eds C. B. Metz and A. Monroy), vol. 1, Academic Press, Orlando, FL, pp. 189–219.

Masui, Y. and Clarke, H. (1979) Oocyte maturation. *Int. Rev. Cytol.*, **57**, 185–282.

Masui, Y. and Markert, C. L. (1971) Cytoplasmic control of nuclear behavior during meiotic maturation of frog oocytes. *J. Exper. Zool.*, **177**, 129–146.

Masui, Y. and Shibuya, E. K. (1987) Development of cytoplasmic activities that control chromosome cycles during maturation of amphibian oocytes, in *Molecular Regulation of Nuclear Events in Mitosis and Meiosis*, (eds R. A. Schlegel, M. S. Halleck and P. N. Rao), Academic Press, Orlando, FL, pp. 1–42.

Masui, Y., Meyerhof, P. G., Miller, M. A. and Wasserman, W. J. (1977) Roles of bivalent cations in maturation and activation of vertebrate oocytes. *Differentiation*, **9**, 49–57.

Matten, W. T. and VandeWoude, G. F. (1994) Cell cycle control and early embryogenesis: *Xenopus laevis* maturation and early embryonic cell cycles. *Semin. Devel. Biol.*, **5**, 173–181.

Mattioli, M. Galeati, G. and Seren, E. (1988) Effect of follicle somatic cells during pig oocyte maturation on egg penetrability and male pronucleus formation. *Gamete Res.*, **20**, 177–183.

Maul, G. G. and Schatten, G. (1986) Nuclear lamins during gametogenesis, fertilization and early development, in *Nucleocytoplasmic Transport*, (eds R. Peters and M. Trendelenburg), Springer-Verlag, New York, pp. 123–134.

Maul, G. G., French, B. T. and Bechtol, K. B. (1986) Identification and redistribution of lamins during nuclear differentiation in mouse spermatogenesis. *Devel. Biol.*, **115**, 68–77.

Mayer, J. F. and Lanzendorf, S. E. (1986) Cryopreservation of gametes and pre-embryos. *In Vitro Fertilization – Norfolk*, (eds H. W. Jones Jr, G. S. Jones, G. D. Hodgen and Z. Rosenwaks), Williams & Wilkins, Baltimore, MD, pp. 260–269.

Mazia, D. (1937) The release of calcium in *Arbacia* eggs on fertilization. *J. Cell Comp. Physiol.*, **10**, 291–304.

Mazia, D. (1961) Mitosis and the physiology of cell division, in *The Cell*, vol. 3, (eds J. Brachet and A. E. Mirsky), Academic Press, New York, pp. 77–412.

Mazia, D. and Ruby, A. (1974) DNA synthesis turned on in unfertilized sea urchin eggs by treatment with NH_4OH. *Exper. Cell Res.*, **85**, 1164–1172.

Mehlmann, L. M. and Kline, D. (1994) Regulation of intracellular calcium in the mouse egg: Calcium release in response to sperm or inositol trisphosphate is enhanced after meiotic maturation. *Biol. Reprod.*, **51**, 1088–1098.

Meijer, L. and Mordet, G. (1994) Starfish oocyte maturation: from prophase to metaphase. *Semin. Devel. Biol.*, **5**, 165–171.

Meizel, S. and Mukerji, S. K. (1975) Proacrosin from rabbit epididymal spermatozoa: Partial purification and initial biochemical characterization. *Biol. Reprod.*, **13**, 83–93.

Merriam, R. W. (1969) Movement of cytoplasmic proteins into nuclei induced to enlarge and initiate DNA or RNA synthesis. *J. Cell Sci.*, **5**, 333–349.

Messenger, S. M. and Albertini, D. F. (1991) Centrosome and microtubule dynamics during meiotic progression in the mouse oocyte. *J. Cell Sci.*, **100**, 289–298.

Metz, C. B. (1967) Gamete surface components and their role in fertilization, in *Fertilization*, vol. 1, (eds C. B. Metz and A. Monroy), Academic Press, New York, pp. 163–236.

Metz, C. B. and Monroy, A. (eds) (1985) *Fertilization*, vols 1–3, Academic Press, New York.

Michod, R. E. and Levin, B. R. (1987) *The Evolution of Sex*, Sinauer Assoc., Sunderland, MA.

Miller, D. J. and Shur, B. D. (1994) Molecular basis of fertilization in the mouse. *Semin. Devel. Biol.*, **5**, 255–264.

Miller, D. J., Macek, M. B. and Shur, B. D. (1992) Complementarity between sperm surface β-1, 4-galactosyltransferase and egg-coat ZP3 mediates sperm–egg binding. *Nature*, **357**, 589–593.

Miller, D. J., Gong, X., Decker, G. and Shur, B. D. (1993) Egg cortical granule N-acetyl glucosaminidase is required for the mouse zona block to polyspermy. *J. Cell Biol.*, **123**, 1431–1440.

Miller, R. L. (1985) Sperm chemo-orientation in the metazoa, in *Biology of Fertilization*, vol. 2, (eds C. B. Metz and A. Monroy), pp. 275–337. Academic Press, New York.

Miller, R. L. (1989) Evidence for the presence of sexual pheromones in free-spawning starfish. *J. Exper. Mar. Biol. Ecol.*, **130**, 205–221.

Millonig, G. (1969) Fine structure analysis of the cortical reaction in the sea urchin egg after normal fertilization and after electric inducation. *J. Submicr. Cytol.*, **1**, 69–84.

Minshull, J., Murray, A., Coleman, A. and Hunt, T. (1991) *Xenopus* oocyte maturation does not require new cyclin synthesis. *J. Cell Biol.*, **114**, 767–772.

Mintz, B. (1964) Synthetic processes and early development in the mammalian egg. *J. Exper. Zool.*, **157**, 85–100.

Mirkes, P. E. (1970) Protein synthesis before and after fertilization in the egg of *Ilyanassa obsoleta*. *Exper. Cell Res.*, **60**, 115–118.

Miyazaki, S. (1988) Inositol 1,4,5-trisphosphate-induced calcium release and guanine nucleotide-binding protein-mediated periodic calcium rises in golden hamster eggs. *J. Cell Biol.*, **106**, 345–353.

Miyazaki, S. (1989) Signal transduction of sperm–egg interaction causing periodic calcium transients in hamster eggs, in *Mechanisms of Egg Activation*, (eds R. Nuccitelli, G. N. Cherr and W. H. Clark Jr), Plenum Press, New York, pp. 231–246.

Miyazaki, S., Yuzaki, M., Nakada, K. *et al.* (1992) Block of Ca^{2+}-wave and Ca^{2+} oscillation by antibody to the inositol 1,4,5-triphosphate receptor in fertilized hamster eggs. *Science*, **257**, 251–255.

Miyazaki, S., Shirakawa, H., Nakada, K. and Honda, Y. (1993) Essential role of the inositol 1,4,5-trisphosphate receptor/Ca^{2+} release channel in Ca^{2+} waves and Ca^{2+} oscillation at fertilization of mammalian eggs. *Devel. Biol.*, **158**, 62–78.

Mohri, T. and Hamaguchi, Y. (1991) Propagation of transient Ca^{2+} increase in sea urchin eggs upon fertilization and its regulation by microinjecting EGTA solution. *Cell Struct. Func.*, **16**, 157–165.

Mohri, H., Hamaguchi, Y., Hamaguchi, M. S. *et al.* (1994) Sperm–egg fusion in the sea urchin is blocked in Mg^{2+}-free seawater. *Zygote*, **2**, 149–157.

Mohri, T., Ivonnet, P. I. and Chambers, E. L. (1995) Effect on sperm-induced activation current and increase of cytosolic Ca^{2+} by agents that modify the mobilization of $[Ca^{2+}]_i$. Heparin and pentosan polysulfate. *Devel. Biol.*, **172**, 139–157.

Moller, C. C. and Wassarman, P. M. (1989) Characterization of a proteinase that cleaves zona pellucida glycoprotein ZP2 following activation of mouse eggs. *Devel. Biol.*, **132**, 103–112.

Monck, J. R. and Fernandez, J. M. (1992) The exocytic fusion pore. *J. Cell Biol.*, **119**, 1395–1404.

Monk, M. M., Boubelik, M. and Lehnert, S. (1987) Temporal and regional changes in DNA methylation in the embryonic, extraembryonic and germ cell lineages during mouse embryo development. *Development*, **99**, 371–382.

Monroy, A. (1957) Adenosine triphosphatase in the mitochondria of unfertilized and newly fertilized sea-urchin eggs. *J. Cell. Comp. Physiol.*, **50**, 73–81.

Monroy, A. (1965) *Chemistry and Physiology of Fertilization*, Holt, Rinehart & Winston, New York.

Moon, R. T., Danilchik, M. V. and Hille, M. (1982) An assessment of the masked messenger hypothesis: Sea urchin egg messenger ribonucleoprotein complexes are efficient templates for *in vitro* protein synthesis. *Devel. Biol.*, **93**, 389–403.

Moon, R. T., Nicosia, R. F., Olsen, C. *et al.* (1983) The cytoskeletal framework of sea urchin eggs and embryos: Developmental changes in the association of messenger RNA. *Devel. Biol.*, **95**, 447–458.

Moore, H. D. M. and Bedford, J. M. (1983) The interaction of mammalian gametes in the female, in *Mechanism and Control of Animal Fertilization*, (ed. J. F. Hartmann), Academic Press, New York, pp. 453–497.

Moore, K. L. and Kinsey, W. H. (1995) Effects of protein tyrosine kinase inhibitors on egg activation and fertilization-dependent protein tyrosine kinase activity. *Devel. Biol.*, **168**, 1–10.

Moore, G., Kopf, G. S. and Schultz, R. M. (1993) Complete mouse egg activation in the absence of sperm by stimulation of an exogenous G protein coupled receptor. *Devel. Biol.*, **159**, 669–678.

Moore, G., Takuya, A., Visconti, P. *et al.* (1994) Roles of heterotrimeric and monomeric G proteins in sperm-induced activation of mouse eggs. *Development*, **120**, 3313–3323.

Moos, J., Visconti, P. E., Moore, G. D. *et al.* (1995) Potential role of mitogen-activated protein kinase in pronuclear envelope assembly and disassembly following fertilization of mouse eggs. *Biol. Reprod.*, **53**, 692–699.

Moos, J., Xu, Z., Schultz, R. M. and Kopf, G. S. (1996) Regulation of nuclear envelope assembly/disassembly by MAP kinase. *Devel. Biol.*, **175**, 358–361.

Moreau, M., Guerrier, P., Dorée, M. and Ashley, C. C. (1978) Hormone-induced release of intracellular Ca^{2+} triggers meiosis in starfish oocytes. *Nature*, **272**, 251–253.

Morisawa, M., Tanimoto, S. and Ohtake, H. (1992) Characterization and partial purification of sperm-activating substance from eggs of the herring, *Clupea palasii*. *J. Exper. Zool.*, **264**, 225–230.

Moriya, M. and Katagiri, C. (1976) Microinjection of toad sperm into oocytes undergoing maturation division. *Devel. Growth Differ.*, **18**, 349–356.

Mortillo, S. and Wassarman, P. M. (1991) Differential binding of gold-labeled zona pellucida glycoprotein mZP2 and mZP3 to mouse sperm membrane components. *Development*, **113**, 141–150.

Morton, D. B. (1975) Acrosomal enzymes: Immunological localization of acrosin and hyaluronidase in ram spermatozoa. *J. Reprod. Fertil.*, **45**, 375–378.

Moser, F. (1939) Studies on cortical layer response to stimulating agents in the *Arbacia* egg. I. Response to insemination. *J. Exper. Zool.*, **80**, 423–471.

Moss, S. B., Donovan, M. J. and Bellvé, A. R. (1987) The occurrence and distribution of lamin proteins during mammalian spermatogenesis and early embryonic development. *Ann. NY Acad. Sci.*, **513**, 74–89.

Motlik, J. and Fulka, J. (1974) Fertilization of pig follicular oocytes cultivated *in vitro*. *J. Reprod. Fertil.*, **36**, 235–237.

Motlik, J., Kopecny, V., Pivko, J. and Fulka, J. (1980) Distribution of proteins labeled during meiotic maturation in rabbit and pig eggs at fertilization. *J. Reprod. Fert.*, **58**, 415–419.

Moy, G. W. and Vacquier, V. D. (1979) Immunoperoxidase localization of bindin during adhesion of sperm to sea urchin eggs. *Curr. Topics Devel. Biol.*, **13**, 31–44.

Moy, G. W., Kopf, G. S., Gache. C. and Vacquier, V. D. (1983) Calcium mediated release of glucanese activity from cortical granules of sea urchin eggs. *Devel. Biol.*, **100**, 267–274.

Murray, A. and Hunt, T. (1993) *The Cell Cycle*, W. H. Freeman, New York.

Myles, D. G. (1993) Molecular mechanisms of sperm–egg membrane binding and fusion in mammals. *Devel. Biol.*, **158**, 35–45.

Myles, D. G. and Primakoff, P. (1984) Localized surface antigens of guinea pig sperm migrate to new regions prior to fertilization. *J. Cell Biol.*, **99**, 1634–1641.

Myles, D. G., Koppel, D. Z., Cowan, A. E. *et al.* (1987) Rearrangement of sperm surface antigens prior to fertilization. *Ann. NY Acad. Sci.*, **513**, 262–273.

Naish, S. J., Perreault, S. D., Foehner, A. L. and Zirkin, B. R. (1987a) DNA synthesis in the fertilizing hamster sperm nucleus: Sperm template availability and egg cytoplasmic control. *Biol. Reprod.*, **36**, 245–253.

Naish, S. J., Perreault, S. D. and Zirkin, B. R. (1987b) DNA synthesis following microinjection of heterologous sperm and somatic cell nuclei into hamster oocytes. *Gamete Res.*, **18**, 109–120.

Navara, C., First, N. and Schatten, G. (1994) Microtubule organization in the cow during fertilization, polyspermy, parthenogenesis and nuclear transfer: The role of the sperm aster. *Devel. Biol.*, **162**, 29–40.

Nebreda, A. R. and Hunt, T. (1993) The c-mos proto-oncogene protein kinase turns on and maintains the activity of MAP kinase, but not MPF, in cell-free extracts of *Xenopus* oocytes and eggs. *EMBO J.*, **12**, 1979–1986.

Newport, J. (1987) Nuclear reconstitution *in vitro*: Stages of assembly around protein-free DNA. *Cell*, **48**, 205–217.

Newport, J. and Spann, T. (1987) Disassembly of the nucleus in mitotic extracts: Membrane vesiculation, lamin disassembly, and chromosome condensation are independent processes. *Cell*, **48**, 219–230.

Newrock, K. M., Alfageme, C. R., Nardi, R. V. and Cohen, L. H. (1977) Histone changes during chromatin remodeling in embryogenesis. *Cold Spring Harbor Symp. Quant. Biol.*, **42**, 421–431.

Nicolson, G. L. and Yanagimachi, R. (1979) Cell surface changes associated with the epididymal maturation of mammalian spermatozoa, in *The Spermatozoon*, (eds D. W. Fawcett and J. M. Bedford), Urban & Schwarzenberg, Baltimore, MD, pp. 187–194.

Nicosia, S. V., Wolf, D. P. and Inoué, M. (1977) Cortical granule distribution and cell surface characteristics in mouse eggs. *Devel. Biol.*, **57**, 56–74.

Nikolopoulou, M., Soucek, D. A. and Vary, J. C. (1985) Changes in the lipid content of boar sperm plasma membranes during epididymal maturation. *Biochem. Biophys. Acta*, **815**, 486–498.

Nishioka, D., Porter, D. C., Trimmer, J. S. and Vacquier, V. D. (1987) Dispersal of sperm surface antigens in the plasma membranes of polyspermically fertilized sea urchin eggs. *Exper. Cell Res.*, **173**, 628–632.

Nogués, C., Ponsà, M., Egozcue, J. and Vidal, F. (1995) Cytogenic studies of oocyte fusion products. *Zygote*, **3**, 27–29.

Nonchev, S. and Tsanev, R. (1990) Protamine-histone replacement and DNA replication in the male mouse pronucleus. *Mol. Reprod. Devel.*, **25**, 72–76.

Nuccitelli, R. (1991) How do sperm activate eggs? *Curr. Topics Devel. Biol.*, **25**, 1–16.

Nuccitelli, R., Kline, D., Busa, W. B. *et al.* (1988) A highly localized activation current yet widespread intracellular calcium increase in the egg of the frog, *Discoglossus pictus*. *Devel. Biol.*, **130**, 120–132.

Nuccitelli, R. Cherr, G. N. and Clark, W. H. Jr (1989) *Mechanisms of Egg Activation*, Plenum Press, New York.

Nuccitelli, R., Yim, D. L. and Smart, T. (1993) The sperm-induced Ca^{2+} wave following fertilization of the *Xenopus* egg requires the production of Ins $(1,4,5)P_3$. *Devel. Biol.*, **158**, 200–212.

Oberdorf, J. A., Head, J. F. and Kaminer, B. (1986) Calcium uptake and release by isolated cortices and microsomes from the unfertilized egg of the sea urchin *Strongylocentrotus drobachiensis*. *J. Cell Biol.*, **102**, 2205–2210.

Oberdorf, J. A., Lebeche, D., Head, J. F. and Kaminer, B. (1988) Identification of a calsequestrin-like protein from sea urchin eggs. *J. Biol. Chem.*, **263**, 6806–6809.

O'Connor, C. M., Robinson, K. R. and Smith, L. D. (1977) Calcium, potassium, and sodium exchange by full-grown and maturing *Xenopus laevis* oocytes. *Devel. Biol.*, **61**, 28–40.

O'Dell, D. S., Ortolani, G. and Monroy, A. (1973) Increased binding of radioactive concanavalin A during maturation of *Ascidia* eggs. *Exper. Cell Res.*, **83**, 408–411.

Ogura, A. and Yanagimachi, R. (1993) Round spermated nuclei injected into hamster oocytes form pronuclei and participate in syngamy. *Biol. Reprod.*, **48**, 219–225.

Ohlendieck, K., Dhume, S. T., Partin, J. S. and Lennarz, W. J. (1993) The sea urchin egg receptor for sperm: Isolation and characterization of the intact, biologically active receptor. *J. Cell Biol.*, **122**, 887–895.

Ohsumi, K. and Katagiri, C. (1991a) Occurrence of H1 subtypes specific to pronuclei and cleavage-stage cell nuclei of anuran amphibians. *Devel. Biol.*, **147**, 110–120.

Ohsumi, K. and Katagiri, C. (1991b) Characterization of the ooplasmic factor inducing decondensation of and protamine removal from toad sperm nuclei: Involvement of nucleoplasmin. *Devel. Biol.*, **148**, 295–305.

Olson, G. E. (1980) Changes in intramembranous particle distribution in the plasma membrane of *Didelphis virginiana* spermatozoa during maturation in the epididymis. *Anat. Rec.*, **197**, 471–488.

Olson, G. E. and Danzo, B. J. (1981) Surface changes in rat spermatozoa during epididymal transit. *Biol. Reprod.*, **24**, 431–443.

Ookata, K., Hisanga, S. I., Okano, T. *et al.* (1992) Relocation and distinct subcellular localization of p34^{cdc2}-cyclin B complex at meiosis reinitiation in starfish oocytes. *EMBO J.*, **11**, 1763–1772.

Oprescu, S. and Thibault, C. (1965) Duplication de l'ADN dans les oeufs de lapine après la fécondation. *Ann. Biol. Anim. Biochem. Biophys.*, **5**, 151–156.

O'Rand, M. G. (1985) Differentiation of mammalian sperm antigens, in *Biology of Fertilization*, vol. 2, (eds C. B. Metz and A. Monroy), Academic Press, New York, pp. 103–119.

Ortolani, G., O'Dell, D. S., Mansueto, C. and Monroy, A. (1975) Surface changes and onset of DNA replication in the *Ascidia* egg. *Exper. Cell Res.*, **96**, 122–128.

Osanai, K., Kyozuka, K.-I., Sato, H. *et al.* (1987) Bioelectric responses of sea urchin eggs inseminated with oyster spermatozoa: A sperm evolved potential without egg activation. *Devel. Biol.*, **124**, 309–315.

Osawa, M., Takemoto, K., Kikuyama, M. *et al.* (1994) Sperm and its soluble extract cause transient increases in intracellular calcium concentration and in membrane potential of sea urchin zygotes. *Devel. Biol.*, **166**, 268–276.

Osman, R. A., Andria, J. L., Jones, A. D. and Meizel, S. (1989) Steroid induced exocytosis: the human sperm acrosome reaction. *Biochem. Biophys. Res. Commun.*, **160**, 828–833.

Otto, J. J. and Schroeder, T. E. (1984) Microtubule arrays in the cortex and near the germinal vesicle of immature starfish oocytes. *Devel. Biol.*, **101**, 274–281.

Otto, J. J., Kane, R. E. and Bryan, J. (1980) Redistribution of actin and fascin in sea urchin eggs after fertilization. *Cell Motil.*, **1**, 31–40.

Paiement, J., Beaufay, H. and Godelaine, D. (1980) Coalescence of microsomal vesicles from rat liver: A phenomenon occurring in parallel with enhancement of the glycosylation activity during incubation of stipped rough microsomes with GTP. *J. Cell Biol.*, **86**, 29–37.

Palacios, M. J., Joshi, H. C., Simerly, C. and Schatten, G. (1993) γ-Tubulin reorganization during mouse fertilization and early development. *J. Cell Sci.* **104**, 383–389.

Palazzo, R. E., Vaisberg, E., Cole, R. W. and Rieder, C. L. (1992) Centriole duplication in lysates of *Spisula solidissima* oocytes. *Science*, **256**, 219–221.

Palermo, G., Joris, H., Devroey, P. and Van Steirteghem, A. C. (1992) Pregnancies after intracytoplasmic injection of single spermatozoon into an oocyte. *Lancet*, **340**, 17–18.

Palermo, G., Joris, H., Derde, M.-P. *et al.* (1993) Sperm characteristics and outcome of human assisted fertilization by subzonal insemination and intracytoplasmic sperm injection. *Fertil. Steril.*, **59**, 826–835.

Palmiter, R. D., Brinster, R. L. and Hammer, R. E. (1982) Dramatic growth of mice that develop from eggs microinjected with metallothionein-growth hormone fusion genes. *Nature*, **300**, 611–615.

Parks, J. E. and Hammerstedt, R. H, (1985) Developmental changes occurring in the lipid of ram epididymal sperm plasma membranes. *Biol. Reprod.*, **32**, 653–668.

Parys, J. B., Sernett, S. W., DeLisle, S. *et al.* (1992) Isolation, characterization, and localization of the inositol 1,4,5-trisphosphate receptor protein in *Xenopus laevis* oocytes. *J. Biol. Chem.*, **267**, 18776–18782.

Parys, J. B., McPherson, S. M., Mathews, L. *et al.* (1993) Presence of inositol 1,4,5-trisphosphate receptor, calreticulin, and calsequestrin in eggs of sea urchins and *Xenopus laevis. Devel. Biol.*, **161**, 466–476.

Pasteels, J. J. (1965) Etude au microscope électronique de la réaction corticale. I. La réaction corticale de fécondation chez *Paracentrotus* et sa chronologie. II. La réaction corticale de l'oeuf vierge de *Sabellaria alveolata. J. Embryol. Exper. Morphol.*, **13**, 327–340.

Patil, J. G. and Khoo, H. W. (1996) Nuclear internalization of foreign DNA by zebrafish spermatozoa and its enhancement by electroporation. *J. Exper. Zool.*, **274**, 121–129.

Paul, M. (1975) Release of acid and changes in light-scattering properties following fertilization of *Urechis caupo* eggs. *Devel. Biol.*, **43**, 299–312.

Paul, M. and Johnston, R. N. (1978) Uptake of Ca^{2+} is one of the earliest responses to fertilization of sea urchin eggs. *J. Exper. Zool.*, **203**, 143–149.

Paules, R. S., Buccione, R., Moschel, R. C. *et al.* (1989) Mouse *mos* protooncogene product is present and functions during oogenesis. *Proc. Natl Acad. Sci, USA*, **86**, 5395–5399.

Paweletz, N. and Mazia, D. (1989) The fine structure of the formation of mitotic poles in fertilized eggs, in *The Cell Biology of Fertilization*, (eds H. Schatten and G. Schatten), Academic Press, New York, pp. 165–187.

Payan, P., Girard, J.-P. and Ciapa, B. (1983) Mechanisms regulating intracellular pH in sea urchin eggs. *Devel. Biol.*, **100**, 29–38.

Peaucellier, G., Veno, P. A. and Kinsey, W. H. (1988) Protein tyrosine phosphorylation in response to fertilization. *J. Biol. Chem.*, **263**, 13806–13811.

Pellicciari, C., Hosokawa, Y., Fukuda, M. and Romanini, M. G. M. (1983) Cytofluorometric study of nuclear sulphydryl and disulphide groups during sperm maturation in the mouse. *J. Reprod. Fertil.*, **68**, 371–376.

Peres, A. (1990) $InsP_3$- and Ca^{2+} release in single mouse oocytes. *FEBS Lett.*, **275**, 213–216.

Perreault, S. D. (1990) Regulation of sperm nuclear reaction during fertilization, in *Fertilization in Mammals*, (eds B. D. Bavister, J. Cummins and E. R. S. Roldan), Serono Symposia, Norwell, MA, pp. 285–296.

Perreault, S. D. (1992) Chromatin remodeling in mammalian zygotes. *Mutation Res.*, **296**, 43–55.

Perreault, S., Zaneveld, L. J. D. and Rogers, B. J. (1979) Inhibition of fertilization in the hamster by sodium aurothiomalate, a hyaluronidase inhibitor. *J. Reprod. Fertil.*, **60**, 461–467.

Perreault, S. D., Naish, S. J. and Zirkin, B. R. (1987) The timing of hamster sperm nuclear decondensation and male pronuclear formation is related to sperm nuclear disulfide bond content. *Biol. Reprod.*, **36**, 239–244.

Perreault, S. D., Barbee, R. R. and Slott, V. L. (1988) Importance of glutathione in the acquisition and maintenance of sperm nuclear decondensing activity in maturing hamster oocytes. *Devel. Biol.*, **125**, 181–186.

Phelps, B. M., Koppel, D. E., Primakoff, P. and Myles, D. G. (1990) Evidence that proteolysis of the surface is an initial step in the mechanism of formation of sperm cell surface domains. *J. Cell Biol.*, **111**, 1839–1847.

Phillips, S. G., Phillips, D. M., Dev, V. G. *et al.* (1976) Spontaneous cell hybridization of somatic cells present in sperm suspensions. *Exper. Cell Res.*, **98**, 429–443.

Philpott, A. and Leno, G. H. (1992) Nucleoplasmin remodels sperm chromatin in *Xenopus* egg extracts. *Cell*, **69**, 759–767.

Philpott, A., Leno, G. H. and Laskey, R. A. (1991) Sperm decondensation in *Xenopus* egg cytoplasm is mediated by nucleoplasmin. *Cell*, **65**, 569–578.

Picheral, B. (1977) La fécodation chez le Triton *Pleurodele*. II. La pénétration des spermatozoides et la réaction locale de l'oeuf. *J. Ultrastruct. Res.*, **60**, 181–202.

Pierce, K. E., Siebert, M. C., Kopf, G. S. *et al.* (1990) Characterization and localization of a mouse egg cortical granule antigen prior to and following fertilization or egg activation. *Devel. Biol.*, **141**, 381–392.

Pillai, M. C., Schields, T. S., Yanagimachi, R. and Cherr, G. N. (1993) Isolation and partial characterization of the sperm motility initiation factor from eggs of the Pacific herring, *Clupea palasii*. *J. Exper. Zool.* **265**, 336–342.

Pines, J. and Hunter, T. (1991) Human cyclins A and B2 are differentially located in the cell and undergo cell cycle-dependent nuclear transport. *J. Cell Biol.*, **115**, 1–17.

Poccia, D. (1986) Remodeling of nucleoproteins during gametogenesis, fertilization and early development. *Int. Rev. Cytol.*, **105**, 1–65.

Poccia, D. (1989) Reaction and remodeling of the sperm nucleus following fertilization, in *The Molecular Biology of Fertilization*, (eds H. Schatten and G. Schatten), Academic Press, New York, pp. 115–135.

Poccia, D. and Green, G. R. (1992) Packaging and unpackaging the sea urchin genome. *TIBS*, **17**, 223–227.

Poccia, D., Krystal, G., Nishioka, D. and Salik, J. (1978) Control of sperm chromatin structure by egg cytoplasm in the sea urchin, in *ICN-UCLA Symposium on Molecular and Cellular Biology: Cell Reproduction, XII*, (eds E. R. Dirksen, D. M. Prescott and C. F. Fox), Academic Press, New York, pp. 197–206.

Poccia, D., Salik, J. and Krystal, G. (1981) Transitions in histone variants of the male pronucleus following fertilization and evidence for a maternal store of cleavage-stage histones in the sea urchin egg. *Devel. Biol.*, **82**, 287–296.

Poccia, D., Wolff, R., Kragh, S. and Williamson, P. (1985) RNA synthesis in male pronuclei of the sea urchin. *Biochem. Biophys. Acta*, **824**, 349–356.

Poccia, D., Simpson, M. V. and Green, G. R. (1987) Histone variant transitions during sea urchin spermatogenesis. *Devel. Biol.*, **121**, 445–453.

Poccia, D., Pavon, W. and Green, G. R. (1990) 6DMAP inhibits chromatin decondensation but not sperm histone kinase in sea urchin male pronuclei. *Exper. Cell Res.*, **188**, 226–234.

Poenie, M. and Epel, D. (1987) Ultrastructural localization of intracellular calcium stores by a new cytochemical method. *J. Histochem. Cytochem.*, **35**, 939–956.

Pollard, T. D. and Craig, S. W. (1982) Mechanisms of actin polymerization. *Trends Biochem. Sci.*, **7**, 55–58.

Pollard, H. B., Burns, A. L. and Rojas, E. (1990) Synexin (annexin VII): a cytosolic calcium-binding protein which promotes membrane fusion and forms calcium channels in artificial and natural membranes. *J. Membr. Biol.*, **17**, 101–112.

Ponce, R. H., Yanagimachi, R., Urch, U. A. *et al.* (1993) Retention of hamster oolemma fusibility with spermatozoa after various enzyme treatments: a search for the molecules involved in sperm–egg fusion. *Zygote*, **1**, 163–171.

Porter, D. C. and Vacquier, V. D. (1986) Phosphorylation of sperm histone H1 is induced by the egg jelly layer in the sea urchin *Strongylocentrotus purpuratus*. *Devel. Biol.*, **116**, 203–212.

Prather, R. S. and Schatten, G. (1992) Construction of the nuclear matrix at the transition from maternal to zygotic control of development in the mouse: An immunocytochemical study. *Mol. Reprod. Devel.*, **32**, 203–208.

Raff, R. A. (1980) Masked messenger RNA and the regulation of protein synthesis in eggs and embryos, in *Cell Biology: A Comprehensive Treatise*, vol. 4, (eds D. M. Prescott and L. Goldstein), Academic Press, New York, pp. 107–136.

Raff, R. A., Greenhouse, G., Gross, K. W. and Gross, P. R. (1971) Synthesis and storage of microtubule proteins by sea urchin embryos. *J. Cell Biol.*, **50**, 516–527.

Rakow, T. L. and Shen, S. S. (1990) Multiple stores of calcium are released in the sea urchin egg during fertilization. *Proc. Natl Acad. Sci. USA*, **87**, 9285–9289.

Ralt, D., Goldenberg, M., Fetterolf, P. *et al.* (1991) Sperm attraction of follicular factor(s) correlates with human egg fertilizability. *Proc. Natl Acad. Sci. USA*, **88**, 2840–2844.

Ralt, D., Manor, M., Cohen-Dayag, A. *et al.* (1994) Chemotaxis and chemokinesis of human spermatozoa to follicular factors. *Biol. Reprod.*, **50**, 774–785.

Ramarao, C. S., Myles, D. G. and Primakoff, P. (1994) Multiple roles for PH-20 and fertilin in sperm–egg interactions. *Semin. Devel. Biol.*, **5**, 265–271.

Rappaport, R. (1971) Cytokinesis in animal cells. *Int. Rev. Cytol.*, **31**, 169–213.

Raven, C. P. (1966) *Morphogenesis: The Analysis of Molluscan Development*, Pergamon Press, Oxford.

Raven, C. P. (1972) Chemical embryology of mollusca, in *Chemical Zoology*, vol. 7, (eds M. Florkin and B. T. Scheer), Academic Press, New York, pp. 155–185.

Rebagliati, M. R., Weeks, D. L., Harvey, R. P. and Melton, D. A. (1985) Identification and cloning of localized maternal RNAs from *Xenopus* eggs. *Cell*, **42**, 769–777.

Rees, B. B., Patton, C., Grainger, J. L. and Epel, D. (1995) Protein synthesis increases after fertilization of sea urchin eggs in the absence of an increase in intracellular pH. *Devel. Biol.*, **169**, 683–698.

Reger, J. F., Fain-Maurel, M. A. and Dadoune, J. P. (1985) A freeze-fracture study on epididymal and ejaculate spermatozoa of the monkey (*Macaca fascicularis*). *J. Submicrosc. Cytol.*, **17**, 49–56.

Reimer, C. L. and Crawford, B. J. (1995) Identification and partial characterization of yolk and cortical granule proteins in eggs and embryos of the starfish, *Pisaster ochraceus. Devel. Biol.*, **167**, 439–457.

Rexroad, C. E. and Pursel, V. G. (1988) Status of gene transfer in domestic animals. *Proc. 11th Intern. Congr. Anim. Reprod. and AI*, **5**, 28–35.

Rime, H. and Ozon, R. (1990) Protein phosphatases are involved in the *in vivo* activation of histone H1 kinase in mouse oocyte. *Devel. Biol.*, **141**, 115–122.

Risley, M. S. (1990) Chromatin organization in sperm, in *Chromosomes: Eukaryotic, Prokaryotic and Viral*, (ed. K. W. Adolph), CRC Press, Boca Raton, FL, pp. 61–85.

Risley, M. S., Einheber, S. and Bumerot, D. A. (1986) Change in DNA topology during spermatogenesis. *Chromosoma*, **94**, 217–227.

Rochwerger, L., Cohen, D. J. and Cuasnicu, P. S. (1992) Mammalian sperm–egg fusion: The rat egg has complementary sites for a sperm protein that mediates gamete fusion. *Devel. Biol.*, **153**, 83–90.

Rodman, T. C., Pruslin, F. H., Hoffmann, H. P. and Allfrey, V. G. (1981) Turnover of basic chromosomal proteins in fertilized eggs: A cytoimmunochemical study of events *in vivo. J. Cell Biol.*, **90**, 351–361.

Roe, J. L., Park, H. R., Strittmatter, W. J. and Lennarz, W. J. (1989) Inhibitors of metalloendoproteases block spiculogenesis in sea urchin primary mesenchyme cells. *Exper. Cell Res.*, **181**, 542–550.

Roldan, E. R. S., Murase, T. and Shi, Q.-X. (1994) Exocytosis in spermatozoa in response to progesterone and zona pellucida. *Science*, **266**, 1578–1581.

Rosati, F., Monroy, A. and De Prisco, P. (1977) Fine structure study of fertilization in the ascidian *Ciona intestinalis. J. Ultrastruct. Res.*, **58**, 261–270.

Rose, D., Thomas, W. and Holm, C. (1990) Segregation of recombined chromosomes in meiosis I requires DNA topoisomerase II. *Cell*, **60**, 1009–1017.

Rosenthal, E. T., Hunt, T and Ruderman, J. V. (1980) Selective translation of mRNA controls the pattern of protein synthesis during early development of the surf clam, *Spisula solidissima. Cell*, **20**, 487–494.

Rosenthal, E. T., Tansey, T. and Ruderman, J. V. (1983) Sequence-specific adenylations and deadenylations accompany changes in the translation of maternal messenger RNA after fertilization in *Spisula* oocytes. *J. Mol. Biol.*, **166**, 309–327.

Rosenwaks, Z. and Muasher, S. J. (1986) Recruitment of fertilizable eggs, in *In Vitro Fertilization – Norfolk*, (eds H. W. Jones Jr, G. S. Jones, G. D. Hodgen and Z. Rosenwaks), Williams & Wilkins, Baltimore, MD, pp. 30–51.

Ruderman, J. V. and Pardue, M. L. (1977) Cell-free translation analysis of messenger RNA in echinoderm and amphibian development. *Devel. Biol.*, **60**, 48–68.

Russell, L., Peterson, R. N. and Freund, M. (1979) Direct evidence for formation of hybrid vesicles by fusion of plasma and outer acrosomal membranes during the acrosome reaction in boar spermatozoa. *J. Exper. Zool.*, **208**, 41–56.

Ryabova, L. V., Betina, M. I. and Vassetzky, S. G. (1986) Influence of cytochalasin B on oocyte maturation in *Xenopus laevis. Cell Diff.*, **19**, 89–96.

Sagata, N., Daar, I., Oskarsson, M. *et al.* (1989a) The product of the *mos* proto-oncogene product as a candidate 'initiator' for oocyte maturation. *Science*, **245**, 643–646.

Sagata, N., Watanabe, N., VandeWoude, G. F. and Ikawa, Y. (1989b) The c-*mos* proto-oncogene is a cytostatic factor responsible for meiotic arrest in verte-brate eggs. *Nature*, **342**, 572–518.

Sano, K. and Kanatani, H. (1980) External calcium ions are requisite for fertiliza-tion of sea urchin eggs by spermatozoa with reacted acrosomes. *Devel. Biol.*, **78**, 242–246.

Santiago, C. L. and Marzluff, W. F. (1989) Changes in gene activity early after fertilization, in *The Molecular Biology of Fertilization*, (eds H. Schatten and G. Schatten), Academic Press, New York, pp. 303–322.

Sardet, C. (1984) The ultrastructure of the sea urchin egg cortex isolated before and after fertilization. *Devel. Biol.*, **105**, 196–210.

Sardet, C., Carré, D., Cosson, M. P. *et al.* P. (1982) Some aspects of fertilization in marine invertebrates. *Prog. Clin. Biol. Res.*, **91**, 185–210.

Sardet, C., Speksnijder, J., Inoué, S. and Jaffe, L. (1989) Fertilization and ooplasmic movements in the ascidian egg. *Development*, **105**, 237–249.

Sardet, C., Gillot, I., Ruscher, A. *et al.* (1992) Ryanodine activates sea urchin eggs. *Devel. Growth Differ.*, **34**, 37–42.

Sargent, T. D. and Raff, R. A. (1976) Protein synthesis and messenger RNA stability in activated, enucleate sea urchin eggs are not affected by actino-mycin D. *Devel. Biol.*, **48**, 327–335.

Sasaki, H. (1984) Modulation of calcium sensitivity by a specific cortical protein during sea urchin egg cortical vesicle exocytosis. *Devel. Biol.*, **101**, 125–135.

Sathananthan, A. H. and Trounson, A. O. (1985) The human pronuclear ovum fine structure of monospermic and polyspermic fertilization *in vitro*. *Gamete Res.*, **12**, 385–398.

Sathananthan, A. H., Osborne, I. K. J., Trounson, A. *et al.* (1991) Centrioles in the beginning of human development. *Proc. Natl Acad. Sci. USA*, **88**, 4806–4810.

Sathananthan, M., MacNamee, M. C., Wick, K. and Matthews, C. D. (1992) Clinical aspects of human embryo cryopreservation, in *A Textbook of In Vitro Fertilization and Assisted Reproduction*, (eds P. R. Brinsden and P. A. Rainsbury), Parthenon Publishing Group, Park Ridge, NJ, pp. 251–263.

Sawada, T. and Schatten, G. (1988) Microtubules in ascidian eggs during meiosis, fertilization and mitosis. *Cell Motil. Cytoskel.*, **9**, 219–230.

Sawada, T. and Schatten, G. (1989) Effects of cytoskeletal inhibitors on ooplasmic segregation and microtubule organization during fertilization and early development in the ascidian *Molgula occidentalis*. *Devel. Biol.*, **132**, 331–342.

Sawada, M. T., Someno, T., Hoshi, M. and Sawada, H. (1992) Participation of 650 kDa protease (20S proteosome) in starfish oocyte maturation. *Devel. Biol.*, **150**, 414–418.

Sawalla, B. J., Moon, R. T. and Jeffrey, W. R. (1985) Developmental significance of a cortical cytoskeletal domain in *Chaetopterus* eggs. *Devel. Biol.*, **11**, 434–450.

Schackmann, R. W. and Shapiro, B. M. (1981) A partial sequence of ionic changes associated with the acrosome reaction of *Strongylocentrotus purpura-tus*. *Devel. Biol.*, **81**, 145–154.

Schackmann, R. W., Eddy, E. M. and Shapiro, B. M. (1978) The acrosome reaction of *Strongylocentrotus purpuratus* sperm. Ion requirements and movements. *Devel. Biol.*, **65**, 483–495.

Schatten, G. (1981) Sperm incorporation, the pronuclear migrations, and their relation to the establishment of the first embryonic axis: Time-lapse video microscopy of the movements during fertilization of the sea urchin *Lytechinus variegatus*. *Devel. Biol.*, **86**, 426–437.

Schatten, G. (1984) The supramolecular organization of the cytoskeleton during fertilization, in *Subcellular Biochemistry*, vol. 1, (ed. D. B. Roodyn), Plenum Press, New York, pp. 359–453.

Schatten, G. (1994) The centrosome and its mode of inheritance: The reduction of the centrosome during gametogenesis and its restoration during fertilization. *Devel. Biol.*, **165**, 299–335.

Schatten, G., Maul, G. G., Schatten, H. *et al.* (1985a) Nuclear lamins and peripheral nuclear antigens during fertilization and embryogenesis in mice and sea urchins. *Proc. Natl Acad. Sci. USA*, **82**, 4727–4731.

Schatten, G., Simerly, C. and Schatten, H. (1985b) Microtubule configurations during fertilization, mitosis and early development in the mouse and the requirement for egg microtubule-mediated motility during mammalian fertilization. *Proc. Natl Acad. Sci. USA*, **82**, 4152–4156.

Schatten, G., Simerly, C., Asai, D. J. *et al.* (1988a) Acetylated α tubulin in microtubules during mouse fertilization and early development. *Devel. Biol.*, **130**, 74–86.

Schatten, G., Simerly, C., Palmer, D. K. *et al.* (1988b) Kinetochore appearance during meiosis, fertilization and mitosis in mouse oocytes and zygotes. *Chromosoma*, **96**, 341–352.

Schatten, H. and Schatten, G. (1986) Motility and centrosomal organization during sea urchin and mouse fertilization. *Cell Motil. Cytoskel.*, **6**, 163–175.

Schatten, H., Schatten, G., Mazia, D. *et al.* (1986) Behavior of centrosomes during fertilization and cell division in mouse oocytes and in sea urchin eggs. *Proc. Natl Acad. Sci. USA*, **83**, 105–109.

Schatten, H., Simerly, C., Maul, G. and Schatten, G. (1989) Microtubule assembly is required for the formation of the pronuclei, nuclear lamin acquisition and DNA synthesis during mouse, but not sea urchin, fertilization. *Gamete Res.*, **23**, 309–322.

Schlegel, R. A., Hammerstedt, R., Cofer, G. P. and Kozarsky, K. (1986) Changes in the organization of the lipid bilayer of the plasma membrane during spermatogenesis and epididymal maturation. *Biol. Reprod.*, **34**, 379–391.

Schlichter, L. C. (1989) Ion channels in *Rana pipiens* oocytes: Change during maturation and fertilization, in *Mechanisms of Egg Activation*, (eds R. Nuccitelli, G. N. Cherr and W. H. Clark Jr), Plenum Press, New York, pp. 89–132.

Schmell, E. D. and Gulyas, B. J. (1980) Ovoperoxidase activity in ionophore treated mouse eggs. II. Evidence for the enzyme's role in hardening the zona pellucida. *Gamete Res.*, **3**, 279–290.

Schmell, E. D., Earles, B. J., Breaux, C. and Lennarz, W. J. (1977) Identification of a sperm receptor on the surface of the eggs of the sea urchin *Arbacia punctulata*. *J. Cell Biol.*, **72**, 35–46.

Schmell, E. D., Gulyas, B. J. and Hedrick, J. L. (1983) Egg surface changes during fertilization and the molecular mechanism of the block to polyspermy, in *Mechanism and Control of Animal Fertilization*, (ed. J. F. Hartmann), Academic Press, New York, pp. 365–413.

Schmidt, T., Patton, C. and Epel, D. (1982) Is there a role for the Ca^{2+} influx during fertilization of the sea urchin egg? *Devel. Biol.*, **90**, 284–290.

Schneider, E. G. (1985) Activation of Na^+-dependent transportation at fertilization in the sea urchin: Requirements of both an early event associated with exocytosis and a later event involving increased energy metabolism. *Devel. Biol.*, **108**, 152–163.

Schroeder, T. E. (1975) Dynamics of the contractile ring, in *Molecules and Cell Movement*, (eds S. Inoué and R. E. Stephens), Raven Press, New York, pp. 305–334.

Schroeder, T. E. (1979) Surface area change at fertilization: Resorption of the mosaic membrane. *Devel. Biol.*, **70**, 306–326.

Schroeder, T. E. (1980) Expression of the prefertilization polar axis in sea urchin eggs. *Devel. Biol.*, **79**, 428–443.

Schroeder, T. E. and Otto, J. J. (1984) Cyclic assembly-disassembly of cortical microtubules during maturation and early development of starfish oocytes. *Devel. Biol.*, **103**, 493–503.

Schuel, H. (1978) Secretory functions of egg cortical granules in fertilization and development. A critical review. *Gamete Res.*, **1**, 299–382.

Schuel, H., Kelly, J. W., Berger, E. R. and Wilson, W. L. (1974) Sulfated acid mucopolysaccharides in the cortical granules of eggs. Effects of quaternary ammonium salts on fertilization. *Exper. Cell Res.*, **88**, 24–30.

Schuetz, A. W. (1975) Induction of nuclear breakdown and meiosis in *Spisula solidissima* oocytes by calcium inophore. *J. Exper. Zool.*, **191**, 433–440.

Schultz, G. A. (1986) Utilization of genetic information in the preimplantation mouse embryo, in *Experimental Approaches to Mammalian Embryonic Development*, (eds J. Rossault and R. Pedersen), Cambridge University Press, New York, pp. 239–265.

SeGall, G. K. and Lennarz, W. J. (1979) Chemical characterization of the component of the jelly coat from sea urchin eggs responsible for induction of the acrosome reaction. *Devel. Biol.*, **71**, 33–48.

Seki, N., Toyama, Y. and Nagano, T. (1992) Changes in the distribution of filipin-sterol complexes in the boar sperm head plasma membrane during epididymal maturation and in the uterus. *Anat. Rec.*, **232**, 221–230.

Shalgi, R. and Phillips, D. M. (1980a) Mechanics of sperm entry in cyclic hamster. *J. Ultrastruct. Res.*, **71**, 154–161.

Shalgi, R. and Phillips, D. M. (1980b) Mechanics of *in vitro* fertilization in the hamster. *Biol. Reprod.*, **23**, 433–444.

Shalgi, R. and Phillips, D. M. (1988) Motility of rat spermatozoa at the site of fertilization. *Biol. Reprod.*, **39**, 1207–1213.

Shalgi, R., Magnus, A., Jones, R. and Phillips, D. M. (1994) Fate of sperm organelles during early embryogenesis in the rat. *Mol. Reprod. Devel.*, **37**, 264–271.

Shapiro, B. M. (1975) Limited proteolysis of some egg components is an early event following fertilization of the sea urchin, *Strongylocentrotus purpuratus*. *Devel. Biol.*, **46**, 88–102.

Shapiro, B. M. (1981) Awakening of the invertebrate egg at fertilization, in *Fertilization and Embryonic Development in Vitro*, (eds L. Mastroianni and J. D. Biggers), Plenum Press, New York, pp. 233–255.

Shapiro, B. M. and Eddy, E. M. (1980) When sperm meets egg: Biochemical mechanisms of gamete interaction. *Int. Rev. Cytol.*, **66**, 257–302.

Shapiro, B. M., Somers, C. E. and Weidman, P. J. (1989) Extracellular remodeling during fertilization, in *The Cell Biology of Fertilization*, (eds H. Schatten and G. Schatten), Academic Press, New York, pp. 251–276.

Shaw, A., Lee, Y.-H., Stout, C. D. and Vacquier, V. D. (1994) The species-specificity and structure of abalone sperm lysin. *Semin. Devel. Biol.*, **5**, 209–215.

Sheehan, M. A., Mills, A. D., Sleeman, A. M. *et al.* (1988) Steps in the assembly of replication-competent nuclei in a cell-free system from *Xenopus* eggs. *J. Cell Biol.*, **106**, 1–12.

Shen, S. S. (1983) Membrane properties and intracellular ion activities of marine invertebrate eggs and their changes during activation, in *Mechanism and Control of Animal Fertilization*, (ed. J. F. Hartmann), Academic Press, New York, pp. 213–267.

Shen, S. S. (1989) Na$^+$–H$^+$ antiport during fertilization of the sea urchin egg is blocked by W-7 but is insensitive to K252 and H-7. *Biochem. Biophys. Res. Commun.*, **161**, 1100–1108.

Shen, S. S. and Buck, W. R. (1990) A synthetic peptide of the pseudosubstrate domain of protein kinase C blocks cytoplasmic alkalinization during activation of the sea urchin egg. *Devel. Biol.*, **140**, 272–280.

Shen, S. S. and Buck, W. R. (1993) Sources of calcium in sea urchin eggs during the fertilization response. *Devel. Biol.*, **157**, 157–169.

Shen, S. S. and Burgart, L. J. (1986) 1,2-Diacylglycerol mimics phorbol 12-myristate 13-acetate activation of the sea urchin egg. *J. Cell Physiol.*, **127**, 330–340.

Shen, S. S. and Steinhardt, R. A. (1978) Direct measurement of the intracellular pH during metabolic derepression of the sea urchin egg. *Nature*, **272**, 253–254.

Sherman, M. I. (1979) Developmental biochemistry of pre-implantation mammalian embryos. *Ann. Rev. Biochem.*, **48**, 443–470.

Shi, Q.-X. and Roldan, E. R. S. (1995) Evidence that a GABA$_A$-like receptor is involved in progesterone-induced acrosomal exocytosis in mouse spermatozoa. *Biol. Reprod.*, **52**, 373–383.

Shih, R. J., O'Connor, C. M., Keem, K. and Smith, L. D. (1978) Kinetic analysis of amino acid pools and protein synthesis in amphibian oocytes and embryos. *Devel. Biol.*, **66**, 172–182.

Shilling, F., Mandel, G. and Jaffe, L. A. (1990) Activation by serotonin of starfish eggs expressing the rat serotonin 1c receptor. *Cell Reg.* **1**, 465–469.

Shilling, F. M., Carroll, D. J., Muslin, A. J. *et al.* (1994) Evidence for both tyrosine kinase and G-protein-coupled pathways leading to starfish activation. *Devel. Biol.*, **162**, 590–599.

Shimizu, T. (1979) Surface contractile activity of the *Tubifex* egg: Its relationship to the meiotic apparatus function. *J. Exper. Zool.*, **208**, 361–378.

Shimizu, T. (1981a) Cortical differentiation of the animal pole during maturation division in fertilized eggs of *Tubifex* (Annelida, Oligochaeta). I. Meiotic apparatus formation. *Devel. Biol.*, **85**, 65–76.

Shimizu, T. (1981b) Cortical differentiation of the animal pole during maturation division in fertilized eggs of *Tubifex* (Annelida, Oligochaeta). II Polar body formation. *Devel. Biol.*, **85**, 77–88.

Shimizu, T. (1988) Localization of actin networks during early development of *Tubifex* embryos. *Devel. Biol.*, **125**, 321–331.

Shirai, H., Hosoya, N., Sawada, T. *et al.* (1990) Dynamics of mitotic apparatus formation and tubulin content during oocyte maturation in starfish. *Devel. Growth Differ.*, **32**, 521–529.

Showman, R. M., Wells, D. E., Anstrom, J. *et al.* (1982) Message-specific sequestration of maternal histone mRNA in the sea urchin egg. *Proc. Natl Acad. Sci. USA,* **79**, 5944–5947.

Shur, B. D. and Hall, N. G. (1982a) Sperm surface galactosyltransferase activities during *in vitro* capacitation. *J. Cell Biol.,* **95**, 567–573.

Shur, B. D. and Hall, N. G. (1982b) A role for mouse sperm surface galactosyltransferase in sperm binding to the egg zona pellucida. *J. Cell Biol.,* **95**, 574–579.

Simerly, C., Balczon, R., Brinkley, B. R. and Schatten, G. (1990) Microinjected kinetochore antibodies interfere with chromosome movement in meiotic and mitotic mouse oocytes. *J. Cell Biol.,* **111**, 1491–1504.

Simerly, C. R., Hecht, N. B., Goldberg, E. and Schatten, G. (1993) Tracing the incorporation of the sperm tail in the mouse zygote and early embryo using an anti-testicular α-tubulin antibody. *Devel. Biol.,* **158**, 536–548.

Simerly, C., Wu, G.-J., Zoran, S. *et al.* (1995) The paternal inheritance of the centrosome, the cell's microtubule-organizing center, in humans, and the implications for infertility. *Nature Med.,* **1**, 47–52.

Simmel, E. B. and Karnofsky, D. A. (1961) Observations on the uptake of tritiated thymidine in the pronuclei of fertilized sand dollar embryos. *J. Biophys. Biochem. Cytol.,* **10**, 59–65.

Singer, S. J. and Nicolson, G. L. (1972) The fluid mosaic model of the structure of cell membranes. *Science,* **175**, 720–731.

Skoblina, M. (1976) Role of karyoplasm in the emergence of capacity of egg cytoplasm to induce DNA synthesis in transplanted sperm nuclei. *J. Embryol, Exper. Morphol.,* **36**, 67–72.

Sluder, G. and Rieder, C. L. (1985) Experimental separation of pronuclei in fertilized sea urchin eggs: Chromosomes do not organize a spindle in the absence of centrosomes. *J. Cell Biol.,* **100**, 897–903.

Sluder, G., Miller, F. J., Lewis, K. *et al.* (1989) Centrosome inheritance in starfish zygotes: Selective loss of the maternal centrosome after fertilization. *Devel. Biol.,* **131**, 567–579.

Sluder, G., Miller, F. J. and Lewis, K. (1993) Centrosome inheritance in starfish zygotes. II. Selective suppression of the maternal centrosome during meiosis. *Devel. Biol.,* **155**, 58–67.

Smith, L. (1989) The induction of oocyte maturation: transmembrane signalling events and regulation of the cell cycle. *Development,* **107**, 685–699.

Smith, L. D. and Ecker, R. E. (1965) Protein synthesis in enucleated eggs of *Rana pipiens. Science,* **150**, 777–779.

Smith, L. D. and Ecker, R. E. (1970) Uterine suppression of biochemical and morphogenetic events in *Rana pipiens. Devel. Biol.,* **22**, 622–637.

Somers, C. E., Battaglia, D. E. and Shapiro, B. M. (1989) Localization and development fate of ovoperoxidase and proteoliaisin, two proteins involved in fertilization envelope assembly. *Devel. Biol.,* **131**, 226–235.

Sorokin, S. P. (1968) Reconstruction of centriole formation and ciliogenesis in mammalian lungs. *J. Cell Sci.,* **3**, 207–230.

Speksnijder, J. E. (1992) The repetitive calcium waves in the fertilized ascidian egg are initiated near the vegetal pole by a cortical pacemaker. *Devel. Biol.,* **153**, 259–271.

Speksnijder, J. E., Sardet, C. and Jaffe, L. F. (1990) The activation wave of calcium in the ascidian egg and its role in ooplasmic segregation. *J. Cell Biol.,* **110**, 1589–1598.

Spudich, A. and Spudich J. A. (1979) Actin in triton-treated cortical preparations of unfertilized and fertilized sea urchin eggs. *J. Cell Biol.*, **82**, 212–226.

Stambaugh, R., Smith, M. and Faltas, S. (1975) An organized distribution of acrosomal proteinase in rabbit sperm acrosomes. *J. Exper. Zool.*, **193**, 119–122.

Standart, N. (1992) Masking and unmasking of material mRNA. *Semin. Devel. Biol.*, **3**, 367–379.

Standart, N., Bray, S. J., George, E. L. *et al.* (1985) The small subunit of ribonucleotide reductase is encoded by one of the most abundant translationally regulated maternal RNAs in clam and sea urchin eggs. *J. Cell Biol.*, **100**, 1968–1976.

Standart, N., Minshull, J., Pines, J. and Hunt, T. (1987) Cyclin synthesis, modification and destruction during meiotic maturation of the starfish oocyte. *Devel. Biol.*, **124**, 248–258.

Stearns, T. and Kirschner, M. (1994) *In vitro* reconstitution of centrosome assembly and function: The central role of γ tubulin. *Cell* , **76**, 623–637.

Stefanini, M., Oura, C. and Zamboni, L. (1969) Ultrastructure of fertilization in the mouse. II. Penetration of sperm into the ovum. *J. Submicrosc. Cytol.*, **1**, 1–23.

Steinhardt, R. A., Epel, D., Carroll, E. J. and Yanagimachi, R. (1974) Is calcium inophore a universal activator for unfertilized eggs? *Nature*, **252**, 41–43.

Steinhardt, R. A., Zucker, R. and Schatten, G. (1977) Intracellular calcium at fertilization in the sea urchin egg. *Devel. Biol.*, **58**, 185–196.

Steinhardt, R. A., Bi, G. and Alderton, J. M. (1994) Cell membrane resealing by a vesicular mechanism similar to neurotransmitter release. *Science*, **263**, 390–393.

Steptoe, P. C. and Edwards, R. G. (1978) Birth after the re-implantation of a human embryo (letter). *Lancet*, **ii**, 366.

Stice, S. L. and Robl, J. M. (1990) Activation of mammalian oocytes by a factor obtained from rabbit sperm. *Mol. Reprod. Devel.*, **25**, 272–280.

Stick, R. and Hausen, P. (1985) Changes in the nuclear lamina composition during early development of *Xenopus laevis*. *Cell*, **41**,191–200.

Stith, B. J., Espinoza, R., Roberts, D. and Smart T. (1994) Sperm increase inositol 1,4,5-trisphosphate mass in *Xenopus laevis* eggs preincubated with calcium buffers or heparin. *Devel. Biol.*, **165**, 206–215.

Storey, B. T., Lee, M. A., Muller, C. *et al.* (1984) Binding of mouse spermatozoa to the zona pellucidae of mouse eggs in cumulus: Evidence that the acrosomes remain substantially intact. *Biol. Reprod.*, **31**, 1119–1128.

Stricker, S. A. and Schatten, G. (1989) Nuclear envelope disassembly and nuclear lamina depolymerization during germinal vesicle breakdown in starfish. *Devel. Biol.*, **135**, 87–98.

Stricker, S., Prather, R., Simerly, C. *et al.* (1989) Nuclear architectural changes during fertilization and development, in *The Cell Biology of Fertilization*, (eds H. Schatten and G. Schatten), Academic Press, New York, pp. 225–250.

Stricker, S. A., Centonze, V. E., Paddock, S. W. and Schatten, G. (1992) Confocal microscopy of fertilization-induced calcium dynamics in sea urchin eggs. *Devel. Biol.*, **149**, 370–380.

Stricker, S. A., Centonze, V. E. and Melendez, R. F. (1994) Calcium dynamics during starfish oocyte maturation. *Devel. Biol.*, **166**, 34–58.

Summers, R. G. and Hylander, B. L. (1975) Species-specificity of acrosome reaction and primary gamete binding in echinoids. *Exper. Cell Res.*, **96**, 63–68.

Surani, M. A. H., Barton, S. C. and Norris, M. L. (1984) Development of reconstituted mouse eggs suggests imprinting of the genome during gametogenesis. *Nature*, **308**, 548–550.

Surani, M. A. H., Barton, S. C. and Norris, M. L. (1986) Nuclear transplantation in the mouse: Heritable differences between parental genomes after activation of the embryonic genome. *Cell*, **45**, 127–136.

Surani, M. A. H., Kothury, R., Allen, N. D. *et al.* (1990) Genome imprinting and development in the mouse. *Development*, **Suppl.**, 89–98.

Suzuki, F. (1990) Morphological aspects of sperm maturation, in *Fertilization in Mammals*, (eds B. D. Bavister, J. Cummins and E. R. S. Roldan), Serono Symposia USA, Norwell, MA, pp. 65–75.

Swann, K. (1993) The soluble sperm oscillogen hypothesis. *Zygote*, **1**, 273–276.

Swann, K. and Whitaker, M. J. (1985) Stimulation of the Na/H exchanger of sea urchin eggs by phorbol ester. *Nature*, **314**, 274–277.

Swann, K. and Whitaker, M. (1986) The part played by inositol trisphosphate and calcium in the propagation of the fertilization wave in sea urchin eggs. *J. Cell Biol.*, **103**, 2333–2342.

Swann, K., McCulloh, D. H., McDougall, A. *et al.* (1992) Sperm-induced currents at fertilization in sea urchin eggs injected with EGTA and neomycin. *Devel. Biol.*, **151**, 552–563.

Swenson, K. I., Farrell, K. M. and Ruderman, J. V. (1986) The clam embryo protein cyclin A induces entry into M phase and the resumption of meiosis in *Xenopus* oocytes. *Cell*, **47**, 861–870.

Swenson, K., Westendorf, J., Hunt, T. and Ruderman, J. (1989) Cyclins and regulation of the cell cycle in early embryos, in *The Molecular Biology of Fertilization*, (eds H. Schatten and G. Schatten), Academic Press, New York, pp. 211–232.

Swezey, R. R. and Epel, D. (1995) The *in vivo* rate of glucose-6-phosphate dehydrogenase activity in sea urchin eggs determined with a photolabile caged substrate. *Devel. Biol.*, **169**, 733–744.

Szöllösi, D. (1965) The fate of sperm middle piece mitochondria in the rat egg. *J. Exper. Zool.*, **159**, 367–378.

Szöllösi, D. and Ozil, J. P. (1991) De novo formation of centrioles in parthenogenetically activated diploidized rabbit embryos. *Biol. Cell*, **72**, 61–66.

Szöllösi, D. and Ris, H. (1961) Observations on sperm penetration in the rat. *J. Biophys. Biochem. Cytol.*, **10**, 275–283.

Szöllösi, D., Calarco, P. and Donahue, R. P. (1972) Absence of centrioles in the first and second meiotic spindles of mouse oocytes. *J. Cell Sci.*, **11**, 521–541.

Szöllösi, D., Czolowska, R., Salytynska, M. S. and Tarkowski, A. K. (1986) Remodeling of thymocyte nuclei in activated mouse oocytes: An ultrastructural study. *Eur. J. Cell Biol.*, **42**, 140–157.

Szöllösi, D., Czolowska, M., Szöllösi, M. and Tarkowski, A. K. (1988) Remodeling of mouse thymocyte nuclei depends on the time of their transfer into activated, homologous oocytes. *J. Cell Sci.*, **91**, 603–613.

Szöllösi, D., Szöllösi, M. S., Czolowska, R. and Tarkowski, A. K. (1990) Sperm penetration into immature mouse oocytes and nuclear changes during maturation: an EM study. *Biol. Cell*, **69**, 53–64.

Szöllösi, M. S., Borusk, E. and Szöllösi, D. (1994) Relationships between sperm nucleus remodeling and cell cycle progression of fragments of mouse parthenogenotes. *Mol. Reprod. Devel.*, **37**, 146–156.

Tachibana, K., Ishiura, M., Uchida, T. and Kishimoto, T. (1990) The starfish egg mRNA responsible for meiosis reinitiation encodes cyclin. *Devel. Biol.*, **140**, 241–252.

Talbot, P. (1985) Sperm penetration through oocyte investments in mammals. *Am. J. Anat.*, **174**, 331–346.

Tan, S. L. and Brinsden, P. R. (1992) Alternative assisted conception techniques, in *A Textbook of In Vitro Fertilization and Assisted Reproduction*, (eds P. R. Brinsden and P. A. Rainsbury), Parthenon Publishing Group, Park Ridge, NJ, pp. 237–250.

Tash, J. S. and Means, A. R. (1982) Regulation of protein phosphorylation and motility of sperm by cyclic adenosine monophosphate and calcium. *Biol. Reprod.*, **26**, 745–763.

Tatone, C., Van Eekelen, C. G. and Colonna, R. (1994) Plasma membrane block to sperm entry occurs in mouse eggs upon parthenogenetic activation. *Mol. Reprod. Devel.*, **38**, 200–208.

Taylor, P. J. and Kredentser, J. V. (1992) Diagnostic and therapeutic laparoscopy and hysteroscopy and their relationship to *in vitro* fertilization, in *A Textbook of In Vitro Fertilization and Assisted Reproduction*, (eds P. R. Brinsden and P. A. Rainsbury), Parthenon Publishing Group, Park Ridge, NJ, pp. 73–92.

Terasaki, M. and Jaffe, L. A. (1991) Organization of the sea urchin egg endoplasmic reticulum and its reorganization at fertilization. *J. Cell Biol.*, **114**, 929–940.

Terasaki, M. and Sardet, C. (1991) Demonstration of calcium uptake and release by sea urchin egg cortical endoplasmic reticulum. *J. Cell Biol.*, **115**, 1031–1037.

Terasaki, M., Henson, J., Begg, D. *et al.* (1991) Characterization of sea urchin egg endoplasmic reticulum in cortical preparations. *Devel. Biol.*, **148**, 398–401.

Tesarik, J. and Kopecny, V. (1989) Nucleic acid synthesis and development of human male pronucleus. *J. Reprod. Fertil.*, **86**, 549–558.

Thadani, V. M. (1979) Injection of sperm heads into immature rat oocyte. *J. Exper. Zool.*, **210**, 161–168.

Thaler, C. D. and Cardullo, R. A. (1995) Biochemical characterization of a glycosylphosphatidylinositol-linked hyaluronidase on mouse sperm. *Biochemistry*, **34**, 7788–7795.

Thibault, C. and Gérard, M. (1973) Cytoplasmic and nuclear maturation of rabbit oocytes *in vitro Ann. Biol. Anim. Biochem. Biophys.*, **13**, 145–156.

Tilney, L. G. (1975a) The role of actin in nonmuscle cell motility, in *Molecules and Cell Movement*, (eds S. Inoué and R. E. Stephens), Raven Press, New York, pp. 339–388.

Tilney, L. G. (1975b) Actin filaments in the acrosomal reaction of *Limulus* sperm. *J. Cell Biol.*, **64**, 289–310.

Tilney, L. G. (1978) The polymerization of actin. V. A new organelle, the actomere, that initiates the assembly of actin filaments in *Thyone* sperm. *J. Cell Biol.*, **77**, 551–564.

Tilney, L. G. (1980) Membrane events in the acrosomal reaction of *Limulus* and *Mytilus* sperm, in *Membrane–Membrane Interactions*, (ed. N. B. Gilula), Raven Press, New York, pp. 59–80.

Tilney, L. G. (1985) The acrosome reaction, in *Biology of Fertilization*, vol. 2, (eds C. B. Metz and A. Monroy), Academic Press, New York, pp. 157–213.

Tilney, L. G. and Jaffe, L. A. (1980) Actin, microvilli and the fertilization cone of sea urchin eggs. *J. Cell Biol.*, **87**, 771–782.

Tilney, L. G., Kiehart, D. P., Sardet, C. and Tilney, M. (1978) Polymerization of actin. IV. Role of Ca^{2+} and H^+ in the assembly of actin and in membrane fusion in the acrosomal reaction of echinoderm sperm. *J. Cell Biol.*, **77**, 536–550.

Tombes, R. M., Simerly, C., Borisy, G. G. and Schatten, G. (1992) Meiosis, egg activation and nuclear envelope breakdown are differentially reliant on Ca^{2+}, whereas germinal vesicle breakdown is Ca^{2+} independent in the mouse oocyte. *J. Cell Biol.*, **117**, 799–811.

Toyoda, Y. and Naito, K. (1990) IVF in domestic animals, in *Fertilization in Mammals*, (eds B. D. Bavister, J. Cummins and E. R. S. Roldan), Serono Symposia USA, Norwell, MA, pp. 335–347.

Trimmer, J. S. and Vacquier, V. D. (1986) Activation of sea urchin gametes. *Ann. Rev. Cell Biol.*, **2**, 1–26.

Trounson, A. (1989a) Fertilization and embryo culture, in *Clinical In Vitro Fertilization*, (eds C. Wood and A. Trounson), Springer-Verlag, New York, pp. 33–50.

Trounson, A. (1989b) Embryo cryopreservation, in *Clinical In Vitro Fertilization*, (eds C. Wood and A. Trounson), Springer-Verlag, New York, pp. 127–142.

Trounson, A. O., Willadsen, S. M. and Rowson, L. E. A. (1977) Fertilization and development capability of bovine follicular oocyte matured *in vitro* and *in vivo* and transferred to the oviducts of rabbits and cows. *J. Reprod. Fertil.*, **51**, 321–327.

Tulsiani, D. R. P., Skudlarek, M. D., Holland, M. K. and Orgebin-Crist, M. C. (1993) Glycosylation of rat sperm plasma membrane during epididymal maturation. *Biol. Reprod.*, **48**, 417–428.

Turner, P. R., Sheetz, M. P. and Jaffe, L. A. (1984) Fertilization increases the polyphosphoinositide content of sea urchin eggs. *Nature*, **310**, 414–415.

Turner, P. R., Jaffe, L. A. and Fein, A. (1986) Regulation of cortical granule exocytosis in sea urchin eggs by inositol 1,4,5-trisphosphate and GTP-binding proteins. *J. Cell Biol.*, **102**, 70–76.

Turner, P. R., Jaffe, L. A. and Primakoff, P. (1987) A cholera-toxin sensitive G-protein stimulates exocytosis in sea urchin eggs. *Devel. Biol.*, **120**, 577–583.

Twigg, J., Patel, R. and Whitaker, M. J. (1988) Translational control of InsP3-induced chromatin condensation during the early cell cycles of sea urchin embryos. *Nature*, **332**, 366–369.

Tyler, A. (1940) Sperm agglutination in the keyhole limpet, *Megathura crenulate*. *Biol. Bull.*, **78**, 159–178.

Uehara, T. and Yanagimachi, R. (1976) Microsurgical injection of spermatozoa into hamster eggs with subsequent transformation of sperm nuclei into male pronuclei. *Biol. Reprod.*, **15**, 467–470.

Uehara, T. and Yanagimachi, R. (1977) Behavior of nuclei of testicular, caput and cauda epididymal spermatozoa injected into hamster eggs. *Biol. Reprod.*, **16**, 315–321.

Ulitzer, N. and Gruenbaum, Y. (1989) Nuclear envelope assembly around sperm chromatin in cell-free preparation from *Drosophilia* embryos. *FEBS Lett.*, **259**, 113–116.

Unsworth, B. R. and Kalenas, M. S. (1975) Changes in ribosome-associated proteins during sea urchin development. *Differentiation*, **3**, 21–27.

Usui, N. and Takahashi, I. (1986) Membrane differentiations in echinoderm spermatozoa before and during the acrosome reaction as revealed by freeze fracture. *J. Ultrastruct. Mol. Struct. Res.*, **96**, 64–76.

Usui, N. and Yanagimachi, R. (1976) Behavior of hamster sperm nuclei incorporated into eggs at various stages of maturation, fertilization and early development: The appearance and disappearance of factors involved in sperm chromatin decondensed in egg cytoplasm. *J. Ultrastruct. Res.*, **57**, 276–288.

Vacquier, V. D. (1975) The isolation of intact cortical granules from sea urchin eggs: Calcium ions trigger granule discharge. *Devel. Biol.*, **43**, 62–74.

Vacquier, V. D. (1979) The fertilizing capacity of sea urchin sperm rapidly decreases after induction of the acrosome reaction. *Devel. Growth Differ.*, **21**, 61–69.

Vacquier, V. D. (1980) The adhesion of sperm to sea urchin eggs, in *The Cell Surface: Mediator of Developmental Processes*, (eds S. Subtelny and N. K. Wessels), Academic Press, New York, pp. 151–168.

Vacquier, V. D. (1981) Dynamic changes of the egg cortex. *Devel. Biol.*, **84**, 1–26.

Vacquier, V. D. and Lee, Y.-H. (1993) Abalone sperm lysin: unusual mode of evolution of a gamete recognition protein. *Zygote*, **1**, 181–196.

Vacquier, V. D. and Moy, G. W. (1977) Isolation of bindin: The protein responsible for adhesion of sperm to sea urchin eggs. *Proc. Natl Acad. Sci. USA*, **74**, 2456–2460.

Vacquier, V. D. and Payne, J. E. (1973) Methods for quantitating sea urchin sperm–egg binding. *Exper. Cell Res.*, **82**, 227–235.

Vacquier, V. D., Tegner, M. J. and Epel, D. (1973) Protease released from sea urchin eggs at fertilization alters the vitelline layer and aids in preventing polyspermy. *Exper. Cell Res.*, **80**, 111–119.

Vacquier, V. D., Porter, D. C., Keller, S. H. and Ankermann, M. (1989) Egg jelly induces the phosphorylation of histone H3 in spermatozoa of the sea urchin *Arbacia punctulata*. *Devel. Biol.*, **133**, 111–118.

Van Blerkom, J. (1977) Molecular approaches to the study of oocyte maturation and embryonic development, in *Immunology of the Gametes*, (eds M. Edidin and M. H. Johnson), Cambridge University Press, Cambridge, pp. 187–202.

Van Blerkom, J. (1979) Molecular differentiation of the rabbit ovum. III. Fertilization-autonomous polypeptide synthesis. *Devel. Biol.*, **72**, 188–194.

Van Blerkom, J. (1991) Microtubule mediation of cytoplasmic and nuclear maturation during the early stages of resumed meiosis in cultured mouse oocytes. *Proc. Natl Acad. Sci. USA*, **88**, 5031–5035.

Van Blerkom, J. (1995) *The Biological Basis of Early Reproductive Facture in the Human: Applications to Medically-assisted Conception and the Treatment of Infertility*, Oxford University Press, Oxford.

Van Blerkom, J. and Motta, P. (1979) *The Cellular Basis of Mammalian Reproduction*. Urban & Schwarzenberg, Baltimore, MD.

Vanderhaeghen, P., Schurmans, S., Vassart, G. and Parmentier, M. (1993) Olfactory receptors are displayed on dog mature sperm cells. *J. Cell Biol.*, **123**, 1441–1452.

Van Meel, F. C. M and Pearson, P. L. (1979) Do human spermatozoa reactivate in the cytoplasm of somatic cells? *J. Cell. Sci.*, **35**, 105–122.

Van Steirteghem, A. C., Nagy, Z., Joris, H. *et al.* (1993) High fertilization and implantation rates after intracytoplasmic sperm injection. *Human Reprod.*, **8**, 1061–1066.

Veeck, L. L. (1986) Morphological estimation of mature oocytes and their preparation for insemination, in *In Vitro Fertilization – Norfolk*, (eds H. W. Jones Jr, G. S. Jones, G. D. Hodgen and Z. Rosenwaks), Williams & Wilkins, Baltimore, MD, pp. 81–93.

Veeck, L. L. and Maloney, M. (1986) Insemination and fertilization, in *In Vitro Fertilization – Norfolk*, (eds H. W. Jones Jr, G. S. Jones, G. D. Hodgen and Z. Rosenwaks), Williams & Wilkins, Baltimore, MD, pp. 168–182.

Verlhac, M.-H., de Pennart, H., Marco, B. *et al.* (1993) MAP kinase becomes stably activated at metaphase and is associated with microtubule-organizing centers during meiotic maturation of mouse oocytes. *Devel. Biol.*, **158**, 330–340.

Vernon, M and Shapiro, B. M. (1977) Binding of concanavalin A to the surface of sea urchin eggs and its alteration upon fertilization. *J. Biol. Chem.*, **252**, 1286–1292.

Vernon, M., Foerder, C., Eddy, E. M. and Shapiro, B. M. (1977) Sequential biochemical and morphological events during assembly of the fertilization membrane of the sea urchin. *Cell*, **10**, 321–328.

Vigers, G. P. A. and Lohka, M. J. (1991) A distinct vesicle population targets membrane and pore complexes to the nuclear envelope in *Xenopus* eggs. *J. Cell Biol.*, **112**, 545–556.

Virji, N., Phillips, D. M. and Dunbar, B. S. (1990) Identification of extracellular proteins in the rat cumulus oophorus. *Mol. Reprod. Devel.*, **25**, 339–344.

Walensky, L. D. and Snyder, S. H. (1995) Inositol 1,4,5-trisphosphate receptors selectively localized to the acrosomes of mammalian sperm. *J. Cell Biol.*, **130**, 857–869.

Walker, J., Dale, M. and Standart, N. (1996) Unmasking mRNA in clam oocytes: Role of phosphorylation of a 3′ UTR masking element-binding protein at fertilization. *Devel. Biol.*, **173**, 292–305.

Ward, W. S. (1993) Deoxyribonucleic acid loop domain tertiary structure in mammalian spermatozoa. *Biol. Reprod.*, **48**, 1193–1201.

Ward, C. R. and Kopf, G. F. (1993) Molecular events mediating sperm activation. *Devel. Biol.*, **158**, 9–34.

Ward, G. E., Brokaw, C. J., Garbers, D. L. and Vacquier, V. D. (1985) Chemotaxis of *Arbacia punctulata* spermatozoa to resact, a peptide from the egg jelly layer. *J. Cell Biol.*, **101**, 2324–2329.

Ward, W. S., Partin, A. W. and Coffey, D. S. (1989) DNA loop domains in mammalian spermatozoa. *Chromosoma*, **98**, 153–159.

Ward, W. S., McNeil, J., deLarca, J. and Lawrence, J. (1996) Localization of three genes in the hook-shaped sperm nucleus by fluorescent *in situ* hybridization. *Biol. Reprod.*, **54**, 1271–1278.

Wassarman, P. M. (1988) Zona pellucida glycoproteins. *Annu. Rev. Biochem.*, **57**, 415–442.

Wassarman, P. M. (1990) Profile of a mammalian sperm receptor. *Development*, **108**, 1–17.

Wassarman, P. M. (1993) Mammalian eggs, sperm and fertilisation: dissimilar cells with a common goal. *Semin. Devel. Biol.*, **4**, 189–197.

Wassarman, P. M., Josefowicz, W. J. and Letourneau, G. E. (1976) Meiotic maturation of mouse oocytes *in vitro*: Inhibition of maturation at specific stages of nuclear progression. *J. Cell Sci.*, **22**, 531–545.

Wassarman. P. M., Bleil, J. D., Cascio, S. M. *et al.* (1981) Programming of gene expression during mammalian oogenesis, in *Bioregulators of Reproduction*, (eds G. Jagiello and H. J. Vogel), Academic Press, New York, pp. 119–150.

Wassarman, P. M., Bleil, J. D., Florman, H. M. *et al.* (1985) The mouse egg's receptor for sperm: What is it and how does it work? *Cold Spring Harbor Symp. Quant. Biol.*, **50**, 11–19.

Wasserman, W. J. and Smith, L. D. (1978) The cyclic behavior of a cytoplasmic factor controlling nuclear membrane breakdown. *J. Cell Biol.*, **78**, R15-R22.

Watanabe, N., VandeWoude, G. F., Ikawa, Y. and Sagata, N. (1989) Specific proteolysis of the c-mos proto-oncogene product by calpain on fertilization of *Xenopus* eggs. *Nature*, **342**, 505–511.

Watanabe, N., Hunt, T., Ikawa, Y. and Sagata N. (1991) The meiotic release on fertilization of *Xenopus* eggs is independent of the inactivation of cytostatic factor (or Mos). *Nature*, **352**, 247–248.

Weber, A., Kubiak, J. Z., Arlinghaus, R. B. *et al.* (1991) c-*mos* proto-oncogene product is partly degraded after release from meiotic arrest and persists during interphase in mouse zygotes. *Devel. Biol.*, **148**, 393–397.

Webster, D. R. and Borisy, G. G. (1988) Microtubules are acetylated in domains which turn over slowly. *J. Cell Sci.*, **92**, 57–65.

Webster, S. D. and McGaughey, R. W. (1990) The cortical cytoskeleton and its role in sperm penetration of the mammalian egg. *Devel. Biol.*, **142**, 61–74.

Weeks, D. L., Bailey, C., Bullock, E. *et al.* (1995) mRNA localized to the animal hemisphere of *Xenopus laevis* oocytes and early embryos and the proteins that they encode (in press).

Weidman, P. J. and Shapiro, B. M. (1987) Regulation of extracellular matrix assembly: *In vitro* reconstitution of a partial fertilization envelope from isolated components. *J. Cell Biol.*, **105**, 561–567.

Weisenberg, R. C. and Rosenfeld, A. C. (1975) *In vitro* polymerization of microtubules into asters and spindles in homogenates of surf clam eggs. *J. Cell Biol.*, **64**, 146–158.

Wessel, G. M. (1995) A protein of the sea urchin cortical granules is targeted to the fertilization envelope and contains an LDL receptor-like motif. *Devel. Biol.*, **167**, 388–397.

Whalley, T., McDougall, A., Crossley, I. *et al.* (1992) Internal calcium release and activation of sea urchin eggs by cGMP are independent of the phosphoinositide signaling pathway. *Mol. Biol. Cell*, **3**, 373–383.

Whitaker, M. J. and Irvine, R. F. (1984) Inositol 1, 4, 5 trisphosphate microinjection activates sea urchin eggs. *Nature*, **312**, 636–639.

Whitaker, M. and Patel, R. (1990) Calcium and cell cycle control. *Development* **108**, 525–542.

Whitaker, M. J. and Steinhardt, R. A. (1981) The relation between the increase in reduced nicotinamide nucleotides and the initiation of DNA synthesis in sea urchin eggs. *Cell*, **25**, 95–103.

Whitaker, M. J. and Steinhardt, R. A. (1982) Ionic regulation of egg activation. *Q. Rev. Biophys.*, **15**, 593–666.

Whitaker, M. J. and Steinhardt, R. A. (1983) Evidence in support of the hypothesis of an electrically mediated fast block to polyspermy in sea urchin eggs. *Devel. Biol.*, **95**, 244–248.

Whitaker, M., Swann, K. and Crossley, I. (1989) What happens during the latent period at fertilization, in *Mechanisms of Egg Activation*, (eds R. Nuccitelli, G. N. Cherr and W. H. Clark Jr), Plenum Press, New York, pp. 157–171.

Wickramasinghe, D. and Albertini, D. F. (1992) Centrosome phosphorylation and the developmental expression of meiotic competence in mouse oocytes. *Devel. Biol.*, **152**, 62–74.

Wickramasinghe, D., Ebert, K. M. and Albertini, D. F. (1991) Meiotic competence acquisition is associated with the appearance of M-phase characteristics in growing mouse oocytes. *Devel. Biol.*, **143**, 162–172.

Wiesel, S. and Schultz, G. A. (1981) Factors which may affect removal of protamine from sperm DNA during fertilization in the rabbit. *Gamete. Res.*, **4**, 25–34.

Wikramanayake, A. H., Uhlinger, K. R., Griffin, F. J. and Clark, W. H. Jr (1992) Sperm of the shrimp *Sicyonia ingentis* undergo a biphase capacitation accompanied by morphological changes. *Devel. Growth Differ.*, **34**, 347–355.

Wilding, M. (1996) Calcium and cell cycle control in early embryos. *Zygote*, **4**, 1–6.

Wildt, D. E. (1990) Potential applications of IVF technology for species conservation, in *Fertilization in Mammals*, (eds B. D. Bavister, J. Cummins and E. R. S. Roldan), Serono Symposia USA, Norwell, MA, pp. 349–364.

Williams, C. J., Schultz, R. M. and Kopf, G. S. (1992) Role of G proteins in mouse egg activation: Stimulatory effects of acetylcholine on the ZP2 to ZP2$_f$ conversion and pronuclear formation in eggs expressing a functional m1 muscarinic receptor. *Devel. Biol.*, **157**, 288–296.

Wilson, E. B. (1925) *The Cell in Development and Heredity*, Macmillan, New York.

Wincek, T. J., Parrish, R. F. and Polakoski, K. L. (1979) Fertilization: A uterine glycosaminoglycan stimulates the conversion of sperm proacrosin to acrosin. *Science*, **203**, 553–554.

Winkler, M. M., Steinhardt, R. A., Grainger, J. L. and Minning, L. (1980) Dual ionic controls for the activation of protein synthesis at fertilization. *Nature*, **287**, 558–560.

Winkler, M. M., Nelson, E. M., Lashbrook, C. and Hershey, J. W. (1985) Multiple levels of regulation of protein synthesis at fertilization in sea urchin eggs. *Devel. Biol.*, **107**, 290–300.

Wolf, D. E. and Voglmayr, J. K. (1984) Diffusion and regionalization in membranes of maturing ram spermatozoa. *J. Cell Biol.*, **98**, 1678–1684.

Wolf, D. E. and Ziomek, C. A. (1983) Regionalization and lateral diffusion of membrane proteins in unfertilized mouse eggs. *J. Cell Biol.*, **96**, 1786–1790.

Wolf, D. E., Kinsey, W., Lennarz, W. and Edidin, M. (1981a) Changes in the organization of the sea urchin egg plasma membrane upon fertilization: Indications from the lateral diffusion rates of lipid-soluble fluorescent dyes. *Devel. Biol.*, **81**, 133–138.

Wolf, D. E., Edidin, M. and Handyside, A. H. (1981b) Changes in the organization of the mouse egg plasma membrane upon fertilization and first cleavage. Indications from the lateral diffusion rates of fluorescent lipid analogs. *Devel. Biol.*, **85**, 195–198.

Wolf, D. P and Hamada, M. (1977) Induction of zonal and egg plasma membrane blocks to sperm penetration in mouse eggs with cortical granule exudate. *Biol. Reprod.*, **17**, 350–354.

Wolf, D. P., Nicosia, S. V. and Hamada, M. (1979) Premature cortical granule loss does not prevent sperm penetration of mouse eggs. *Devel. Biol.*, **71**, 22–32.

Wolf, R. (1978) The cytaster, a colchicine-sensitive migration organelle of cleavage nuclei in an insect egg. *Devel. Biol.*, **62**, 464–472.

Wood, C. and Trounson, A. (1989) *Clinical In Vitro Fertilization*, Springer-Verlag, London.

Wooten, M. W., Voglmayr, J. K. and Wrenn, R. W. (1987) Characterization of cAMP-dependent protein kinase and its endogenous substrate proteins in ram testicular, cauda epididymal and ejaculated spermatozoa. *Gamete Res.*, **16**, 57–68.

Wright, S. J. and Longo, F. J. (1988) Sperm nuclear enlargement in fertilized hamster eggs is related to meiotic maturation of the maternal chromatin. *J. Exper. Zool.*, **247**, 155–165.

Wright, S. J. and Schatten, G. (1990) Teniposide, a topoisomerase II inhibitor, prevents chromosome condensation and separation but not decondensation in fertilized surf clam, *Spisula solidissima* oocytes. *Devel. Biol.*, **142**, 224–232.

Wright, G., Wiker, S., Elsner, C. *et al.* (1990) Observations on the morphology of human zygotes, pronuclei and nucleoli and implications for cryopreservation. *Human Reprod.*, **5**, 109–115.

Wu, J. T. and Chang, M. C. (1973) Reciprocal fertilization between the ferret and short-tailed weasel with special reference to the development of ferret eggs fertilized by weasel sperm. *J. Exper. Zool.*, **183**, 281–290.

Wyrick, R. E., Nishihara, T. and Hedrick, J. L. (1974) Agglutination of jelly coat and cortical granule components and the block to polyspermy in the amphibian *Xenopus laevis. Proc. Natl Acad. Sci. USA*, **71**, 2067–2071.

Xu, Z., Kopf, G. S. and Schultz, R. M. (1994) Involvement of inositol 1,4,5-trisphosphate-mediated Ca^{2+} release in early and late events of mouse egg activation. *Development*, **120**, 1851–1859.

Yamada, H. and Hirai, S. (1984) Role of contents of the germinal vesicle in male pronuclear development and cleavage of starfish oocytes. *Devel. Growth Differ.*, **26**, 479–487.

Yamamoto, K. and Yoneda, M. (1983) Cytoplasmic cycle in meiotic division of starfish oocytes. *Devel. Biol.*, **96**, 166–172.

Yamashita, M., Onozato, H., Nakanishi, T. and Nagahama, Y. (1990) Breakdown of the sperm nuclear envelope is a prerequisite for male pronucleus formation: Direct evidence from the gynogenetic crucian carp, *Carassius auratus langsdorffii. Devel. Biol.*, **137**, 155–160.

Yanagimachi, R. (1981) Mechanisms of fertilization in mammals, in *Fertilization and Embryonic Development in Vitro*, (eds L. Mastroianni and J. D. Biggers), Plenum Press, New York, pp. 81–182.

Yanagimachi, R. (1982) Requirement of extracellular calcium ions for various stages of fertilization and fertilization related phenomena in the hamster. *Gamete Res.*, **5**, 323–344.

Yanagimachi, R. (1988) Sperm–egg fusion. *Curr. Topics Memb. Trans.*, **32**, 3–43.

Yanagimachi, R. (1994) Mammalian Fertilization, in *The Physiology of Reproduction*, 2nd edn, (eds E. Knobil and J. D. Neill), Raven Press, New York, pp189–317.

Yanagimachi, R. and Nicolson, G. L. (1976) Lectin-binding properties of hamster egg zona pellucida and plasma membrane during maturation and preimplantation development. *Exper. Cell Res.*, **100**, 249–257.

Yanagimachi, R. and Noda, Y. D. (1970a) Ultrastructural changes in the hamster sperm head during fertilization. *J. Ultrastruct. Res.*, **31**, 465–485.

Yanagimachi, R. and Noda, Y. D. (1970b) Electron microscopic studies of sperm incorporation into the golden hamster egg. *Amer. J. Anat.*, **128**, 429–462.

Yanagimachi, R. and Suzuki, F. (1985) A further study of lysolecithin-mediated acrosome reaction of guinea pig spermatozoa. *Gamete Res.*, **11**, 29–40.

Yanagimachi, R., Nicolson, G. L., Noda, Y. D. and Fujimoto, M. (1973) Electron microscopic observations of the distribution of acidic anionic residues on hamster spermatozoa and eggs before and during fertilization. *J. Ultrastruct. Res.*, **43**, 344–353.

Yanagimachi, R., Kamiguchi, Y., Sugawara, S. and Mikano, K. (1983) Gametes and fertilization in the Chinese hamster. *Gamete Res.*, **8**, 97–117.

Yanagimachi, R., Cherr, G. N., Pillai, M. C. and Baldwin, J. D. (1992) Factors controlling sperm entry into the micropyles of salmonid and herring eggs. *Devel. Growth Differ.*, **34**, 447–461.

Yanagimachi, R., Pillai, M. and Cherr, G. N. (1993) Sperm guidance factors in the micropyle area of fish eggs. *J. Reprod. Devel.*, **39** (Suppl.), 19–20.

Yasumasu, I., Tazawa, E. and Fujiwara, A. (1975) Glycolysis in the eggs of the echiuroid, *Urechis unicinctus* and the oyster, *Crassostrea gigas*. Rate-limiting steps and activation at fertilization. *Exper. Cell Res.*, **93**, 166–174.

Yew, N., Mellini, M. L. and VandeWoude, G. F. (1992) Meiotic initiation in *Xenopus* by the mos protein. *Nature*, **355**, 649–652.

Yim, D., Opresco, L. K., Wiley, H. S. and Nuccitelli, R. (1994) Highly polarized EGF receptor tyrosine kinase activity initiates egg activation in *Xenopus*. *Devel. Biol.*, **162**, 41–55.

Yllera-Fernandez, M. D. M., Crozet, N. and Ahmed-Ali, M. (1992) Microtubule distribution during fertilization in the rabbit. *Mol. Reprod. Devel.*, **32**, 221–276.

Yoneda, M. Ikeda, M. and Washitani, S. (1978) Periodic change in the tension at the surface of activated non-nucleate fragments of sea urchin eggs. *Devel. Growth Differ.*, **20**, 329–336.

Young, R. J. (1979) Rabbit sperm chromatin is decondensed by thiol-induced proteolytic activity not endogenous to its nucleus. *Biol. Reprod.*, **20**, 1001–1004.

Young, E. M. and Raff, R. A. (1979) Messenger ribonucleoprotein particles in developing sea urchin embryos. *Devel. Biol.*, **72**, 24–40.

Young, R. J. and Sweeney, K. (1978) Mammalian ova and one-cell embryos do not incorporate phosphate into nucleic acids. *Eur. J. Biochem.*, **91**, 111–117.

Yu, S. F. and Wolf, D. P. (1981) Polyspermic mouse eggs can dispose of supernumerary sperm. *Devel. Biol.*, **82**, 203–210.

Zamboni, L. (1971) *Fine Morphology of Mammalian Fertilization*, Harper & Row, New York.

Zamboni, L. (1972) Fertilization in the mouse, in *Biology of Mammalian Fertilization and Implantation*, (eds K. S. Moghissi and E. S. E. Hafez), Charles C. Thomas, Springfield, IL, pp. 213–262.

Zamboni, L., Mishell, D. R., Bell, J. H. and Baca, M. (1966) Fine structure of the human ovum in the pronuclear stage. *J. Cell Biol.*, **30**, 579–600.

Zeng, Y., Clark, E. N. and Florman, H. M. (1995) Sperm membrane potential: Hyperpolarization during capacitation regulates zona pellucida-dependent acrosomal secretion. *Devel. Biol.*, **171**, 554–563.

Zeng, Y., Oberdorf, J. A. and Florman, H. M. (1996) pH regulation in mouse sperm: Identification of Na^+-, CL^-, and HCO_3^- dependent and arylaminobenzoate-dependent regulatory mechanisms and characterization of their roles in sperm capacitation. *Devel. Biol.*, **173**, 510–520.

Zimmerberg, J. (1988) Fusion in biological and model membranes: Similarities and differences, in *Molecular Mechanisms of Membrane Fusion*, (eds S. Ohki, D. Doyle, T. D. Flanagan *et al.*), Plenum Press, New York, pp. 181–195.

Zimmerman, A. M. and Forer, A. (1981) *Mitosis/Cytokinesis*, Academic Press, New York.

Zimmerman, A. M. and Zimmerman, S. (1967) Action of colcemid in sea urchin eggs. *J. Cell Biol.*, **34**, 483–448.

Zirkin, B. R. and Chang, T. S. K. (1977) Involvement of endogenous proteolytic activity in thiol-induced release of DNA template restrictions in rabbit sperm nuclei. *Biol. Reprod.*, **17**, 131–137.

Zirkin, B. R., Chang, T. S. K. and Heaps, J. (1980) Involvement of an acrosin-like proteinase in the sulfhydryl-induced degradation of rabbit sperm nuclear protamine. *J. Cell Biol.*, **85**, 116–121.

Zirkin, B. R., Soucek, D. A., Chang, T. S. K. and Perreault, S. D. (1985) *In vitro* and *in vivo* studies of mammalian sperm nuclear decondensation. *Gamete Res.*, **11**, 349–365.

Zirkin, B. R., Perreault, S. D. and Naish, S. J. (1989) Formation and function of the male pronucleus during mammalian fertilization, in *The Molecular Biology of Fertilization*, (eds H. Schatten and G. Schatten), Academic Press, New York, pp. 91–114.

Zuccotti, M., Yanagimachi, R. and Yanagimachi, H. (1991) The ability of hamster oolemma to fuse with spermatozoa: Its acquisition during oogenesis and loss after fertilization. *Development*, **112**, 143–152.

Zuccotti, M., Piccinelli, A., Rossi, P. G. *et al.* (1995) Chromatin organization during mouse oocyte growth. *Mol. Reprod. Devel.*, **41**, 479–485.

Zucker, R. S. and Steinhardt, R. A. (1978) Prevention of the cortical reaction in fertilized sea urchin eggs by injection of calcium-chelating ligands. *Biochem. Biophys. Acta*, **541**, 459–466.

Zucker, R. S., Steinhardt, R. A. and Winkler, M. M. (1978) Intracellular calcium release and the mechanisms of parthenogenetic activation of the sea urchin egg. *Devel. Biol.*, **65**, 285–295.

Index